Computational Biology

Volume 32

Endorsed by the *International Society for Computational Biology*, the *Computational Biology* series publishes the very latest, high-quality research devoted to specific issues in computer-assisted analysis of biological data. The main emphasis is on current scientific developments and innovative techniques in computational biology (bioinformatics), bringing to light methods from mathematics, statistics and computer science that directly address biological problems currently under investigation.

The series offers publications that present the state-of-the-art regarding the problems in question; show computational biology/bioinformatics methods at work; and finally discuss anticipated demands regarding developments in future methodology. Titles can range from focused monographs, to undergraduate and graduate textbooks, and professional text/reference works.

More information about this series at http://www.springer.com/series/5769

Fabricio Alves Barbosa da Silva ·
Nicolas Carels · Marcelo Trindade dos Santos ·
Francisco José Pereira Lopes
Editors

Networks in Systems Biology

Applications for Disease Modeling

Editors
Fabricio Alves Barbosa da Silva
Scientific Computing Program (PROCC)
Oswaldo Cruz Foundation
Rio de Janeiro, Brazil

Nicolas Carels
CDTS
Oswaldo Cruz Foundation
Rio de Janeiro, Brazil

Marcelo Trindade dos Santos
Department of Computational Modeling
National Laboratory of Scientific Computing
Petrópolis, Rio de Janeiro, Brazil

Francisco José Pereira Lopes
Graduate Program in Nanobiosystems
Federal University of Rio de Janeiro
Duque de Caxias, Rio de Janeiro, Brazil

ISSN 1568-2684 ISSN 2662-2432 (electronic)
Computational Biology
ISBN 978-3-030-51864-6 ISBN 978-3-030-51862-2 (eBook)
https://doi.org/10.1007/978-3-030-51862-2

This Springer imprint is published by the registered company Springer Nature Switzerland AG
The registered company address is: Gewerbestrasse 11, 6330 Cham, Switzerland

Foreword

Since the early 60s, when Prigogine puts forward the emergent properties of complex chemical reaction networks in non-equilibrium and non-linear conditions, researchers around the world try to apply such ideas to describe the highly complex biological networks. Especially, a greater effort has been made since the beginning of the twenty-first century due to the development of omics sciences, when a huge set of global data on biological systems became available. It also appears that these technical improvements in sequencing and detecting biological molecules have propelled classical biological studies in a much more quantitative way. Furthermore, massive and universal computational resources available now have contributed crucially to what we can call a "new" model-driven quantitative biology. In this context, representation, visualization, analysis, and modeling of the topological and dynamic properties of the complex biological networks are in order.

This volume shows relevant aspects of this new area, with contributions mainly from Brazilian groups, trying to describe gene expression, metabolic, and signaling networks, as well as the brain functioning and epidemiological models. In the following chapters, complex networks are explored not only from the point of view of inference but also from dynamics and time evolution, pointing out the emerging properties of biological systems. In the first part, theoretical and computational analysis of complex biological networks is reviewed, involving visualization, inference, topological, and differential analysis, as well as modeling time evolution and sensitivity of biological processes. In the second part, the emphasis is on the application of these methods to investigate infectious and degenerative diseases, including cancer, aimed at a better understanding of the evolution of diseases and searching for relevant pharmacological targets and biomarkers.

Sophisticated mathematical and computational tools are necessary to understand the intricate processes occurring in biological networks at different levels, from the regulation of gene expression and metabolic networks into each cell, as well as signaling at the intracellular level and between cells and organisms. A broad view of modeling of these processes is presented by the authors, pointing out the importance of this approach in the rational design of new drugs, innovative gene, and immune therapies, and also in advancing the concept of personalized medicine.

Although recognized as being in the early stages, the power of this approach is well demonstrated in this volume. In fact, we can imagine much broader and new applications in this area, especially related to the new trends arising from the current revolution in information technology, such as the promising resources of the "internet of things." Also, surprising is the number and quality of Brazilian groups involved in this area, showing a very promising evolution of research in our country.

Finally, we can point out that the current pandemic of the new coronavirus, which required rapid and accurate responses, mobilized, as expected, scientists from around the world who have largely employed tools like those discussed here, emphasizing once again the importance of these contributions.

Paulo Mascarello Bisch
Institute of Biophysics "Carlos Chagas Filho"
Federal University of Rio de Janeiro
Rio de Janeiro, Brazil

Preface

In the last decades, we have witnessed a transition from descriptive biology to a systemic understanding of biological systems that was possible due to the impressive progress in high-throughput technologies. The wave of data produced by these technologies is tremendous and offered an opportunity for big data as well as mathematical and computational modeling to take off. Systems Biology is a rapidly expanding field and comprises the study of biological systems through mathematical modeling and analysis of large volumes of biological data. Now we testify exciting times where sciences integrate one another for the benefit of solving specific problems. Of course, medical sciences do not escape this trend, and we have to follow these developments for participating and translating them into medical applications as well as for transmitting them to the next generations. Indeed, the transmission of knowledge on cutting-edge developments in System Biology was the purpose of the III International Course on Theoretical and Applied Aspects of Systems Biology, held in Rio de Janeiro in July 2019, whose contributions are now translated into the present book.

This book presents current research topics on biological network modeling, as well as its application in studies on human hosts, pathogens, and diseases. The chapters were written by renowned experts in the field. Some topics discussed in-depth here include networks in systems biology, computational modeling of multidrug-resistant bacteria, and systems biology of cancer. It is intended for researchers, advanced students, and practitioners of the field. Chapters are research-oriented and present some of the most recent results related to each topic.

This book is organized into two main sections: Biological Networks and Methods in Systems Biology and Disease and Pathogen Modeling. Although the whole book is made of contributions from researchers with a clear commitment to applied sciences, the first part brings chapters where the more fundamental aspects of biological networks in systems biology are addressed. The remaining chapters on the second part of the book deal with the application of such fundamentals in disease modeling.

We take the opportunity to acknowledge the Brazilian Coordination for Improvement of Higher-Level Personnel (CAPES), FIOCRUZ's Vice-Presidency of Education, Information and Communication, the National Laboratory for Scientific Computing, and the Computational and Systems Biology Graduate Program of IOC/FIOCRUZ that made possible this event to occur with their financial and logistics support. Finally, we cannot emphasize enough how thankful we are for all authors contributing to the book for their dedication and generosity. A special thanks to Professor Paulo Bisch for writing such a splendid foreword that much honored us and contributed to further elevate this book.

Rio de Janeiro, Brazil — Fabricio Alves Barbosa da Silva
Rio de Janeiro, Brazil — Nicolas Carels
Petrópolis, Brazil — Marcelo Trindade dos Santos
Duque de Caxias, Brazil — Francisco José Pereira Lopes

Contents

Part II Disease and Pathogen Modeling

Part I
Biological Networks and Methods in Systems Biology

Chapter 1
Network Medicine: Methods and Applications

Italo F. do Valle and Helder I. Nakaya

Abstract The structure and function of biological systems are determined by a complex network of interactions among cell components. Network medicine offers a toolset for us to systematically explore perturbations in biological networks and to understand how they can spread and affect other cellular processes. In this way, we can have mechanistic insights underlying diseases and phenotypes, evaluate gene function in the context of their molecular interactions, and identify molecular relationships among apparently distinct phenotypes. These tools have also enabled the interpretation of heterogeneity among biological samples, identification of drug targets and drug repurposing as well as biomarker discovery. As our ability to profile biological samples increases, these network-based approaches are fundamental for data integration across the genomic, transcriptomic, and proteomic sciences. Here, we review and discuss the recent advances in network medicine, exploring the different types of biological networks, several methods, and their applications.

Keywords Network medicine · Graph theory · High-throughput technologies

1.1 Introduction

High-throughput technologies such as next-generation sequencing, mass spectrometry, and high-dimensional flow cytometry have revolutionized medicine. By providing the molecular and cellular profile of patients, these technologies can help physicians into their medical decisions. For instance, the analysis of whole-genome

I. F. do Valle
Center for Complex Network Research, Department of Physics, Northeastern University, 11th Floor, 177 Huntington Avenue, Boston, MA 02115, USA

H. I. Nakaya (✉)
Department of Clinical and Toxicological Analyses, School of Pharmaceutical Sciences, University of São Paulo, São Paulo, Brazil
e-mail: hnakaya@usp.br

Scientific Platform Pasteur USP, São Paulo, Brazil

Av. Prof. Lúcio Martins Rodrigues, 370, block C, 4th floor, São Paulo, SP 05508-020, Brazil

F. A. B. da Silva et al. (eds.), *Networks in Systems Biology*, Computational Biology 32,
https://doi.org/10.1007/978-3-030-51862-2_1

sequencing allows the identification of mutations associated with a disease or a response to treatment. It is also possible to measure the activity of tens of thousands of genes, proteins, and metabolites to find the set of markers capable of predicting a medical outcome. Another example is the analysis of DNA methylation patterns in liquid biopsies that can reveal the presence of tumors in the early stages of the disease [1]. However, just having this comprehensive catalog of a patient's genes and biological components is often not sufficient to understand the mechanisms of human diseases.

Network medicine studies the interactions among molecular components to better understand the pathogenesis of a disease. The underlying idea is that a cell can be thought as networks of interacting biomolecules, and a disease is can be seen as a "malfunctioning" in one or more regions in human biological networks [2]. A mutation that affects the correct functioning of a single protein will interfere not only with the function of that specific protein, but also with the proper functioning of many other proteins that are connected to it. Network medicine uses graph theory to analyze how networks behave in the context of a disease and one of its aims is to identify the key players related to the disease.

In this chapter, we will describe the types of biological networks and the main methods of analysis utilized in network medicine. We will also address the techniques that identify subnetworks and gene modules associated with human diseases. Finally, we will show how drug treatment can affect the network behavior.

1.1.1 Basic Concepts in Graph Theory

A network (or a graph) is a catalog of a system's components often called nodes or vertices and the interactions between them, called links or edges. In biological networks, nodes represent biomolecules, such as proteins, metabolites, and genes, while links represent different types of biological interactions between them, such as physical binding, enzymatic reaction, or transcriptional regulation. Link networks can be directed, like in the interaction where a transcription factor activates a given target gene, or undirected, where the interaction is bidirectional, like in a physical interaction between two proteins.

A key property of a node is its degree, which is equal to its total number of connections. Depending on the network, it can represent the number of proteins a given protein binds to or the number of reactions a given metabolite participates in. For directed networks, such as regulatory networks, the degree can be differentiated in outgoing degree, the number of nodes it points to, or incoming degree, and the number of nodes that point to it. The degree distribution of a network, which gives the probability $P(k)$ that a selected node has exactly k links, is important to understand how the network works. For example, a peaked degree distribution indicates that a system has a characteristic degree from which most of the nodes do not highly deviate from (Fig. 1.1a, c). By contrast, most networks found in nature, are characterized by a power-law degree distribution, which means that most nodes have a few interactions

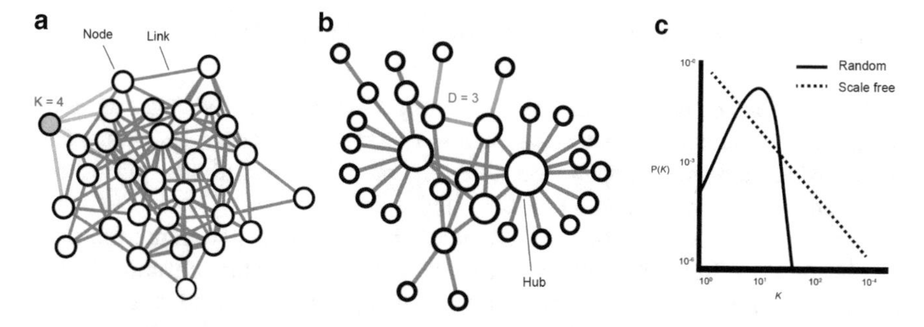

Fig. 1.1 Example of a random non-scale-free network (**a**) and of a scale-free network (**b**), together with a schematic representation of their degree distributions (**c**). A node with degree K = 4 is highlighted in green, and a shortest path of length D = 3 is highlighted in blue

and that these coexist with a few highly connected nodes, the hubs, that hold the whole network together (Fig. 1.1b, c). Networks with this property are usually referred to as scale-free network. These are typical of several real-world systems, and this degree distribution implies important properties of these systems' behavior, such as the high robustness against accidental node failures [3].

Complex networks are also often characterized by the small-world property [4]. This means that any pair of nodes can be connected by relatively short paths. In biological networks, this property indicates that, for example, most proteins (or metabolites) are only a few interactions (or reactions) from any other protein (metabolite) [5–7]. Therefore, perturbing the state of a given node can affect the activity of several others in their vicinity.

1.2 Biological Networks

Cells are comprised of complex webs of molecular interactions between cell components [8]. These interactions form complex networks or interactomes, and many experimental approaches have been developed to completely map them. These approaches include (1) curation of existing data available in the literature (literature curation), (2) computational predictions based on different information such as sequence similarity and evolutionary conservation, and (3) systematic and unbiased high-throughput experimental strategies applied at the scale of whole genomes or proteomes. The networks derived by each of these methods have their own biases and limitations that should be carefully taken into account during computational analysis. Here, we discuss a few examples of biological networks and their respective properties.

1.2.1 Protein–Protein Interaction (PPI) Networks

PPIs are undirected networks in which nodes represent proteins and edges represent a physical interaction between two proteins. Two main methodologies are used for large-scale interaction mapping: yeast-two hybrid (Y2H) and affinity purification followed by mass spectrometry (AP/MS) (Fig. 1.2).

In the Y2H technique, a transcription factor is split into its two components: the binding domain (BD), which binds to the DNA sequence, and the activation domain (AD), which activates the transcription. DNA recombinant tools are used to create chimeric proteins in which one protein of interest (prey) is fused to the transcription factor BD, while the other protein of interest (bait) is fused to the AD. If the prey and bait proteins physically interact, the transcription factor is reconstructed and is then able to activate the transcription of a reporter gene, which will create an indicator that the interaction occurred (Fig. 1.2a). In AP/MS, a protein of interest (bait) is purified from a cell lysate (often referred to as pull-down), and co-purified proteins (preys) are identified through mass spectrometry (Fig. 1.2b). Mappings derived from Y2H contain physical interactions, while AP/MS ones contain co-complex information—that is, the interactions can be either physical or indirect.

Several high-throughput mappings have been used to map protein interactomes in humans and model organisms. The most recent efforts for human interactomes include the Human Reference Interactome (HuRI) [9], mapped by Y2H, and BioPlex2.0 [10], and mapped using AP/MS. Several databases with literature-curated PPIs are available, and a few efforts have been made to produce high-quality interactomes derived from literature-curated data [11, 12].

Literature-curated PPIs are inherently biased toward heavily studied proteins: most interactions occur among genes characterized by many publications and their network is depleted of interactions for proteins with few or no publications [13].

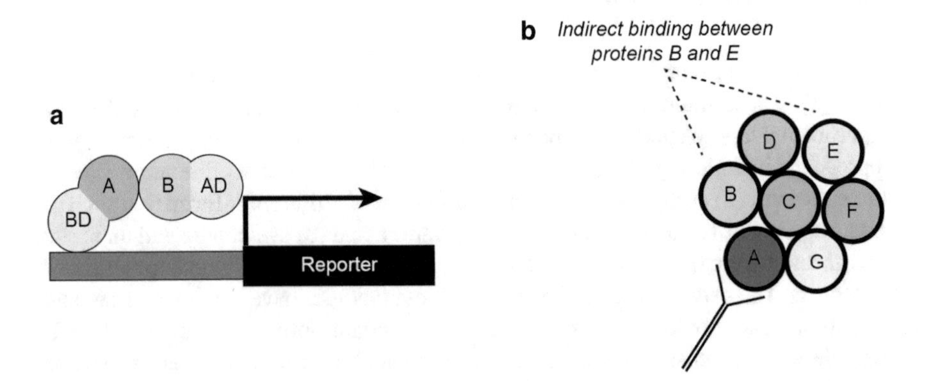

Fig. 1.2 Schematic representations of the techniques used to detect protein–protein interactions: **a** Yeast-two hybrid and **b** affinity purification of protein complexes. BD: DNA-binding domain, AD: transcription activation domain, A-B hypothetic proteins

The analysis of such networks may lead to incorrect conclusions, such as the previously reported correlation between number of interaction partners (degree) and gene essentiality [9, 13].

Small overlaps are observed among protein interactomes, even those derived by unbiased and systematic studies, which can be partially attributed to the different properties of their respective experimental strategies. For example, PPIs have different binding affinities, which may or may not be in the range of detectability for that specific method. Other factors might include fusion constructs, washing buffer, and protein expression in the cell.

1.2.2 Gene Regulatory Networks

In gene regulatory networks, nodes are transcription factors (TFs) (and/or miRNAs) and their targets, and directed edges exist between TFs (miRNA) and their targets. The most common approach for detection of regulatory interactions is chromatin immunoprecipitation (ChIP-seq)-based approaches: DNA-binding proteins are cross-linked with the DNA, an antibody is used to immunoprecipitate the protein of interest, and DNA sequencing strategies are used to identify the genomic regions where the protein binds to. The human regulatory network derived in this way by the ENCODE project revealed important features of cellular regulation: hierarchical organization of TFs in which top-level factors more strongly influence expression, while middle-level TFs co-regulate several targets. These properties avoid information-flow bottlenecks and allow the presence of feed-forward network motifs. It was also possible to observe stronger evolutionary and allele-specific activity of the most connected network components [14].

Other strategies also take advantage of DNAse-Seq to identify regions that can be occupied by TFs and then identify these TFs by their binding motifs for that particular genomic region. The mapping of the regulatory networks across 41 cell lines using this technique has revealed that networks are markedly cell-specific and even TFs that are expressed across cells of a given lineage show distinctive regulatory roles in the different cells [15].

Several approaches have been developed for reverse engineering cellular networks from gene expression data. Most of these methods are based on the notion of similarity among co-expression across different experimental conditions. Methods based on measures based on correlation, mutual information and graphical models (including Bayesian networks) identify undirected edges between nodes by capturing probabilistic dependences of different kinds [16].

1.2.3 Metabolic Networks

Metabolic networks attempt to describe biochemical reactions for a particular cell or organism. In most representations, nodes are metabolites and edges are the reactions that convert one metabolite into another. In this case, edges can be directed or undirected, depending on the reversibility of the reaction. Other representations are also possible, such as nodes as metabolites and edges representing co-participation in the same biochemical reactions.

Network reconstruction involves manual curation of literature data describing experimental results on metabolic reactions as well as predicted reactions derived from orthologous enzymes experimentally characterized in other species [17].

1.2.4 Genetic Interaction Networks

Genetic interactions (GIs) are functional relationships between genes, and they can be classified into positive and negative interactions. In negative GIs, the observed fitness by mutating a pair of genes at the same time (double mutants) is worse than what is expected when mutating genes individually (single mutants) (Fig. 1.3a). In positive GIs, a gene mutation can mitigate the effect caused by another mutation, and the double mutant is healthier than most sick of the single mutants (Fig. 1.3b). Mapping GIs allows us to understand the mechanisms underlying robustness of

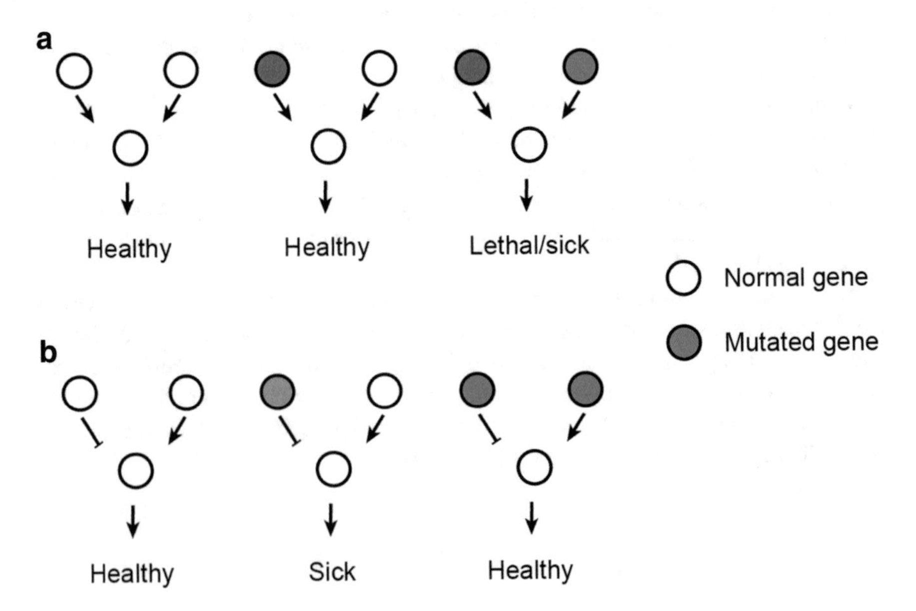

Fig. 1.3 Schematic representation of negative (**a**) and positive (**b**) genetic interactions

biological systems and how compensatory mechanisms emerge after perturbations. Recent technological advances have made possible the systematic mapping of genetic interactions in yeast, *C. elegans* and human cells. These maps can be represented as networks in which nodes are genes and an edge exists between genes that have high similarity in their genetic interaction profiles. Genetic interaction networks have enabled the understanding of the hierarchical dependencies of cell functions, identification of functionally related processes and pleiotropic genes [18–20].

1.2.5 Pathogen–Host Interactomes

Pathogens have complex mechanisms to perturb host intracellular networks to their advantage and the understanding of parasite–host interactomes could provide important insights for the development of treatment strategies. For instance, it has been observed that pathogen's proteins preferentially target hubs in human and plant interactomes [21, 22]. Systematic maps capturing viral–host protein–protein interactions have been obtained for Epstein–Barr virus [23], hepatitis C virus [23], herpesviruses [24], influenza [25], HIV [26], and others [27]. Other pathogen–host interactomes have been assembled or predicted for bacteria [28], fungi [29], worms, such as *Schistosoma mansoni* [30], and several protozoans, such as *Leishmania* [30], *Plasmodium* [30–32], and *Trypanosoma* [30].

1.3 Biological Networks for Functional Annotations of Proteins and Complexes

The network neighborhood of a protein reflects several of its properties: cellular localization, biological, and molecular function. Therefore, the most basic assumption is that proteins that are close to each other and/or share many neighbors in the interactomes are more likely to have a similar function. This "guilt-by-association" principle underlies many network-based methods for protein function prediction.

An example of this principle can be observed in the recent demonstration that PPI networks can be used to predict protein subcellular localization [9]. The authors showed that extracellular vesicle (EV) proteins form a significant subnetwork in the PPI network. The interaction partners of the EV subnetwork already included many proteins with established roles in EV biogenesis and cargo recruitment, and the other interaction partners with unknown subcellular localization were ranked as potential EV proteins based on the number of interactions they shared with the EV subnetwork. Experimental validation demonstrated that candidate proteins were indeed related to extracellular vesicle functions [9].

Several network-based indexes attempt to quantify the size of neighborhood that is shared between two proteins—these are often referred to in network science as node similarity indexes. Common similarity indices for pairs of nodes take into account the number of shared interaction between the nodes normalized by their total number of interactions (Jaccard index), or by the smallest degree of either node (Simpson index) or by the product of the individual node degrees (geometric and cosine indexes) [33]. Uncharacterized proteins can be ranked based on their similarity indexes with proteins of known function, and the high-ranking proteins can be annotated to the same function. Exploiting the same principle on genetic interaction (GI) networks has been shown to be very effective for the discovery of functional complexes, control, and regulatory strategies, as well as unrecognized biosynthetic pathways [34].

Other approaches for the prediction of protein function take into account the full topology of the network [35]. Flow-based approaches consider each protein annotated to a given function as the source of a "functional flow". After simulating the spread of this flow over time through the network, each unannotated protein is assigned a score that is proportional to the amount of flow it received during the simulation [36, 37]. A recent distance metric based on network diffusion that is able to capture similarities based on multiple paths in the network has been shown to provide finer grained distinctions when transferring functional annotation in PPI networks [38]. Other approaches integrate PPI network data with high-throughput biological data, creating functional networks for the predictions [39].

More recently, algorithms based on network embedding have also been used for the prediction of protein function. In classic versions of these approaches, matrix factorization leads to a representation of network nodes as vectors in a low-dimensional space while preserving the neighborhood similarity between nodes [40]. A network embedding algorithm has been applied in a multi-layer network, where each layer represents protein interactions in a different tissue, to provide predictions of cellular function that take into account tissue-specific protein functions [41].

1.4 Biological Networks and Diseases

1.4.1 Disease Genes and Subnetworks

Biological networks provide us with a unique opportunity to study disease mechanisms in a holistic manner. The interconnectivity between cellular components—genes, proteins, and metabolites—implies that the effect of specific perturbations, like mutations, will spread through the network to areas not originally affected. Biological networks provide us with the context for a given gene or protein which is essential in determining the phenotypic effects of perturbations [2].

Protein–protein interaction networks have been extensively used to study disease mechanisms. It has been observed that proteins genetically associated to a given disease tend to be colocalized in a given neighborhood of the network, forming a connected subgraph, often referred to as *disease module*. Thus, the disease module indicates a region in the network that, if perturbed, leads to the disease phenotype

[2]. In asthma, for example, out of 129 asthma-related genes, 37 formed a connected subgraph, or disease module [42]. In order to measure whether the disease module could have emerged by chance, 129 genes were randomly selected from the network and the size of the largest connected component formed by these genes is registered. This process is repeated through several iterations, usually 1,000 times, producing a null distribution. This null distribution can be then used as reference to compare the module size observed from the disease genes, providing an empirical p-value: the proportion of random iterations that produced a module size equal or greater than the real observation [42]. The biological significance of the disease modules can be verified by the fact that the asthma module contained proteins related to immune response and pathways involved in other immune-related disorders [42]. The asthma module also resulted in enriched with differentially expressed genes from normal and asthmatic fibroblast cells treated with an asthma-specific drug and close evaluation of the module revealed GAB1 signaling pathway as an important modulator in asthma [42]. In summary, this represents the simplest approach for disease module discovery: size of the largest connected component (LCC) of the subgraph formed by disease proteins, and strategy that was later demonstrated to work in hundreds of other diseases rather than asthma only [43].

However, the discovery and detection of disease modules can be often challenging, since a large proportion of the disease-associated proteins remains unknown, as well as many of the possible protein–protein interactions remain to be discovered. These limitations result in modules that are often fragmented, limited in size, and only partially describing the underlying disease mechanisms.

Several methods have been proposed for the discovery of disease-associated proteins and subnetworks. Some methods are based on the "guilt-by-association" principle and exploit the network proximity of diseased genes [44, 45]. Other methods explore the global structural and topological properties of PPI networks to identify disease-related subnetworks or disease modules. For example, some methods are based on the principle of network diffusion or random walk [37, 46]. In these methods, disease genes are starting points (seeds) of a random walker that moves from node to node along the links of a network. After a given number of iterations, the frequency in which the nodes are visited converges is used to rank highly visited subnetworks. Examples of methods based on this principle are HotNet [47] and HotNet2 [48] algorithms that aim to find modules of somatic mutations in cancer and modules of common variants in complex diseases.

The DIseAse Module Detection (DIAMOnD) algorithm introduced the concept of connectivity significance [49]. In this method, disease genes are mapped in the PPI network and, at each step of an iterative process, the node most significantly connected to the disease genes is added to the module. The connectivity significance is based on a hypergeometric test that assigns a p-value to the proteins that share more connections with the seed proteins than expected by chance [49]. However, as a cautionary note, we highlight the fact that statistical tests that assume independence of observations are not appropriate for networks [50], and degree-preserving network randomization strategies could provide better statistical support for the same network-based principles [51, 52]. The comparison of different types of similar algorithms (i.e., node-ranking) suggests that each one has its strengths and weakness and their application might depend on the specific use case [53].

Another class of methods for module identification is based on the principle that nodes related to the same disease or function are more densely connected to each other than expected by change (i.e., high modularity). A recent study compared different types of module identification algorithms in different biological networks [54]. Again, results showed that methods from different categories can achieve comparable performances complementary to each other [54].

Different networks can provide very different predictive performances when used with disease module identification algorithms. A comparison of different network sources indicates that the size of the network can improve performance and outweigh the detrimental effects of false positives [55]. It also showed that parsimonious composite networks, which only include edges that are also observed in other networks, can also increase performance and efficiency [55].

Integration of phenotypic data can also help in the identification of disease genes and disease modules. Caceres and Paccanaro [57] recently proposed an approach that uses disease phenotype similarity [56], to define a prior probability distribution over disease genes on the interactome. Subsequently, a semi-supervised learning method establishes a prioritization ordering for all genes in the interactome. The important advantage of this method is that it provides predictions of disease genes even for diseases with no known genes. Their method can also be used to retrieve disease modules [57].

1.4.2 Disease Networks

The interconnectivity between cellular components also implies that different diseases might be connected by the same underlying molecular mechanisms. To map these disease–disease relationships, Goh et al. [58] created the diseasome, a network in which nodes represent diseases; two diseases are linked if they share common genes, and these links are labeled by the number of gene-causing mutations that are shared. The network representation of disease interrelationships provides a global perspective, offering the possibility to identify patterns and principles not readily apparent from the study of individual disorders.

However, different diseases could share disease pathways and processes while not having any causal gene in common. Therefore, methods have been developed for measuring the network proximity (or overlap) between disease modules in the interactome. The S_{AB} measure compares the shortest distances between proteins within each disease (A and B, for example), to the shortest distance between A-B protein pairs [43]. It was shown that overlapping disease modules ($S_{AB} < 0$) share several pathobiological properties: the respective disease proteins have similar functions and show higher co-expression across tissues, while the diseases have similar symptoms and higher risk of co-occurrence in patients [43].

It is also possible to integrate clinical information to map and understand how different diseases are related. Disease networks can map the level of disease co-occurrence, or comorbidity, from Electronical Health Records (EHR). In these networks, the nodes are usually disease codes used in clinical practice, such as ICD-9 and ICD-10, and the links are defined by a statistical measure of co-occurrence. Examples of co-occurrence measures are the phi-coefficient, a correlation measure for binary variables, and the relative risk, the ratio between observed co-occurrence and random expectation [59]. Strategies based on such maps were able to reveal comorbidities that were demographically modulated in a given population (e.g., diseases more frequently co-occurring in black males) [60], the impact of age and sex on disease comorbidities [61, 62], and to reveal temporal patterns of disease progression [63–66]. For example, the study of a disease network was able to identify patterns of disease trajectories significantly associated with sepsis mortality, which started from three major points: alcohol-abuse, diabetes, and cardiovascular diagnoses [64].

The human symptoms—disease network—that connected diseases that showed symptom similarity, indicated that strong associations in symptom similarity also reflect common disease genes and PPIs [67]. It also indicated that diseases with diverse clinical manifestations also showed diversity in their underlying mechanisms [67].

Disease networks will improve as more molecular and phenotypic data will become available for a larger number of diseases. They represent a global reference map for clinicians to better visualize and understand disease interrelationships. They might reveal principles for better treatment and prevention, as well as offer a mechanism-driven approach for the development of new disease classification guidelines [68].

1.5 Biological Networks and Drugs

As the study of biological networks reveals significant insights into the systemic organization of cellular mechanisms, they provide a powerful platform where we can study the interplay between drugs and diseases and identify emergent properties not apparent when single molecules are studied in isolation [69]. Biological networks have been applied for the discovery of new targets, characterization of mechanism of action, identification of drug repurposing strategies, and for prediction of drug safety and toxicity.

PPI networks have been extensively used to study drug targets and their relationship with disease proteins. The targets of most drugs tend to form connected subgraphs within the PPI that are significantly larger than expected by change and most compound targets are characterized by significantly shorter path lengths between their associated targets [70]. Additionally, drug targets tend to be significantly proximal, in the network, to the proteins of the diseases for which they are indicated [71, 72]. These network proximity measures take into account the shortest path lengths among drug targets and proteins, and the statistical significance is evaluated by comparing the observed distance to the distance obtained from random sets of proteins, while preserving the size and degree of the original sets. Guney et. al. [72] applied these principles to study the proximity between all possible pairs among 238 drugs and 78 diseases, showing that the proximity between drug targets and disease proteins provided a good discriminating performance for distinguishing known drug–disease pairs (i.e., with clinical use) from unknown ones [72].

These observations suggest that network-based methods could aid in the identification of drugs that could be reused for conditions different from their intended indications. In particular, it has been observed that drugs often target regions and pathways that are shared across multiple conditions [73]. Potential therapeutic interventions targeting the common pathologic processes of Type 2 Diabetes and Alzheimer's Diseases were revealed by first identifying pathways proximal to the disease modules and then ranking pathways targeted by drugs using topological information from the protein interactome [73]. In another example, transcriptomic data integrated with a protein–protein interaction network were used to identify molecular pathways shared across different tumor types, revealing therapeutic candidates that could eventually be repurposed for the treatment of a whole group of tumors [74]. Using the network proximity between drug targets and disease proteins, Cheng et al. [76] identified hydroxychloroquine, a drug indicated for rheumatoid arthritis, as a potential therapeutic intervention for coronary artery disease [75]. The authors analyzed data from over 220 million patients in healthcare databases to demonstrate that patients who happened to be prescribed for hydroxychloroquine had indeed lower risk of being diagnosed for coronary heart disease later in their lives. The study provided additional in vitro data suggesting that the mechanism of action for this association might involve hydroxychloroquine's anti-inflammatory effects on endothelial cells [75].

Complex diseases tend to be associated with multiple proteins and drugs often work by targeting several proteins besides their primary target. Consequently, several approaches attempt to develop and predict drugs that target multiple proteins, as well as to identify new drug combination strategies. Recent analysis of drug targets in PPIs showed that targets are clustered in specific network neighborhoods and proximity among targets of drug pairs also correlates with chemical, biological, and clinical similarities of the corresponding drugs [70, 76]. It also showed that for a drug–pair combination to be effective, the drug targets of both drugs should overlap with the disease module of the disease for which the treatment is intended for, while not overlapping with each other [76]. Based on these principles, a network proximity method showed good accuracy on the discrimination of approved hypertensive drug combinations, and it outperformed traditional cheminformatics and bioinformatics approaches [76]. These observations also agree with experimental data evaluating morphology perturbations caused by drug combinations in cell lines [70].

In a recent application of neural networks on graphs, a multi-layer representation of protein–protein, drug–protein, and drug–drug (with links representing side effects) interaction networks was used to predict side effects with improved performance over previous methods [77]. In contrast with previous methods, this method could not only predict a probability/strength score of a drug interaction, but could also identify which exact side effect would result from the interaction.

Acknowledgements We would like to thank Alberto Paccanaro for his valuable inputs.

Funding HIN is supported by CNPq (313662/2017-7) and the São Paulo Research Foundation (FAPESP; grants 2018/14933-2, 2018/21934-5, and 2013/08216-2).

References

1. Shen SY, Singhania R, Fehringer G, Chakravarthy A, Roehrl MHA, Chadwick D et al (2018) Sensitive tumour detection and classification using plasma cell-free DNA methylomes. Nature 563(7732):579–583
2. Barabási A-L, Gulbahce N, Loscalzo J (2011) Network medicine: a network-based approach to human disease. Nat Rev Genet 12(1):56–68
3. Reka A, Jeong H, Barabási A-L, Albert R, Jeong H, Barabási A-L (2000) Error and Attack Tolerance of Complex Networks. Nature 406:378–381
4. Watts DJ, Strogatz SH (1998) Collective dynamics of "small-world" networks. Nature 393(6684):440–442
5. Jeong H, Tombor B, Albert R, Oltvai ZN, Barabasi A-L, The large-scale organization of metabolic networks. Nature
6. Fell DA, Wagner A (2000) The small world of metabolism. Nat Biotechnol 18(11):1121–1122
7. Jeong H, Mason SP, Barabási AL, Oltvai ZN (2001) Lethality and centrality in protein networks. Nature 411(6833):41–42
8. Vidal M, Cusick ME, Barabasi A-L (2011) Interactome networks and human disease. Cell 144(6):986–998
9. Luck K, Kim D-K, Lambourne L, Spirohn K, Begg BE, Bian W, et al (2019) A reference map of the human protein interactome. bioRxiv 605451.
10. Huttlin EL, Bruckner RJ, Paulo JA, Cannon JR, Ting L, Baltier K et al (2017) Architecture of the human interactome defines protein communities and disease networks. Nature [Internet] 545(7655):505–509
11. Das J, Yu H (2012) HINT: high-quality protein interactomes and their applications in understanding human disease. BMC Syst Biol 6
12. Alonso-López D, Campos-Laborie FJ, Gutiérrez MA, Lambourne L, Calderwood MA, Vidal M et al (2019) APID database: redefining protein-protein interaction experimental evidences and binary interactomes. Database (i):1–8
13. Luck K, Sheynkman GM, Zhang I, Vidal M (2017) Proteome-scale human interactomics. Trends Biochem Sci 42(5):342–354
14. Gerstein MB, Kundaje A, Hariharan M, Landt SG, Yan KK, Cheng C et al (2012) Architecture of the human regulatory network derived from ENCODE data. Nature [Internet] 489(7414):91–100
15. Neph S, Stergachis AB, Reynolds A, Sandstrom R, Borenstein E, Stamatoyannopoulos JA (2012) Circuitry and dynamics of human transcription factor regulatory networks. Cell [Internet] 150(6):1274–1286
16. Wang YXR, Huang H (2014) Review on statistical methods for gene network reconstruction using expression data. J Theor Biol [Internet] 362:53–61
17. Mo ML, Palsson BØ (2009) Understanding human metabolic physiology: a genome-to-systems approach. Trends Biotechnol [Internet] 27(1):37–44

18. Costanzo M, VanderSluis B, Koch EN, Baryshnikova A, Pons C, Tan G, et al (2016) A global genetic interaction network maps a wiring diagram of cellular function. Science (80-) [Internet] 353(6306):aaf1420–aaf1420
19. Baryshnikova A, Costanzo M, Myers CL, Andrews B, Boone C (2013) Genetic interaction networks: toward an Understanding of Heritability. Annu Rev Genomics Hum Genet 14(1):111–133
20. Boucher B, Jenna S (2013) Genetic interaction networks: better understand to better predict. Front Genet 4:1–16
21. Ahmed H, Howton TC, Sun Y, Weinberger N, Belkhadir Y, Mukhtar MS (2018) Network biology discovers pathogen contact points in host protein-protein interactomes. Nat Commun 9(1):2312
22. Calderwood MA, Venkatesan K, Xing L, Chase MR, Vazquez A, Holthaus AM et al (2007) Epstein-Barr virus and virus human protein interaction maps. Proc Natl Acad Sci U S A 104(18):7606–7611
23. de Chassey B, Navratil V, Tafforeau L, Hiet MS, Aublin-Gex A, Agaugué S et al (2008) Hepatitis C virus infection protein network. Mol Syst Biol 4:230
24. Uetz P, Dong Y-A, Zeretzke C, Atzler C, Baiker A, Berger B et al (2006) Herpesviral protein networks and their interaction with the human proteome. Science [Internet] 311(5758):239–242
25. Shapira SD, Gat-Viks I, Shum BO V, Dricot A, de Grace MM, Wu L et al (2009) A physical and regulatory map of host-influenza interactions reveals pathways in H1N1 infection. Cell 139(7):1255–1267
26. Jäger S, Gulbahce N, Cimermancic P, Kane J, He N, Chou S et al (2011) Purification and characterization of HIV-human protein complexes. Methods 53(1):13–19
27. Mendez-Rios J, Uetz P (2010) Global approaches to study protein-protein interactions among viruses and hosts. Future Microbiol 5(2):289–301
28. Penn BH, Netter Z, Johnson JR, Von Dollen J, Jang GM, Johnson T et al (2018) An Mtb-human protein-protein interaction map identifies a switch between host antiviral and antibacterial responses. Mol Cell 71(4):637–648.e5
29. Remmele CW, Luther CH, Balkenhol J, Dandekar T, Müller T, Dittrich MT (2015) Integrated inference and evaluation of host–fungi interaction networks. Front Microbiol 4:6
30. Cuesta-Astroz Y, Santos A, Oliveira G, Jensen LJ (2019) Analysis of predicted host–parasite interactomes reveals commonalities and specificities related to parasitic lifestyle and tissues tropism. Front Immunol 13:10
31. Aditya R, Mayil K, Thomas J, Gopalakrishnan B (2010) Cerebral malaria: insights from host-parasite protein-protein interactions. Malar J 9:1–7
32. Wuchty S (2011) Computational prediction of host-parasite protein interactions between P. falciparum and H. sapiens. Borrmann S (ed). PLoS One 6(11):e26960
33. Bass JIF, Diallo A, Nelson J, Soto JM, Myers CL, Walhout AJM (2013) Using networks to measure similarity between genes: Association index selection. Nat Methods 10(12):1169–1176
34. Collins SR, Miller KM, Maas NL, Roguev A, Fillingham J, Chu CS et al (2007) Functional dissection of protein complexes involved in yeast chromosome biology using a genetic interaction map. Nature 446(7137):806–810
35. Sharan R, Ulitsky I, Shamir R (2007) Network-based prediction of protein function. Mol Syst Biol 3(88):1–13
36. Nabieva E, Jim K, Agarwal A, Chazelle B, Singh M (2005) Whole-proteome prediction of protein function via graph-theoretic analysis of interaction maps. Bioinformatics 21(SUPPL. 1):302–310
37. Cowen L, Ideker T, Raphael BJ, Sharan R (2017) Network propagation: a universal amplifier of genetic associations. Nat Rev Genet 18(9):551–562
38. Cao M, Zhang H, Park J, Daniels NM, Crovella ME, Cowen LJ et al (2013) Going the distance for protein function prediction: a new distance metric for protein interaction networks. PLoS ONE 8(10):1–12

39. Warde-Farley D, Donaldson SL, Comes O, Zuberi K, Badrawi R, Chao P et al (2010) The GeneMANIA prediction server: biological network integration for gene prioritization and predicting gene function. Nucleic Acids Res 38(SUPPL. 2):214–220
40. Nelson W, Zitnik M, Wang B, Leskovec J, Goldenberg A, Sharan R (2019) To embed or not: Network embedding as a paradigm in computational biology. Front Genet 10:1–11
41. Zitnik M, Leskovec J (2017) Predicting multicellular function through multi-layer tissue networks. Bioinformatics 33(14):i190–i198
42. Sharma A, Menche J, Chris Huang C, Ort T, Zhou X, Kitsak M et al (2014) A disease module in the interactome explains disease heterogeneity, drug response and captures novel pathways and genes in asthma. Hum Mol Genet 24(11):3005–3020
43. Menche J, Sharma A, Kitsak M, Ghiassian SD, Vidal M, Loscalzo J et al (2015) Disease networks. Uncovering disease-disease relationships through the incomplete interactome. Science [Internet] 347(6224):1257601
44. Guney E, Oliva B (2012) Exploiting protein-protein interaction networks for genome-wide disease-gene prioritization. PLoS One 7(9)
45. Yin T, Chen S, Wu X, Tian W (2017) GenePANDA-a novel network-based gene prioritizing tool for complex diseases. Sci Rep 7:1–10
46. Köhler S, Bauer S, Horn D, Robinson PN (2008) Walking the interactome for prioritization of candidate disease genes. Am J Hum Genet 82(4):949–958
47. Vandin F, Upfal E, Raphael BJ (2011) Algorithms for detecting significantly mutated pathways in cancer. J Comput Biol 18(3):507–522
48. Leiserson MDM, Vandin F, Wu H-T, Dobson JR, Eldridge J V, Thomas JL et al (2015) Pancancer network analysis identifies combinations of rare somatic mutations across pathways and protein complexes. Nat Genet [Internet] 47(2):106–114
49. Ghiassian SD, Menche J, Barabási AL (2015) A DIseAse MOdule Detection (DIAMOnD) algorithm derived from a systematic analysis of connectivity patterns of disease proteins in the human interactome. PLoS Comput Biol 11(4):1–21
50. Croft DP, Madden JR, Franks DW, James R (2011) Hypothesis testing in animal social networks. Trends Ecol Evol 26(10):502–507
51. Iorio F, Bernardo-Faura M, Gobbi A, Cokelaer T, Jurman G, Saez-Rodriguez J (2016) Efficient randomization of biological networks while preserving functional characterization of individual nodes. BMC Bioinfor 17(1):542
52. Farine DR (2017) A guide to null models for animal social network analysis. Methods Ecol Evol 8(10):1309–1320
53. Hill A, Gleim S, Kiefer F, Sigoillot F, Loureiro J, Jenkins J et al (2019) Benchmarking network algorithms for contextualizing genes of interest. PLoS Comput Biol 15(12):1–14
54. Choobdar S, Ahsen ME, Crawford J, Tomasoni M, Fang T, Lamparter D et al (2019) Assessment of network module identification across complex diseases. Nat Methods 16(9):843–852
55. Huang JK, Carlin DE, Yu MK, Zhang W, Kreisberg JF, Tamayo P et al (2018) Systematic evaluation of molecular networks for discovery of disease genes. Cell Syst. 6(4):484-495.e5
56. Caniza H, Romero AE, Paccanaro A (2015) A network medicine approach to quantify distance between hereditary disease modules on the interactome. Sci Rep 5:1–10
57. Cáceres JJ, Paccanaro A (2019) Disease gene prediction for molecularly uncharacterized diseases. PLoS Comput Biol 15(7):1–14
58. Goh K-I, Cusick ME, Valle D, Childs B, Vidal M, Barabási A-L (2007) The human disease network. Proc Natl Acad Sci U S A 104(21):8685–8690
59. Fotouhi B, Momeni N, Riolo MA, Buckeridge DL (2018) Statistical methods for constructing disease comorbidity networks from longitudinal inpatient data. Appl Netw Sci 3(1)
60. Hidalgo CA, Blumm N, Barabási AL, Christakis NA (2009) A dynamic network approach for the study of human phenotypes. PLoS Comput Biol 5(4)
61. Chmiel A, Klimek P, Thurner S. Spreading of diseases through comorbidity networks across life and gender. New J Phys 16(11):115013
62. Kalgotra P, Sharda R, Croff JM (2017) Examining health disparities by gender: A multimorbidity network analysis of electronic medical record. Int J Med Inform 108:22–28

63. Jensen AB, Moseley PL, Oprea TI, Ellesøe SG, Eriksson R, Schmock H et al (2014) Temporal disease trajectories condensed from population-wide registry data covering 6.2 million patients. Nat Commun 5:1–10
64. Beck MK, Jensen AB, Nielsen AB, Perner A, Moseley PL, Brunak S (2016) Diagnosis trajectories of prior multi-morbidity predict sepsis mortality. Sci Rep 6:1–9
65. Giannoula A, Gutierrez-Sacristán A, Bravo Á, Sanz F, Furlong LI (2018) Identifying temporal patterns in patient disease trajectories using dynamic time warping: a population-based study. Sci Rep 8(1):1–14
66. Jeong E, Ko K, Oh S, Han HW (2017) Network-based analysis of diagnosis progression patterns using claims data. Sci Rep 7(1):1–12
67. Zhou X, Menche J, Barabási A-L, Sharma A, Zhou X (2914) Human symptoms–disease network. Nat Commun 5
68. Dozmorov MG (2018) Disease classification: from phenotypic similarity to integrative genomics and beyond. Brief Bioinform 1–12
69. Loscalzo J, Barabási A-L, Silverman EK (2017) Network medicine. Harvard University Press
70. Caldera M, Müller F, Kaltenbrunner I, Licciardello MP, Lardeau CH, Kubicek S et al (2019) Mapping the perturbome network of cellular perturbations. Nat Commun [Internet] 10(1)
71. Wang RS, Loscalzo J (2016) Illuminating drug action by network integration of disease genes: a case study of myocardial infarction. Mol Biosyst 12(5):1653–1666
72. Guney E, Menche J, Vidal M, Barabási A-L (2016) Network-based in silico drug efficacy screening. Nat Commun 7(1):10331
73. Aguirre-Plans J, Piñero J, Menche J, Sanz F, Furlong LI, Schmidt HHHW et al (2018) Targeting comorbid diseases via network endopharmacology. Pharmaceuticals [Internet] 11(61)
74. do Valle ÍF, Menichetti G, Simonetti G, Bruno S, Zironi I, Durso DF et al (2018) Network integration of multi-tumour omics data suggests novel targeting strategies. Nat Commun [Internet] 9(1):4514
75. Cheng F, Desai RJ, Handy DE, Wang R, Schneeweiss S, Barabási AL et al (2018) Network-based approach to prediction and population-based validation of in silico drug repurposing. Nat Commun 9(1):1–12
76. Cheng F, Kovács IA, Barabási A-L (2019) Network-based prediction of drug combinations. Nat Commun 10(1):1197
77. Zitnik M, Agrawal M, Leskovec J (2018) Modeling polypharmacy side effects with graph convolutional networks. Bioinformatics 34(13):i457–i466

Chapter 2
Computational Tools for Comparing Gene Coexpression Networks

Vinícius Carvalho Jardim, Camila Castro Moreno, and André Fujita

Abstract The comparison of biological networks is a crucial step to better understanding the underlying mechanisms involved in specific experimental conditions, such as those of health and disease or high and low concentrations of an environmental element. To this end, several tools have been developed to compare whether network structures are "equal" (in some sense) across conditions. Some examples of computational methods include DCGL, EBcoexpress, DiffCorr, CoDiNA, DiffCoEx, coXpress, DINGO, DECODE, dCoxS, GSCA, GSNCA, CoGA, GANOVA, and BioNetStat. We will briefly describe these algorithms and their advantages and disadvantages.

Keywords Network science · Differential network analysis · Coexpression network · Systems biology · Network theory

2.1 Introduction

To understand complex systems, we need to consider the interactions between their elements. A graph is a useful tool for studying these systems due to the plasticity of network models for interpreting biological problems. In a biological context, network vertices can represent system elements such as proteins, metabolites, genes, among other examples. In coexpression networks, vertices represent genes, while edges represent coexpression between gene pairs.

V. C. Jardim · C. C. Moreno
Interdepartmental Bioinformatics Program, Institute of Mathematics and Statistics - University of São Paulo, São Paulo, Brazil
e-mail: vinicius.jardim.carvalho@usp.br

C. C. Moreno
e-mail: camila.moreno@usp.br

A. Fujita (✉)
Department of Computer Science, Institute of Mathematics and Statistics - University of São Paulo, São Paulo, Brazil
e-mail: andrefujita@usp.br

F. A. B. da Silva et al. (eds.), *Networks in Systems Biology*, Computational Biology 32,
https://doi.org/10.1007/978-3-030-51862-2_2

We define coexpression as the statistical dependence (correlation) between the expression values of two genes. Correlation measures how coordinated the variation of expression values of two genes are in same condition samples (obtained from microarray or RNA-seq analysis). In this chapter, we use the terms *Conditions* or *experimental conditions* as synonyms of experimental treatments, such as of high, mean, and low temperatures or clinical status, such as healthy versus cancer tissues. Usually, correlation allows us to infer whether two genes belong to the same metabolic pathway or biological process. However, it does not imply that one variable influences another. Therefore, the edges that represent the correlations have no direction, constructing undirected networks.

Changes in correlations (edges) between conditions are of interest to many studies. In some cases, the aim is to verify whether the environment or genome variations affect the relationship between genes. Considering that each network represents an experimental condition, to achieve this goal, we need effective means to compare these networks. The scientific community has developed several strategies to accomplish this task, with approaches ranging from verifying the edge's existence in differing conditions to network model comparisons.

The most used correlation measure in coexpression network studies is the Pearson correlation. However, the non-parametric Spearman correlation is also frequently used since it does not demand the assumption of normality and is not limited to only detecting linear correlations. Other strategies use mutual entropy and Bayesian inference to define coexpression between genes [1].

Beyond the choice of correlation methods, it is also vital to select the threshold for a given correlation to become an edge. In this sense, we commonly use two main kinds of techniques. The most used is the hard threshold. It works as a cut-off value to remove correlations that are below a defined value (correlation threshold) or with a predetermined level of significance (p-value threshold). Another strategy is the soft threshold proposed in WCGA paper [2]. The soft threshold ponders (or rescales) the correlation values according to a power value β. This threshold technique works by powering the correlation to a β value: the higher values increase and the lower ones decrease, therefore highlighting the most relevant correlations. At the soft threshold, the network remains complete without edge removal. Once parameters for constructing networks are defined, those such as coexpression criteria and threshold technique, we can compare the resulting networks in many ways.

2.2 Network Comparison Methods

Many studies apply network analysis to compare different experimental conditions. One way is to quantify and compare the structural features of networks such as presence or absence of edges or the number of connections of a vertex [3, 4]. Other strategies look for edges that are exclusive of a condition [5] or identify a differential network resulting from the combination of differential expression analysis (DE) and differential coexpression (DC) [6]. Despite these methods being useful, they do not take into account the intrinsic fluctuations of data to compare the networks, leading to erroneous conclusions about differences between experimental conditions.

Some statistical techniques give reliability to significant differences in network comparison [7]. For instance, data permutation and resampling techniques allow the association of a p-value (probability of significance) to each pair of genes [7] or to compare entire networks [8]. Many tools approach the need to compare networks between different conditions. Here, we present and discuss some of these differential network analysis tools. All the tools presented in this chapter are summarized in Fig. 2.1.

2.2.1 Edge Comparison

Some tools test whether edges (or a group of them) are statistically different between two or more conditions. They usually return, as a result, a list of differentially coexpressed edges. Sometimes, these methods also return a list of enriched vertices belonging to these edges.

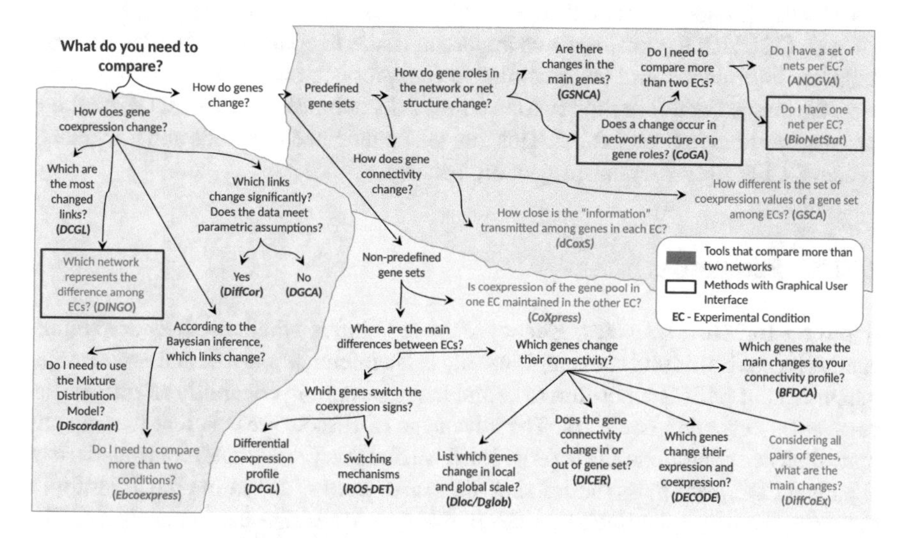

Fig. 2.1 Decision tree of differential coexpression analysis tools. Each tool answers a specific question about the data set. Then, the user has to follow the questions to choose the ideal method. The methods that have a Graphical User Interface (GUI) to perform the analyses are outlined with a dark blue square. The red color indicates that the method compares more than two networks. The blue background color indicates the edge comparison tools; the yellow background indicates the untargeted vertex comparison tools and the green background indicates the targeted vertex comparison tools

2.2.1.1 Diffcorr

To compare two biological conditions, Fukushima (2013) checks whether each edge occurs under both conditions [9]. **Diffcorr** applies a direct and straightforward strategy to define differentially coexpressed edges based on correlations transformed by the Fisher Z-scores method [10]. An advantage of Diffcorr is that it is possible to test each correlation, allowing the user to examine in detail the changes between conditions. One disadvantage is the high number of tests performed, incurring the problem of multiple tests. This method is implemented in the R language and is available in the SourceForge platform (https://sourceforge.net/projects/diffcorr/).

2.2.1.2 DCGL

Liu et al. [11] proposed a method to verify which edges are distinct between two conditions after checking the vertices associated with them. **DCGL** performs the Differential Coexpression Enrichment (DCe) analysis, which applies the limit fold change (LFC) model over each pair of edges in both conditions. This method returns a list of differential coexpression links (DCLs). Moreover, based on DCLs and a binomial model, DCGL selects a set of differentially coexpressed genes (DCGs). According to Yu et al. [12], DCGL considers two important issues to compare networks: the gene neighbor information and the quantitative coexpression change information. However, the comparisons are limited to two networks only. We present DCGL methods based on vertices in Sect. 2.2.2.2. This tool is implemented in R code and is available in cran DCGL (https://cran.r-project.org/package=DCGL).

2.2.1.3 Ebcoexpress

Based on Bayesian statistics, **Ebcoexpress** [13] infers whether edges are significantly changed. This tool uses empirical Bayesian inference and a nested expectation-maximization (NEM) algorithm to estimate the posterior probability of differential correlation between gene pairs. The advantage of EBcoexpress is that it compares more than two conditions and to provide a false discovery rate (FDR) controlled list of significant DC gene pairs minimizing the loss of power. Regarding the algorithm's run-time, there is a restriction on the number of genes that can be analyzed. The authors recommend 10,000 genes as a limit. Besides, it must be verified if genes have high correlations among them and remove these highly correlated genes pairs to avoid false-positive detections. EBcoexpress is implemented in R code and is available in the Bioconductor repository (https://bioconductor.org/packages/EBcoexpress/).

2.2.1.4 Discordant

Similar to Ebcoexpress (Sect. 2.2.1.3), the **Discordant** tool [14] also uses empirical Bayesian inference and the expectation-maximization technique to estimate the posterior probability and identify differential coexpression of gene pairs. According to Siska et al. [14], Discordant fits a mixture distribution model based on Z-scores of correlations. This technique allows Discordant to detect more types of differential coexpression scenarios than EBcoexpress. It also outperforms the Ebcoexpress method in computational time and accuracy [15]. To reduce the computational time, it assumes that the expression levels of gene pairs are independent and bivariate distributed. However, this assumption is not biologically probable. Discordant is implemented in R and is available in the Bioconductor repository (http://bioconductor.org/packages/discordant/).

2.2.1.5 DGCA

DGCA [16] identifies sets of genes as differentially correlated. It classifies differential correlation into nine possible scenarios. As does Diffcorr, DGCA also applies the transformed correlations by the Fisher Z-scores method [10]. However, DGCA differs from the existing differential correlation approaches since it calculates the FDR of differential correlation using nonparametric sample permutation and calculates the average difference in correlation between one gene and a gene set across two conditions. The permutation tests also minimize parametric assumptions. One disadvantage of the DGCA methodology is that it can compare only two experimental conditions at a time. DGCA is implemented in R code and is available in the CRAN repository (https://cran.r-project.org/package=DGCA).

2.2.1.6 DINGO

The last edge comparison tool detects differentially coexpressed edges by the use of a Gaussian Graphical Model (GGM). **DINGO** [17] calculates a global component (graph), composed of common edges between two conditions. Based on this global component, the algorithm determines specific local components for each condition. It attributes a score to each edge, determining how altered they are between conditions. Then, DINGO selects the edges that have significantly altered scores. Finally, the tool returns a single network that has significantly altered edges between the studied conditions. DINGO estimates conditional dependencies for each group. The newer version of DINGO, iDINGO, is an R package and is available in the CRAN repository (https://cran.r-project.org/web/packages/iDINGO/).

2.2.2 *Untargeted Vertex Comparison*

Instead of comparing edges among conditions, other tools look for subsets of genes that are differentially coexpressed between different conditions. It is possible to classify vertex comparison methods into two subgroups according to the applied strategy: untargeted and targeted (adapted from [18]). Untargeted approaches search for non-predefined gene sets. This strategy is based on grouping genes into modules according to their coexpression status under the compared conditions. We present targeted approaches in Sect. 2.2.3.

2.2.2.1 coXpress

The first untargeted method presented here is **coXpress** [19]. This methodology detects a gene set that is highly correlated in one condition and tests whether the other condition maintains the same genes in the strongly connected group. Based on hierarchical clustering, it groups the vertices in one condition and calculates a t-statistic for this group. A gene set is differentially coexpressed between two conditions when t is statistically significant in one condition, but not in the other. It also detects which gene pairs changed their correlation among networks and can compare more than two experimental conditions. The t-statistic allows coXpress to state whether the formation of a group is a random process. However, it considers that each gene belongs to only one group, as opposed to what actually occurs in biological systems, where genes generally participate in more than one process. The R package for this method is available at the coXpress website (http://coxpress.sourceforge.net/).

2.2.2.2 DCGL

The method proposed by Liu et al. [11] identifies differentially coexpressed genes using the differential coexpression profile (DCp) method [12]. Unlike the DCe method presented in Sect. 2.2.1.2, DCp is based on the coexpression profile of each vertex with all other vertices. It measures whether the average coexpression of a gene with its neighbors changes between conditions. Besides DCp and DCe, the authors implemented three other methods of gene connectivity measures to perform differential coexpression analysis: log-ratio of connections (LRC), average specific connection (ASC), and weighted gene coexpression network analysis (WGCNA). According to the authors, DCp and DCe detect if the coexpression of genes pairs changes from a positive to a negative sign, while other methods are focused on gene connectivity. All five comparison algorithms mentioned are limited to comparing two networks. This method is implemented as an R package available in the CRAN repository (http://cran.r-project.org/web/packages/DCGL/index.html).

2.2.2.3 DiffCoEx

Tesson et al. [18] proposed a method based on the dissimilarity matrix between two correlation matrices of each experimental condition. **DiffCoEx** performs the Topological Overlap Method (TOM) [2] on the dissimilarity matrix resulting in a list of altered genes. Then, it enriches this list according to biological pathways. Furthermore, DiffCoEx does not need to detect a coexpressed module in one condition to verify if this module is coherent in another; instead, it determines the differential coexpression based only on the dissimilarity matrix. Aside from this, DiffCoEx also has a similar algorithm (based on the dissimilarity matrix) that allows comparing more than two networks. This method is implemented in R and is available on the DiffCoEx website (https://rdrr.io/github/ddeweerd/MODifieRDev/man/diffcoex.html).

2.2.2.4 DICER

Amar et al. [20] state that the main differences among biological conditions occur more frequently between modules of genes than within them. Thus, **DICER** [20] classifies a set of genes as differentially coexpressed (DC) if its set of altered correlations fits in at least one of the following two scenarios: the DC cluster and Meta-module. The DC cluster is a gene set in which genes correlations are statistically different between experimental conditions. Meta-modules are the pairs of gene sets that are highly correlated within the gene sets and have high dissimilarity between them comparing two experimental conditions. The differentiation of these two scenarios allows the user to know which kinds of differences (DC cluster or Meta-module) the system has between conditions. DICER is implemented in Java code and is freely available for download at the DICER website (http://acgt.cs.tau.ac.il/dicer/).

2.2.2.5 DCloc and DCglob

Bockymayr et al. [21] developed two untargeted algorithms, **DCloc** and **DCglob**, that identify differential correlation patterns by comparing the local or global structure of correlation networks. The construction of networks from correlation structures requires fixing a correlation threshold. Instead of a single cutoff, the algorithms systematically investigate a series of correlation thresholds and permit the detection of different kinds of correlation changes at the same level of significance: great changes of a few genes and moderate changes of many genes. Using random subsampling and cross-validation methods, DCloc and DCglob identify accurate lists of differentially correlated genes. The codes to run each function are in R code and are available in additional files of the article (https://www.ncbi.nlm.nih.gov/pmc/articles/PMC3848818/).

2.2.2.6 BFDCA

BFDCA (Bayes Factor Approach for Differential Coexpression Analysis) [22] aims to detect gene sets that possess different distributions of gene coexpression profiles between two different conditions. It first estimates the differential coexpression of gene pairs based on Bayes factors. Then, it infers DC modules with higher Bayes factor edges and selects significant DC gene pairs based on the vertex and edge importance. BFDCA provides a relatively small number of gene pairs, which can lead to a high-accuracy classifier [22]. One limitation of BFDCA is the small sample problem: the method needs enough samples to estimate the hyperparameters. BFDCA is implemented in a comprehensive R package freely available for download at this website (http://dx.doi.org/10.17632/jdz4vtvnm3.1).

2.2.2.7 ROS-DET

Kayano et al. [23] consider that other approaches have problems under three real cases regarding experimental data: 1) when there are outliers, 2) when there are expression values with a tiny range and 3) when there is a small number of samples. The authors proposed the **ROS-DET** (RObust Switching mechanisms DETector), a detector of switching mechanisms. This switch is the alteration of the correlation signal between two conditions. The ROS-DET overcomes these three problems while keeping the computational complexity of current approaches. ROS-DET is implemented in shell script and is available on this website (https://www.bic.kyoto-u.ac.jp/pathway/kayano/ros-det.htm).

2.2.2.8 DECODE

Lui et al. [24] proposed to combine differential expression (DE) and coexpression (DC) analysis. **DECODE** identifies characteristics not detected through DC or DE approaches alone. This tool combines both strategies and performs a Z-test to select significant differences in coexpression between two conditions [24]. The main advantage of DECODE is that it combines both strategies, which allows it to detect differential coexpression scenarios not detected by other methods. DECODE is implemented in R code and is available in the CRAN repository (http://cran.r-project.org/web/packages/decode/index.html.).

2.2.3 Targeted Vertex Comparison

Following the classification cited in Sect. 2.2.2 (adapted from [18]), targeted approaches compare predefined genes modules according to previous knowledge about the studied system.

2.2.3.1 dCoxS

dCoxS is a targeted strategy because it compares and tests whether predefined gene sets are differentially coexpressed between experimental conditions [25]. It verifies if the Interaction Score (IS) between two gene groups changes among conditions. dCoxS calculates the relative entropy among genes to build the coexpression network. Considering that adjacency matrices represent the networks, the distance between them is measured by the correlation method. dCoxS is unique in that it applies entropy as coexpression measure and correlation as a distance measure between two adjacency matrices. It is implemented in R and is available on this website (http://www.snubi.org/publication/dCoxS/index.html.).

2.2.3.2 GSCA

Choi et al. [26] proposed a method to perform a network comparison based on the distance between adjacency matrices. **GSCA** constructs the adjacency matrices by calculating the correlation measure and compares them using Euclidean distances. If the distance is statistically significant, tested by the permutation samples technique, the two networks are classified as differentially coexpressed. Besides this, the GSCA method has a generalization for more than two conditions: for this, GSCA calculates the average of pairwise distances between correlation matrices. As GSCA does not correct the comparisons for multiple tests, comparing many experimental conditions could lead to false-positive results. The GSCA package is implemented in R and is available on the GSCA website (https://www.biostat.wisc.edu/ kendzior/GSCA/).

2.2.3.3 GSNCA

The comparison of network structures is a strategy employed by this and the following three tools. Rahmatallah et al. [27] state that the significant changes in a system occur at the most critical vertices of the network. Based on this statement, the **GSNCA** tool tests whether there are differences between the vertex's weight vectors given by the eigenvector centrality. Centrality is the weight of a vertex according to its position in a network. The eigenvector centrality determines a vertex centrality by the centralities of its neighbors pondered by the strength of the connections [28]. In other words, the method tests whether the most critical vertices of the network (higher eigenvector centralities) change between conditions. However, this method only compares two experimental conditions. The GSNCA implementation in R is available upon request from the authors.

2.2.3.4 CoGA

Santos et al. (2015) [29] proposed a statistical method to compare the graph spectrum of correlation networks. The spectrum of a graph is the probability distribution of eigenvalues of the adjacency matrix, which represents the network, also called spectral distribution. Based on this measure, **CoGA** (Coexpression Graph Analyzer) tests the equality between the spectral distributions of two networks. CoGA also compares two networks by other structural measures, such as spectral entropy (the entropy of spectral distribution), centralities, clustering coefficient, and distribution of vertex degrees. However, as GSNCA, this method is also restricted to the comparison of only two conditions. CoGA was implemented in R code. It is available on the CoGA website (https://www.ime.usp.br/ suzana/coga/) and can be used with graphical interface features to perform the analysis easily.

2.2.3.5 ANOGVA

To solve the problem of being limited to comparing only two experimental conditions, Fujita et al. [30] generalized CoGA statistics (Sect. 2.2.3.4). The **ANOGVA** (ANalysis Of Graph VAriability) compares graph populations through spectral distributions using the Kullback-Leibler divergence, much like an ANOVA method for graphs. This tool is useful for comparing two or more sets of networks, such as functional brain networks, where each sample has one network and, consequently, each experimental condition has many networks. However, this method does not compare experimental conditions that have only one coexpression network each. ANOGVA is implemented in R code and is available in the package statGraph (http://www.ime.usp.br/ fujita/software.html).

2.2.3.6 BioNetStat

Finally, **BioNetStat** [31] generalizes the graph spectrum comparison performed by CoGA (Sect. 2.2.3.4), without the necessity of graph populations, as ANOGVA. Also, it compares networks by spectral entropy, vertex centralities, clustering coefficient, and degree distribution. BioNetStat also performs statistical tests for each vertex (centralities and clustering coefficient), highlighting which vertices differ among networks. As we have other methods that compare the spectral distribution of networks (CoGA Sect. 2.2.3.4 and ANOGVA Sect. 2.2.3.5), BioNetStat has a restriction for the number of genes. A high number of genes—over 5,000 —slows the algorithm since it has to find the eigenvalues of the adjacency matrix, which is time-consuming for larger data sets. BioNetStat is an R package and is available in the Bioconductor repository (https://bioconductor.org/packages/BioNetStat/). It is also possible to perform this analysis behind a Graphical User Interface.

2.3 Conclusion

As was shown, there are many methods based on a wide range of strategies to compare coexpression networks. Beyond just coexpression, it is also possible to determine the correlation between metabolites and protein concentrations, applying all methodologies mentioned above to metabolites or protein networks. Unfortunately, most of these methods are only readily usable for those who have some prior knowledge in programming, specifically in the R language. For this reason, researchers and developers should provide a graphical user interface for their methods, both improving the reach of these tools and increasing the number of data analysis tools available for nonprogramming scientists.

Acknowledgements This work was partially supported by FAPESP (2018/21934-5, 2019/03615-2), CAPES (Finance Code 001), CNPq (303855/2019-3), Alexander von Humboldt Foundation, Newton Fund, and The Academy of Medical Sciences.

References

1. Santos SDS, Takahashi DY, Nakata A, Fujita A (2013) Brief Bioinform **15**(6):906. https://doi.org/10.1093/bib/bbt051
2. Langfelder P, Horvath S (2008) BMC Bioinform 9:559. https://doi.org/10.1186/1471-2105-9-559
3. Caldana C, Degenkolbe T, Cuadros-Inostroza A, Klie S, Sulpice R, Leisse A, Steinhauser D, Fernie AR, Willmitzer L, Hannah MA (2011) Plant J 67(5):869. https://doi.org/10.1111/j.1365-313X.2011.04640.x
4. Weston DJ, Karve AA, Gunter LE, Jawdy SS, Yang X, Allen SM, Wullschleger SD (2011) Plant, Cell Environ **34**(9):1488. https://doi.org/10.1111/j.1365-3040.2011.02347.x
5. Zhang H, Yin T (2016) Tree Genet Genomes 12(3):61. https://doi.org/10.1007/s11295-016-1016-9
6. Sun SY, Liu ZP, Zeng T, Wang Y, Chen L (2013) Scientific reports 3:2268. https://doi.org/10.1038/srep02268, http://www.pubmedcentral.nih.gov/articlerender.fcgi?artid=3721080&tool=pmcentrez&rendertype=abstract
7. Walley AJ, Jacobson P, Falchi M, Bottolo L, Andersson JC, Petretto E (2012) Int J 36(1):137. https://doi.org/10.1038/ijo.2011.22.Differential
8. Hochberg U, Degu A, Toubiana D, Gendler T, Nikoloski Z, Rachmilevitch S, Fait A (2013) BMC Plant Biol 13(1):184. https://doi.org/10.1186/1471-2229-13-184, http://dx.plos.org/10.1371/journal.pone.0115581://www.biomedcentral.com/1471-2229/13/184
9. Fukushima A (2013) Gene 518(1):209. https://doi.org/10.1016/j.gene.2012.11.028
10. Fisher RA (1915) Biometrika 10(4):507. www.jstor.org/stable/2331838
11. Liu BH, Yu H, Tu K, Li C, Li YX, Li YY (2010) Bioinformatics 26(20):2637. https://doi.org/10.1093/bioinformatics/btq471
12. Yu H, Liu BH, Ye ZQ, Li C, Li YX, Li YY (2011) BMC Bioinform 12(1):315. https://doi.org/10.1186/1471-2105-12-315, http://www.pubmedcentral.nih.gov/articlerender.fcgi?artid=3199761&tool=pmcentrez&rendertype=abstract
13. Dawson JA, Ye S, Kendziorski C (2012) Bioinformatics 28(14):1939. https://doi.org/10.1093/bioinformatics/bts268
14. Siska C, Bowler R, Kechris K (2015) Bioinformatics 32(5):690. https://doi.org/10.1093/bioinformatics/btv633

15. Singh AJ, Ramsey SA, Filtz TM, Kioussi C (2018) Cell Mol Life Sci 75(6):1013. https://doi.org/10.1007/s00018-017-2679-6
16. McKenzie AT, Katsyv I, Song WM, Wang M, Zhang B (2016) BMC Syst Biol 10(1):106. https://doi.org/10.1186/s12918-016-0349-1, http://www.ncbi.nlm.nih.gov/pubmed/27846853%5Cnwww.pubmedcentral.nih.gov/articlerender.fcgi?artid=PMC5111277
17. Ha MJ, Baladandayuthapani V, Do KA (2015) Bioinformatics 31(21):3413. https://doi.org/10.1093/bioinformatics/btv406
18. Tesson BM, Breitling R, Jansen RC (2010) BMC Bioinform 11(1):497. https://doi.org/10.1186/1471-2105-11-497, http://bmcbioinformatics.biomedcentral.com/articles/
19. Watson M (2006) BMC Bioinform 7:509. https://doi.org/10.1186/1471-2105-7-509, http://eutils.ncbi.nlm.nih.gov/entrez/eutils/elink.fcgi?dbfrom=pubmed&id=17116249&retmode=ref&cmd=prlinks%5Cnpapers3://publication/
20. Amar D, Safer H, Shamir R (2013) PLoS Comput Biol 9:3. https://doi.org/10.1371/journal.pcbi.1002955
21. Bockmayr M, Klauschen F, Györffy B, Denkert C, Budczies J (2013) BMC Syst Biol 7(1):78. https://doi.org/10.1186/1752-0509-7-78, http://www.ncbi.nlm.nih.gov/pubmed/23945349%5Cnwww.biomedcentral.com/1752-0509/7/78
22. Wang D, Gu J (2016) Quant Biol 4(1):58. https://doi.org/10.1007/s40484-016-0063-4
23. Kayano M, Takigawa I, Shiga M, Tsuda K, Mamitsuka H (2011) Nucl Acids Res 39(11):1. https://doi.org/10.1093/nar/gkr130
24. Lui TWH, Tsui NBY, Chan LWC, Wong CSC, Siu PMF, Yung BYM (2015) BMC Bioinform 16(1):182. https://doi.org/10.1186/s12859-015-0582-4, http://bmcbioinformatics.biomedcentral.com/articles/
25. Cho SB, Kim J, Kim JH (2009) BMC Bioinform 10(1):109. https://doi.org/10.1186/1471-2105-10-109
26. Choi Y, Kendziorski C (2009) Bioinformatics. https://doi.org/10.1093/bioinformatics/btp502
27. Rahmatallah Y, Emmert-Streib F, Glazko G (2014) Bioinformatics 30(3):360. https://doi.org/10.1093/bioinformatics/btt687
28. Bonacich P (1972) Sociol Methodol 4:176. https://www.jstor.org/stable/270732
29. Santos SDS, De Almeida Galatro TF, Watanabe RA, Oba-Shinjo SM, Marie SKN, Fujita A (2015) PLoS ONE. https://doi.org/10.1371/journal.pone.0135831
30. Fujita A, Vidal MC, Takahashi DY (2017) Front Neurosci 11. https://doi.org/10.3389/fnins.2017.00066, http://journal.frontiersin.org/article/10.3389/fnins.2017.00066/full
31. Jardim VC, Santos SSS, Fujita A, Buckeridge MS (2019) Front Genet 10(1). https://doi.org/10.3389/fgene.2019.00594

Chapter 3
Functional Gene Networks and Their Applications

Hong-Dong Li and Yuanfang Guan

Abstract One essential goal in functional genomics is to understand the functions and functional interactions of genes. The functional interaction between genes can happen in many ways and at different molecular levels, including co-expression, protein–protein interaction, shared sequence motif, etc. The functional interaction supported by such heterogeneous genomic data can be integrated into functional gene networks (FGNs) based on machine learning approaches. In FGNs, a node represents a gene and the edge indicates the probability that two genes co-function in the same pathway or biological process. By addressing the functional difference between isoforms generated from the same gene, recent efforts have focused on building FGNs at the finer isoform level, i.e., functional isoform networks (FINs). In this chapter, we will first present an introduction to FGNs and describe how heterogeneous genomic data can be integrated to build the network by machine learning approaches. We will then describe the refinement of FGNs from global networks to tissue-specific networks and from gene level to isoform level. Finally, we will describe and discuss the applications of FGNs in predicting gene functions and disease genes.

3.1 Introduction

In the post-genomic era, more and more high-throughput biological data, such as protein–protein interaction data and gene co-expression, etc., have been generated. These high-throughput data may solve the problem of large-scale gene function annotation. How to effectively utilize and integrate these high-throughput data is challenging. Functional gene networks (FGNs) represent a powerful method for

H.-D. Li (✉)
Hunan Provincial Key Lab on Bioinformatics, School of Computer Science and Engineering, Central South University, Changsha 410083, Hunan, People's Republic of China
e-mail: hongdong@csu.edu.cn

Y. Guan
Department of Computational Medicine and Bioinformatics, University of Michigan, Ann Arbor, MI 48109, USA

F. A. B. da Silva et al. (eds.), *Networks in Systems Biology*, Computational Biology 32,
https://doi.org/10.1007/978-3-030-51862-2_3

modeling the functional interaction among genes by synthesizing different high-throughput biological data [1–7]. In FGNs, a node is a gene, and the edge represents the probability that the two connected genes co-function in the same biological process.

Troyanskaya et al. [6] proposed a Bayesian-based network approach for building FGNs. This network combines evidence from different functional genomic data such as microarray gene expression to predict whether two proteins are functionally related, namely, being involved in the same biological process or pathway. FGNs are often constructed by integrating different functional genomic data such as gene co-expression data, protein–protein interaction data, protein docking data, gene–phenotype annotation data, chemical and genetic perturbation data, micro-RNA target profiles data, etc. The gold standard of functionally related gene pairs is constructed from Gene Ontology biological process terms and Kyoto Encyclopedia of Genes and Genomes (KEGG) pathways.

There are two types of networks. The Bayesian network [8] constructed by all the genes that do not distinguish tissues can be called the tissue-naive network or the global network, which means there is no correlation between genes and tissues in the network. However, the precise function of genes often depends on their organizational environment, while human diseases are caused by the disordered interaction of specific processes of tissue and cell lineage. Overcoming the influence of the above factors is the key to our comprehending of tissue-specific gene functions and gene disease associations [9]. Therefore, it is necessary to construct tissue-specific network to address tissue specificity [10]. The main difference between global network and tissue-specific network lies in the different construction methods of gold standard for positive and negative samples (gene pairs). In the construction of global functional gene networks, the construction of gold standard only uses Gene Ontology biological process terms (GO term) [11] and KEGG pathways [12], that is, to judge whether a gene pair is annotated by the same biological process. And let G_A and G_B represent two genes. A gene pair G_A–G_B contains two genes, if there are co-annotated by the same biological process in the GO term, then the gene pair is a positive sample, and if there are not co-annotated then the gene pair is a negative sample. However, in the construction of the tissue-specific network, the above step is only to complete the gene function mapping. Because of the need to build tissue-specific, it also needs a step of gene-tissue mapping to select genes expressed in the tissue of interest. We select tissue-specific genes by annotating widely expressed genes. The positive and negative samples can be selected according to different methods, such as taking tissue-specific gene pairs with common annotation as positive samples and other gene pairs as negative samples. The Bayesian approach proposed by Troyanskaya et al. has been applied to build FGNs for humans and other model organisms including mouse, elegans, rats, etc. Alternatively, spliced isoforms from the same gene may have different functions and functional interactions. The isoform-level functional interaction could not be modeled by functional networks at the gene level. By addressing this limitation, FGNs have been extended to the isoform level and functional isoform networks (FINs) have been established [13].

In this chapter, we will first describe the computational method for building FGNs and introduce the resources of FGNs that have been constructed for humans and other model organisms and their applications to disease/trait gene prediction. Second, we will describe FGNs at the isoform level and build functional isoform networks on different organisms.

3.2 Functional Networks at the Gene Level

3.2.1 The Bayesian Approach for Building Functional Gene Networks

Gene function prediction and disease gene prediction are important areas in functional genomics. Constructing FGNs by integrating heterogeneous functional genomic data is a promising way for studying gene functions and disease genes. Troyanskaya et al. proposed to build FGNs by integrating a variety of genomic data through Bayesian network classifiers [6]. The construction of functional gene networks can be generally divided into three components (Fig. 3.1): (1) feature matrix construction, (2) compilation of gold standard of functionally related (positive) and unrelated (negative) gene pairs based on gene function annotation databases such as Gene Ontology (GO) biological process terms and KEGG pathways, and (3) building FGNs with the Bayesian integration approach.

Feature matrix construction. We need to perform data pre-processing on the input data. The input data includes gene co-expression data, protein–protein interaction data, protein docking data, gene–phenotype annotation data, chemical and genetic perturbation data, micro-RNA target profiles data, etc. For example, when processing gene co-expression data, for each gene expression datasets, we can calculate the Pearson correlation coefficient (PCC) of gene pairs in all gene expression datasets. Since the distribution of PCCs produced by different gene expression sets is different, it may not be comparable. Therefore, Fisher's z transformation is used to normalize the PCCs in each dataset, so that PCCs are approximately normally distributed and comparable in the dataset [14]. Then, we used the pre-processed data to build the feature matrix.

Compilation of gold standard of functionally related (positive) and unrelated (negative) gene pairs. FGNs are constructed with supervised machine learning methods. So, positive and negative samples (gene pairs in this case) need to be compiled. A pair of genes is positive if the two genes are co-annotated in the same biological process or pathway. It is negative if the two genes do not participate in a biological process or pathway. GO biological process terms and KEGG pathways are often used to construct gold standard.

Building FGNs with the Bayesian integration approach. In this step, the feature matrix that characterizes gene pairs and the gold standard are integrated to build FGNs. For a gene pair G_A–G_B, the probability that the two genes co-function in the

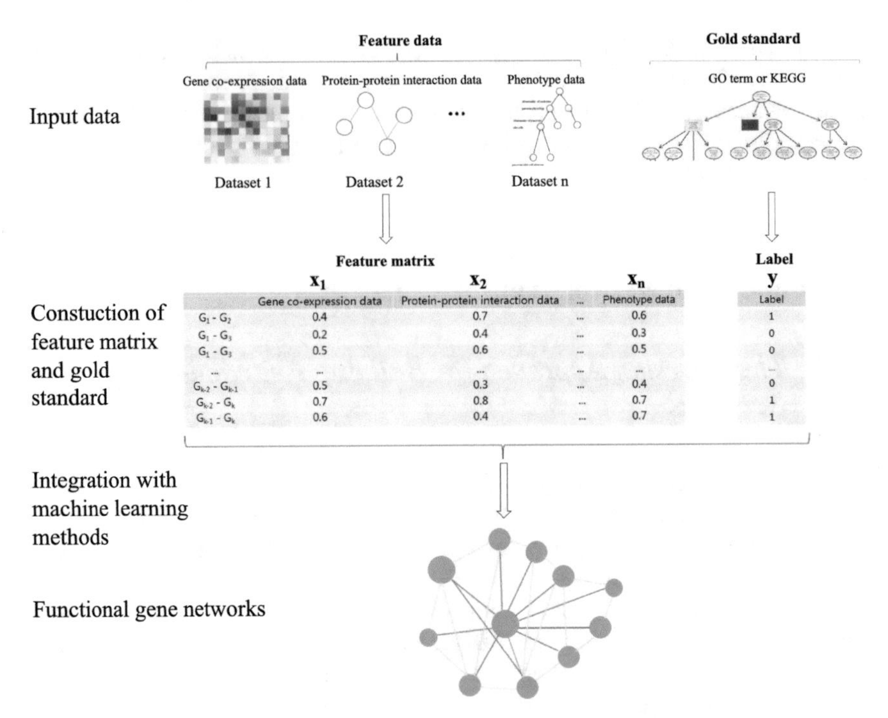

Fig. 3.1 The schematic for building functional gene networks (FGNs). The feature is constructed from various functional genomic data such as gene co-expression data, protein–protein interaction and phenotype annotation data, etc. Functional annotation data such as Gene Ontology terms and KEGG pathways are used to generate gold standard of functionally related (positive) and (unrelated) gene pairs. If a pair of genes G_A–G_B is co-annotated in a GO term or a KEGG pathway, the label is 1; it is 0 otherwise. A machine learning approach, which is often Bayesian network in this context, is applied to train a model. The model is then used to predict the co-functional probability for all pairs of genes, resulting in an FGN. In the network, a node represents a gene and the weight of the edge represents the probability that the two connected genes co-function in the same biological process or pathway

same biological process can be calculated with the Bayesian formula,

$$P(y = 1|x_1, x_2, x_3, \ldots, x_n) = \frac{P(y = 1)\prod_{i=1}^{n} P(x_i|y = 1)}{C},$$

where $P(y = 1)$ is the prior probability that a pair of genes is positive, $P(x_i|y = 1)$, i = 1, 2, 3,…, n is, the probability for the observed value x_i conditioned that the gene pair is positive, and C is a constant. $P(y = 1|x_1, x_2, x_3, \ldots, x_n)$ is the predicted posterior probability that a gene pair is functionally related [14]. In FGNs, the posterior co-functional probability (CFP) is used as the weight of the edge.

3.2.2 *Description of Established Functional Gene Networks*

3.2.2.1 Functional Gene Networks for Humans

FGNs can be used to study the association among genes and the relationship between genes and diseases [15]. A series of studies have contributed to the construction of FGNs for humans. The established FGNs include global functional gene networks (tissue-naive network) and tissue-specific networks.

The construction of the global FGNs for humans is mainly based on the Bayesian network method named MAGIC as proposed in [6]. Huttenhower et al. [10] used the regularized Bayesian method to integrate protein function information, the relationship between diseases, genes and pathways, and the interaction between biological processes to build a global human FGN. The FGN can be used in several ways. For example, the authors showed that protein functions can be predicted with the FGN. As another example, putative functional gene modules could be extracted from the network by detecting tightly connected clusters. One limitation is that the above method does not consider tissue specificity of functional relationship among genes.

However, the exact function of a gene often depends on its tissue environment, while the human disease is caused by the disordered interaction of specific processes of tissue and cell lineage [16–20]. The global human FGNs constructed by the above methods do not consider the relationship between gene expression and tissue environment, so it is not capable of accurately predicting gene functions [10, 19].

Greene et al. built tissue-specific networks for 144 human tissues and cell types with the Bayesian approach. The global FGN assumes that the functional relationship of genes across all tissues remains unchanged. In contrast, tissue-specific networks can capture the functional relationship among genes in different tissues [9]. GIANT mainly uses four kinds of tissue-specific data: protein physical interactions, JASPAR [21] transcription site data, micro-RNA target profiles from the Molecular Signatures Database (MSigDB [22]), and gene expression data from the NCBI Gene Expression Omnibus (GEO) database [23]. Tissue-specific networks were used to quantify molecular interactions between diseases and analyze tissue-specific functional associations. This paper focused on Parkinson's disease and created a functional disease map based on the substantia nigra network [24]. And many documented disease associations were observed. It also verified the tissue-specific molecular response of vascular cells to IL-1β (IL1B) [25] stimulation, which validated that the network can respond to tissue-specific gene functions and response to pathway disturbances. This paper provided a portal (https://hb.flatironinstitute.org/gene) for functional gene networks through multi-gene query, network visualization, including NetWAS analysis tools and downloadable networks. The NetWAS method consists of an organization network classifier that learns network connection patterns related to a target phenotype and predicts genes across the genome [9, 26].

Many of the gene expression data in public databases were generated from cancerous and cell line samples [9, 27]. Including such data for building gene networks may bias the functional relationship among genes. However, the gene expression data are not often curated for building networks. To address this limitation, Li et al. [14] performed an intensive hand curation of non-cancer and brain-specific gene expression data from humans, mice, and rats, and used these data to build a brain-specific functional gene network, called BaiHui. BaiHui uses four kinds of functional genomic data: 213 non-cancerous brain-specific gene expression datasets, protein–protein interaction, protein docking, and gene–phenotype annotation data. BaiHui was then constructed by integrating these genomic features and the gold standard of gene pairs derived from GO biological process terms and KEGG pathways with the Bayesian approach.

3.2.2.2 Functional Gene Networks for Mice

Guan et al. [28] built global FGNs for the mice and then mined the network for phenotype prediction based on machine learning approaches. First, the authors gathered various sources of genomic data to construct a global FGN for the mouse, including protein–protein physical interactions, phylogenetic data, homologous functional relationship predictions in yeast, and expression and tissue localization data. Each input dataset was transformed to pairwise feature values. The gold standard was generated from the Mouse Genome Informatics (MGI) database [29]. The positive samples are defined as pairs of genes known to interact or functionally related. The negative samples are defined as pairs of genes not known to interact or functionally unrelated. With the obtained gold standard and input feature data, a global FGN was built for the mouse through a Bayesian network.

Then, the authors applied machine learning approaches on the basis of FGNs to generate hypotheses of genotype–phenotype associations. Training data of gene–phenotype associations were derived from the mammalian phenotype [30] (MP) ontology annotations. A total of 1157 different phenotypes were analyzed. If any allele is annotated to the phenotype under investigation or its descendent, the gene was considered as positive examples. All other genes were treated as negative examples. The first method considered is support vector machine (SVM) classification. For each sample (genes) in the training data, its co-functional probabilities with all the positive genes were used as features for SVM modeling. Then the trained SVM was used to predict all genes for their association with each phenotype. The predictions for each phenotype were computationally evaluated by bootstrapping. The results showed that FGN-based SVMs are promising for predicting phenotypes. Further, the top predicted associations of *Timp2* and *Abcg8* with bone mineral density were experimentally validated. The results showed that FGNs were capable of identifying potential disease genes that have not been identified by quantitative genetics screens.

However, genetic diseases often present tissue-specific pathologies [18, 19, 31, 32]. Therefore, tissue-specific functional information is essential for drug discovery, and diagnosis, biomarker identification. Hence, tissue-specific functional networks are more accurate to identify the phenotype-associated genes of mouse. Guan et al. [33] proposed a conceptual progress to address tissue specificity of FGNs. For tissue-specific expression, the data are derived from mouse Gene eXpression Database

[34] (GXD), including in situ hybridization data, RT-PCR data, and immunohistochemistry data. These data reflect only the presence or absence of genes in a given tissue. The data for 107 tissues were used for tissue-specific analysis. Tissue-specific gold standard of gene pairs was generated from GXD. For each tissue, specific gold standard positives were defined as pairs of co-annotated genes that are also coexpressed in the same tissue. Negatives were defined as pairs of genes which were coexpressed in the tissue but were not functionally related. Gene expression data [35–37] across tissues were interrogated, and different data sources were integrated using the Bayesian approach to generate tissue-specific FGNs. A total of 107 tissue-specific FRNs were generated. Then, a cross-network comparison metric was developed to identify significantly changed genes between networks. SVM classifiers used these networks as input to predict phenotype-related genes. These tissue-specific networks were shown to be more robust and accurate than global networks for mining phenotype-related genes. Based on the networks, they identified candidate genes associated with male fertility specifically (*Mbyl1*).

3.2.2.3 Functional Gene Networks for Other Model Species

Troyanskaya et al. [6] developed the Multisource Association of Genes by Integration of Clusters (MAGIC), an accurate and efficient gene function annotation tool. MAGIC is a general Bayesian framework that integrates heterogeneous high-throughput biological data and related expert knowledge, leading to more biologically accurate gene grouping and gene function prediction. Based on the fact that pairwise data from non-expression biological data cannot be simply added to microarray clustering methods, *S. cerevisiae* protein–protein interactions and transcription factor binding sites data with clustering analyses of a stress-response microarray dataset were used to verify the utility of MAGIC's integrated analysis of heterogeneous biological data. To avoid double-counting, MAGIC implements a probability weighting strategy on different data sources, which can be obtained through relevant expert knowledge. For each pair of predicted proteins or functional relationships, MAGIC calculates the posterior probability and determines whether or not it is involved in the same biological process (the definition of biological process is consistent with the Gene Ontology), and allows users to set thresholds to change the rigor of the prediction. Considering the relative sparseness of non-microarray experimental data, some independent assumptions made by MAGIC are unlikely to affect the results, and the different basic principles of the data will also make their combinations robust to functional inference. MAGIC is good at solving the problem of identifying an "ideal" clustering algorithm on microarray data and provides great flexibility and simplicity to easily include new methods and data sources from different organisms. Due to the statistical and biological constraints of quantitative genetics approaches, a large number of disease-related genes may be lost. Network-based methods can help identify these missing genes.

Yao et al. [38] developed disease-associated Quantitative Unbiased Estimation across Species and Tissues (diseaseQUEST), a method that systematically combines calculation methods with experimental methods, as well as combining experimental tools in model organisms. By combining human quantitative genetics with in silico functional network representations of model-organism biology, high-throughput behavioral analysis can be conducted with large-scale human genetics research to further understand complex diseases, and systematically list disease gene candidates in functionally conserved processes and pathways. These integrated networks encapsulate the functional relationships of specific tissue or cell types that represent direct or indirect synergies between genes in biological pathways. The tissue-specific networks also reflect the associations between diverse tissues and complex diseases, which are critical to the accuracy and interpretation of disease gene prediction. By using *C. elegans* a model system and generating a tissue-specific functional network in a semi-supervised manner, diseaseQUEST can accurately identify disease candidate genes across organ systems by combining only human genome-wide association studies [39] (GWAS). Using experimental tools for high-throughput behavioral testing in *C. elegans* to rapidly determine disease-related genes and quickly screen for age-related diseases, diseaseQUEST identified 45 candidate Parkinson's disease (PD) genes from 13,255 individual worms. These candidate genes appear to be related to human PD biology, enriching genes that are significantly differentially expressed in the substantia nigra of PD patients. Some genes have been experimentally verified to be related to age-dependent motor deficits that reflect the clinical symptoms of PD. In addition, knocking out the best candidate gene bcat-1 and encoding a branched-chain amino acid transferase resulted in C. elegans spastic-like "curls" and exacerbated dopaminergic neuron degeneration induced by α-synuclein-mediated, and reduced BCAT1 expression in the brains of PD patients [38]. Modularity and scalability are the main advantages of diseaseQUEST, which allows researchers to choose different models or diseases based on disease correlation and existing experimental conditions, such as entorhinal cortex-specific network in mice can be combined with Alzheimer's disease GWAS.

3.3 Functional Gene Networks at the Isoform Level

3.3.1 Alternative Splicing

Alternative splicing (AS) [40, 41] refers to the process of generating different mRNA isoforms from one pre-mRNA through selecting different combinations of splice sites. AS is an important mechanism for regulating gene expression and producing proteome diversity. The splicing process is regulated by the interaction of multiple cis-acting sequences and trans-acting factors, including multiple splicing factors including SR and hnRNP family proteins. There are three major types of AS: (1) intron retention; (2) retention or excision of alternative exons; and (3) shifts in $3'$ and $5'$ splice sites that result in exon growth or shorten. The effect of AS on protein structures is also diverse, such as the increase or decrease of one to hundreds of amino acids in the polypeptide chain.

AS is one of the mechanisms that regulate gene expression at the RNA level. A gene produces multiple transcript isoforms through alternative splicing. Each different transcript isoform encodes a protein with a different structure and function, which have specific expressions and functions in different tissues or at different stages of cell differentiation and development. Therefore, AS is an important mechanism for regulating gene expression at the post-transcriptional RNA level. AS is also an important mechanism to increase proteome diversity from a relatively simple genome. The diversity of the proteome is compatible with the complexity of multicellular higher organisms. From the analysis of gene distribution patterns involved in AS, AS mostly occurs on genes involved in complex processes such as signal transduction and expression regulation, such as receptors, signaling pathways (apoptosis), transcription factors, and so on. AS is of great significance for the precise regulation of individual differentiation and development as well as some key physiological processes such as apoptotic and cell excitation.

3.3.2 Methods for Building Functional Isoform Networks

The main challenge for isoform network modeling is the lack of ground-truth functionally related isoform pairs. Moreover, biological functions and pathways are recorded at the gene level rather than at the isoform level. Therefore, the classical method for establishing a gene-level network cannot be directly applied to splice isoforms. Li et al. proposed a single-instance bag multiple-instance learning (SIB-MIL) algorithm based on Bayesian network to build an isoform network [42]. Multi-instance learning (MIL) [43] is a form of semi-supervised learning in which the training data has only incomplete knowledge on the labels. Specifically, instances in MIL are grouped into a set of bags. The labels of the bags are provided, but the labels of instances in the bags are unknown. Here, the isoform pairs are the instances. MIL formulates a gene pair into a bag of multiple isoform pairs with potentially different probabilities of functional relevance. Functionally related gene pairs are called positive bags, in which at least one of the isoform pairs must be functionally related. Functionally unrelated gene pairs are called negative bags, and none of the isoform pairs have no functional relationship. It attempts to learn a classification function that can predict the truly functionally related isoform pairs, called "witnesses" of the positive bags. The algorithm is detailed in the following.

Firstly, a set of isoform pairs in the positive gene bag are selected as witnesses for the initial model. Due to the fact that the isoform pairs in single-instance positive gene pair bags must be positive, all instances in these bags are labeled as type 1 (functionally related), and the negative bags are marked as type 0 (functionally unrelated). Secondly, an initial Bayesian network classifier is built based on the current witness set and the negative isoform pairs. The instance with the highest probability score is then selected as the "witness" and labeled as a class 1. For each negative bag, only the highest scoring instance is marked as class 0. When the cross-validation performance no longer changes, we have a final classifier based on isoform, which will be used to predict isotype networks.

This paper demonstrates that isoform networks can reveal functional differences of the isoforms of the same gene, which can help deepen our understanding of gene functions and functional relationships and may provide useful information about diseases caused by alternative splicing. Compared to traditional gene networks, isoform networks provide higher resolution views of the functional relationship among genes [42].

3.3.3 Functional Isoform Networks for Humans

Functional isoforms networks of highest connected isoforms (HCIs) can be an efficient and accurate isoform function annotation approach in databases. Li et al. [13] gathered various sources of heterogeneous genomic data as inputs of network, including heart, kidney, and liver of human, totaling 940 RNA-seq samples. Gold standard was obtained from Kyoto Encyclopedia of Genes and Genomes (KEGG) pathways and branch annotations of GO terms. They try to avoid situations that annotations are too specific or too general. Only term/pathways including at least five and fewer than three hundred annotated genes were considered. If gene pairs are co-annotated to a particular term/pathways, gene pairs were regarded as gold standard positives. Because there is no gold standard for negative gene pairs, negative gene pairs are randomly produced from the entire genome space. However, isoform networks constructing faced challenge. Because there is lack of functional annotation, networks cannot use Bayesian networks and supervised learning methods directly. To solve this problem, they used multiple-instance learning [43] (MIL) to predict isoform-level networks. MIL integrates functional annotation data of gene level and genomic features of isoform level.

The integrated network (AUC = 0.62) revealed that the prediction accuracy is higher than a single feature network (AUC = 0.53). The pairs of transcript isoforms (ABCC3, RBM34, ERBB2, and ANXA7) shown that functional isoform networks of the same gene have large differences. From the network, 6157 HCIs were identified from multi-isoform genes. The transcript-level expression of HCIs was significantly higher than non-highest connected isoforms (NCIs). The differential expression of HCIs and NCIs was observed. In functional isoform networks, HCIs have the strongest interactions and are complementary to the major or dominant isoforms.

3.3.4 Functional Isoform Networks for Mice

In studying of alternative splicing events, the major mammalian species utilized is the laboratory mouse. There are significant differences between isoform FRNs and

Tseng et al. [44] work in computational methods and research contents. The isoform FRNs, described the probability of splice isoforms that participate in the same GO term or KEGG pathway, has recently been studied in the mouse.

Li et al. [45] gathered various sources of heterogeneous genomic data as inputs of network, including encompassing 11 RNA-seq datasets, 52 exon array datasets, 1 protein docking dataset, and 1 amino acid composition dataset, totaling 175 RNA-seq datasets. They used Bayesian network-based MIL to build isoform-level networks. MIL integrates functional annotation data of gene level and genomic features of isoform level. The transcript-level expression of 65% of the HCIs was significantly higher than NCIs. At the protein level, these HCIS were highly overlapped with the expressed splice variants, according to proteomic data of eight different normal tissues.

Li et al. [42] constructing network used different sources of heterogeneous genomic data. Data included RNA-seq data, exon array, protein docking, and pseudo-amino acid composition. The resulting data size is not sufficient to serve as the gold standard of isoform networks. Therefore, Bayesian network-based MIL algorithm was used to solve the prediction problem of isoform functional network. The authors applied a high-resolution view of FRNs at isoform level to reveal diverse functions of isoforms of the same gene (such as *Anxa6*).

Kandoi and Dickerson [46] used different RNA-Seq datasets of tissue-specific and sequence information as the input of tissue-specific FRNs. There is lack of isoform-level gold standard. If single-isoform genes were co-annotated to GO term, BioCyc pathways, KEGG pathways and protein–protein interactions, the genes were regarded as positive pair. And the randomly selected negative pairs of isoforms were assumed to decrease bias. They followed a leave-one-tissue-out strategy and trained a random forest model to predict the functional network. Therefore, the isoform-level functional network prediction problem was transformed into a simple supervised learning task. The network revealed tissue-specific functional differences for the isoforms of the same gene.

3.4 Conclusion

Functional gene networks (FGNs) represent an important model to characterize the complex functional interaction among genes. FGNs are often constructed by integrating various high-throughput biological data. Based on tissues FGNs are built on, FGNs can be divided into global (tissue-naive) and tissue-specific ones. Tissue-specific networks can more accurately capture the interaction between genes. By addressing the functional diversity of isoforms generated from the same gene through alternative splicing, FGNs have been extended to the isoform level and functional isoform networks have been established. Functional gene networks have been shown promising in predicting gene functions and disease genes. It is expected that more accurate networks will be constructed in the future and the networks will find more applications.

References

1. Myers CL et al (2005) Discovery of biological networks from diverse functional genomic data. Genome Biol 6(13):R114
2. Chen Y, Xu D (2004) Global protein function annotation through mining genome-scale data in yeast Saccharomyces cerevisiae. Nucleic Acids Res 32(21):6414–6424
3. Jiang T, Keating AE (2005) AVID: an integrative framework for discovering functional relationships among proteins. BMC Bioinfor 6(1):136
4. Jansen R et al (2003) A Bayesian networks approach for predicting protein-protein interactions from genomic data. Science 302(5644):449–453
5. Lee I et al (2004) A probabilistic functional network of yeast genes. Science 306(5701):1555–1558
6. Troyanskaya OG et al (2003) A Bayesian framework for combining heterogeneous data sources for gene function prediction (in *Saccharomyces cerevisiae*). Proc Natl Acad Sci 100(14):8348–8353
7. Myers CL, Troyanskaya OG (2007) Context-sensitive data integration and prediction of biological networks. Bioinformatics 23(17):2322–2330
8. Pearl J (2014) Probabilistic reasoning in intelligent systems: networks of plausible inference. Elsevier
9. Greene CS et al (2015) Understanding multicellular function and disease with human tissue-specific networks. Nat Genet 47(6):569
10. Huttenhower C et al (2009) Exploring the human genome with functional maps. Genome Res 19(6):1093–1106
11. Ashburner M et al (2000) Gene ontology: tool for the unification of biology. Nat Genet 25(1):25–29
12. Kanehisa M, Goto S (2000) KEGG: kyoto encyclopedia of genes and genomes. Nucleic Acids Res 28(1):27–30
13. Li H-D et al (2015) Functional networks of highest-connected splice isoforms: from the chromosome 17 human proteome project. J Proteome Res 14(9):3484–3491
14. Li H-D et al (2019) BaiHui: cross-species brain-specific network built with hundreds of hand-curated datasets. Bioinformatics 35(14):2486–2488
15. Franke L et al (2006) Reconstruction of a functional human gene network, with an application for prioritizing positional candidate genes. Am J Human Gene 78(6):1011–1025
16. D'Agati DV (2008) The spectrum of focal segmental glomerulosclerosis: new insights. Current Opin Nephrol Hyperten 17(3):271–281
17. Cai JJ, Petrov DA (2010) Relaxed purifying selection and possibly high rate of adaptation in primate lineage-specific genes. Genome Biol Evolut 2:393–409
18. Lage K et al (2008) A large-scale analysis of tissue-specific pathology and gene expression of human disease genes and complexes. Proc Natl Acad Sci 105(52):20870–20875
19. Winter EE et al (2004) Elevated rates of protein secretion, evolution, and disease among tissue-specific genes. Genome Res 14(1):54–61
20. Lee Y-S et al (2018) Interpretation of an individual functional genomics experiment guided by massive public data. Nat Methods 15(12):1049–1052
21. Portales-Casamar E et al (2010) JASPAR 2010: the greatly expanded open-access database of transcription factor binding profiles. Nucleic Acids Res 38(suppl_1):D105–D110
22. Subramanian A et al (2005) Gene set enrichment analysis: a knowledge-based approach for interpreting genome-wide expression profiles. Proc Natl Acad Sci 102(43):15545–15550
23. Barrett T et al (2012) NCBI GEO: archive for functional genomics data sets—update. Nucleic Acids Res 41(D1):D991–D995
24. Forno LS (1988) The neuropathology of Parkinson's disease. In: Progress in Parkinson research. Springer, pp 11–21
25. Kofler S et al (2005) Role of cytokines in cardiovascular diseases: a focus on endothelial responses to inflammation. Clin Sci 108(3):205–213

26. Wong AK et al (2018) GIANT 2.0: genome-scale integrated analysis of gene networks in tissues. Nucleic Acids Res 46(W1):W65–W70
27. Hu J et al (2010) Computational analysis of tissue-specific gene networks: application to murine retinal functional studies. Bioinformatics 26(18):2289–2297
28. Guan Y et al (2010) Functional genomics complements quantitative genetics in identifying disease-gene associations. PLoS Computat Biol 6(11)
29. Bult CJ et al (2008) The Mouse Genome Database (MGD): mouse biology and model systems. Nucleic Acids Res 36(suppl_1):D724–D728
30. Smith CL et al (2005) The Mammalian phenotype ontology as a tool for annotating, analyzing and comparing phenotypic information. Genome Biol 6(1):R7
31. Goh K-I et al (2007) The human disease network. Proc Natl Acad Sci 104(21):8685–8690
32. Chao EC, Lipkin SM (2006) Molecular models for the tissue specificity of DNA mismatch repair-deficient carcinogenesis. Nucleic Acids Res 34(3):840–852
33. Guan Y et al (2012) Tissue-specific functional networks for prioritizing phenotype and disease genes. PLoS Computat Biol 8(9)
34. Smith CM et al (2007) The mouse gene expression database (GXD): 2007 update. Nucleic Acids Res 35(suppl_1):D618–D623
35. Su AI et al (2004) A gene atlas of the mouse and human protein-encoding transcriptomes. Proc Natl Acad Sci 101(16):6062–6067
36. Zhang W et al (2004) The functional landscape of mouse gene expression. J Biol 3(5):21
37. Siddiqui AS et al (2005) A mouse atlas of gene expression: large-scale digital gene-expression profiles from precisely defined developing C57BL/6J mouse tissues and cells. Proc Natl Acad Sci 102(51):18485–18490
38. Yao V et al (2018) An integrative tissue-network approach to identify and test human disease genes. Nat Biotechnol 36(11):1091–1099
39. Liu JZ et al (2010) A versatile gene-based test for genome-wide association studies. Am J Human Genet 87(1):139–145
40. Graveley BR (2001) Alternative splicing: increasing diversity in the proteomic world. Trends Genet 17(2):100–107
41. Modrek B, Lee C (2002) A genomic view of alternative splicing. Nat Genet 30(1):13–19
42. Li H-D et al (2016) A network of splice isoforms for the mouse. Scienti Report 6:24507
43. Maron O, Lozano-Pérez T (1998) A framework for multiple-instance learning. Adv Neural Infor Process Syst 570–576
44. Tseng Y-T et al (2015) IIIDB: a database for isoform-isoform interactions and isoform network modules. BMC Genom S10. Springer
45. Li HD et al (2014) Revisiting the identification of canonical splice isoforms through integration of functional genomics and proteomics evidence. Proteomics 14(23–24):2709–2718
46. Kandoi G, Dickerson JA (2019) Tissue-specific mouse mRNA isoform networks. Scienti Reports 9(1):1–24

Chapter 4
A Review of Artificial Neural Networks for the Prediction of Essential Proteins

Kele Belloze, Luciana Campos, Ribamar Matias, Ivair Luques, and Eduardo Bezerra

Abstract Identifying essential proteins is vital to understanding the minimum requirements for maintaining life. The correct identification of essential proteins contributes to guide the diagnosis of diseases and to identify new drug targets. The continuous advances of experimental methods contribute to generate and accumulate gene essentiality data that facilitates computational methods. Machine learning methods focused on predicting essential proteins have gained much traction in recent years. Among them, we can highlight artificial neural networks. Most of these methods make use of sequence and network-based features. In this chapter, we initially present a background related to artificial neural networks, which encompasses different neural network architectures. Then, we review the research papers that used artificial neural networks as a machine learning method for predicting essential proteins.

Keywords Artificial neural networks · Essential proteins · Systematic mapping of the literature

4.1 Introduction

Essential genes correspond to the minimum set of critical genes responsible for its reproduction and survival. Essential genes play a significant role in survival and development of an organism since the identification of these genes is crucial to understand the minimum requirements for maintaining life [34]. Proteins are products of gene

K. Belloze (✉) · R. Matias · I. Luques · E. Bezerra
Centro Federal de Educação Tecnológica Celso Suckow da Fonseca—CEFET/RJ, Rio de Janeiro, Brazil
e-mail: kele.belloze@cefet-rj.br

E. Bezerra
e-mail: ebezerra@cefet-rj.br

L. Campos
Universidade Federal de Juiz de Fora—UFJF, Juiz de Fora, Brazil
e-mail: luciana.campos@ice.ufjf.br

F. A. B. da Silva et al. (eds.), *Networks in Systems Biology*, Computational Biology 32,
https://doi.org/10.1007/978-3-030-51862-2_4

expressions, which are vital material and the functional units of living organisms. Essential proteins are those that act in essential biological functions for organisms to grow and multiply. The research on essential proteins is very attractive and has theoretical and practical values. The correct identification of essential proteins contributes in an important way in order to understand the main biological processes of an organism at molecular level, which is useful to guide disease diagnosis and also identify new drug targets [27].

Essential proteins can be identified by biological experiments such as single-gene knockout [30], transposon mutagenesis [48], RNA Interference [8], and more recently the CRISPR (clustered regularly interspaced short palindromic repeats) gene-editing technology [13]. Although experimental methods are quite accurate, they are expensive and time-consuming. Considering the restrictions in these biological experiments, several computational methods to identify essential proteins have been proposed.

Computational methods are becoming more critical in the study of essential proteins because they can save a lot of time and effort. The first computational methods developed were based on comparative genomics, in which annotations of essentiality of genes or proteins are transferred between species through homology mappings. Subsequently, new methods began using data from essential genes that started to become available in public databases such as Database of Essential Genes (DEG) [47] and Online Gene Essentiality database (OGEE) [7]. The genome-scale data from high-throughput technologies in recent years also supports these methods. In this case, the aim is to study essential genes characteristics and develop machine learning methods for essential proteins prediction [31].

The machine learning models applied to the identification of essential proteins are classifiers, whose construction and training are based on the known characteristics of essential and non-essential proteins. Once fitted, such classifiers can be applied to a set of unlabeled proteins to predict their essentiality. Machine learning methods normally used to fit such classifiers include Artificial Neural Networks (ANN), Support Vector Machines (SVM), Random Forests (RF), Decision Trees (DT), Naïve Bayes (NB), Logistic Regression (LR), Bayesian Networks (BN), and others.

In order to develop a machine learning model for predicting essential proteins, a four-step procedure is usually applied. The first step refers to selecting the appropriate features to build the datasets for training and testing the classification models. The second step refers to the classifier training stage that uses the training dataset gathered in the previous step. Once the classifier is trained, the third involves conducting an evaluation of its predictive performance, using a previously held out validation dataset in step one. Finally, in the last step, the classifier is used to make predictions on proteins whose essentiality is unknown [28].

A wide range of features is associated with gene essentiality. Generally, features can be divided into two groups: sequence-based features and context-dependent features [29, 34, 43]. Except for sequence-based features, which can be obtained directly from gene or protein sequences, the others require pre-computed experimental data [31]. The features of both groups can be combined in order to improve the prediction performance. However, different combinations of features may be necessary, according to each studied organism. Besides, the combination of features can make the labor more complex [34]. Selecting the most suitable features for the studied organism to predict essential proteins is still an issue that needs further research [29].

Sequence-based features are information about the gene or protein sequence. As an example, we can cite GC content (gene sequence with a high GC content is considered more stable) [37], amino acid usage (frequencies of the 20 amino acids in each gene) [37], codon usage (codon frequencies in all genes) [21], and protein length (essential genes compared to non-essential genes have a significantly higher proportion of large and small proteins compared to medium-sized proteins) [16].

Context-dependent features include domain properties (a function of protein domains or combinations of domains) [11], protein localization (essential proteins exist in the cytoplasm with a higher proportion compared to non-essential proteins) [33], Gene Ontology (GO) (the GO terms related to cell localization and biological process indicate reliable predictors of essential genes) [1], gene expression (genes whose levels of expression are higher and more stable under certain conditions are more likely to be essential) [20], and protein–protein interaction network (genes or their protein products are connected. Essential genes tend to be more highly connected when compared to non-essential genes in a protein–protein interaction network. Network topology features such as Degree Centrality (DC), Betweenness Centrality (BC), Closeness Centrality (CC), Eigenvector Centrality (EC), and Subgraph Centrality (SC) have been used for essential proteins identification) [2].

In this manuscript, we present background about artificial neural networks and a review of the research papers that applied such technique as a machine learning method for predicting essential proteins. The choice for this type of method is due to its ability to learn non-linear and complex relationships between inputs and outputs. Added to this, the observation in the literature of recent publications on the topic shows us ways for future works.

This chapter is organized as follows. In Sect. 4.2, we frame the problem of predicting protein essentiality as a classification problem. We also present basic concepts related to artificial neural networks, as well as descriptions of different neural network architectures. In Sect. 4.3, we describe the systematic mapping of the literature, in which we focus on publications that report on the use of artificial neural network models to predict essential proteins. The works highlighted in the systematic mapping are described emphasizing the features of the input data for the classification models, the neural network architectures adopted, and the performance evaluation measures of the models. Section 4.4 presents our final considerations.

4.2 Background

4.2.1 Essentiality as a Classification Problem

In a general supervised learning setting, we assume the availability of a sample taken from a large (potentially infinite) population. We call this sample a *training set* and denote it as $\{\mathbf{x}^{(i)}\}_{i=1}^{m}$, where each $\mathbf{x}^{(i)}$ is a training example. Each training example $\mathbf{x}$ is vector of features and is associated with a label y. The term *label* is generally used for an element belonging to a finite set of classes $\{1, 2, \ldots, C\}$ and target for a real number, or a more complex structure (i.e., vector, matrix, graph). The goal of a supervised learning algorithm is to learn how to predict y from $\mathbf{x}$.

The problem of predicting essential proteins can be framed as a supervised learning problem. In this case, each element $\mathbf{x}$ in the training set is a representation of a protein. Furthermore, $C = 2$ and $y \in \{0, 1\}$ (0 for labeling non-essential proteins, and 1 for labeling essential ones).

The problem of essentiality is a particularly challenging classification task for two main reasons. First, the available training datasets for such a task are very unbalanced, that is, a vast majority of the training examples is associated to label 0 (non-essential proteins). It is known that this can dramatically skew the predictive performance of classifiers, introducing a prediction bias for the majority class [26]. Second, the structure of each training example is complex: each $\mathbf{x}$ is a variable-length sequence of symbols (each symbol corresponding to a particular amino acid present in the protein represented by $\mathbf{x}$). Besides, the length distribution of proteins is also very skewed.

4.2.2 Artificial Neural Networks

An Artificial Neural Network (ANN) is a computational model loosely based on the biological brain that aims to process information in parallel. The transmission of a signal from one neuron to another inside a biological brain is a complex chemical process (which is far from being well understood), in which specific substances are released by the transmitting neuron. The effect is an increase or a decrease in the electrochemical potential in the body of the receiving cell. If this potential reaches the cell's activation limit, a pulse or action of fixed power and duration is sent to other neurons. The neuron is then said to be *active*. Similar to its biological counterpart, an ANN has a system of *artificial neurons* (also known as *processing units* or simply *units*). Each unit performs a computation based on the activations of other units to which it is connected.

The emergence of ANNs can be attributed to the complexity of implementing systems for solving hard recognition problems based on heuristic rules. As a way to overcome this difficulty, ANN models are fitted in such a way as to receive training data as input and, through successive observations, learn to recognize the underlying patterns. The goal is to generate a model capable of making inferences about data not used during the training phase, but which have similar patterns.

The task of training ANNs is an iterative process that can be divided into two steps: forward and backward pass. In the forward pass, the ANN receives input data, generating an output. This output is then used to compute the value of a cost function. In the backward pass, the generated output is compared with the expected output and the backpropagation algorithm is used to compute the gradients of the cost function with relation to all the weights for every neuron in the ANN [35]. These gradients are finally used to update the parameters (i.e., the weights) of the ANN in the direction that minimizes the value of the cost function. This minimization process commonly uses some variant of the gradient descent method [40].

The remaining of this section describes some neural network architectures commonly used in essential protein identification.

4.2.2.1 Multilayer Perceptrons

The simplest possible ANN is made up of a single artificial neuron. Networks of this nature are quite limited, because they only solve binary decision processes (i.e., in which there are two possible outputs) and linearly separable (e.g., Boolean AND and OR functions). On the other hand, it is possible to build more complex networks (i.e., capable of modeling non-linear separable decision processes) through a procedure of composing simpler computational blocks organized in layers. Multilayer Perceptrons (MLP), also known as vanilla neural networks, are an old family of artificial neural networks in which neurons are organized in layers, with connections between neurons in consecutive layers. Connections between units are weighted by real values called *weights*. Figure 4.1 illustrates an artificial neuron, which is the basic building block of most artificial neural network families, MLPs included.

As in the human brain, neurons are the basic processing unit of an ANN. An artificial neuron receives a number of inputs, and each input has an associated weight.

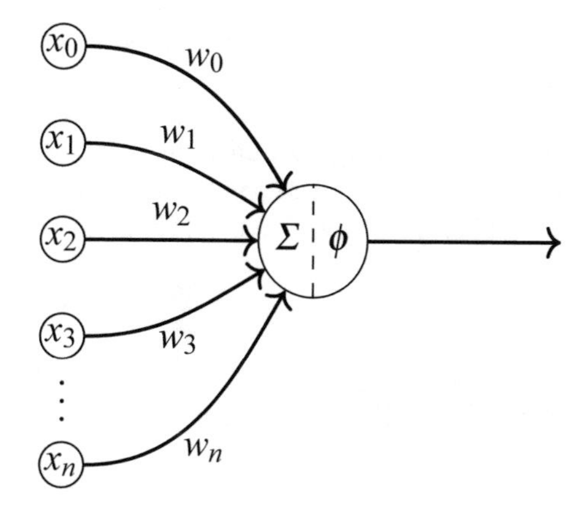

Fig. 4.1 Two simple operations are conducted inside an artificial neuron: preactivation (weighted sum of the input) and activation (a non-linear transformation applied to the result generated by the preactivation)

The neuron then first computes its preactivation, that is, the inner product between the input and weight vectors, as shown in Eq. 4.1. In this equation, n is the number of inputs received by the neuron, x_j is the jth input received, and w_j is the weight associated to this input. Notice that the weight vector has a special component w_0 named bias, which is associated to a fixed input $x_0 = 1$.

$$z = \sum_{j=0}^{n} w_j x_j \tag{4.1}$$

The value of z is then passed as input to the second part of the computation conducted inside a neuron, which uses an activation function. An activation function ϕ is some non-linear function used to provide the neuron with the ability to capture complex patterns in its input. Indeed, it can be shown that a neuron without an non-linear activation function is not able to build non-linear a decision boundary. Sigmoid Logistic (Eq. 4.2a), Hyperbolic Tangent (Eq. 4.2b), and Rectified Linear Unit (ReLU) (Eq. 4.2c) are examples of commonly used activation functions.

$$\phi(z) = \frac{1}{1 + e^{-z}} \tag{4.2a}$$

$$\phi(z) = \frac{e^{z} - e^{-z}}{e^{z} + e^{-z}} \tag{4.2b}$$

$$\phi(z) = \max(0, z) \tag{4.2c}$$

On an MLP, artificial neurons are stacked on layers, and layers are put in a sequence to compose the network architecture. Figure 4.2 schematically illustrates how the artificial neurons are organized in an MLP. There are three types of layers in an MLP: one input layer (the input features representing some object), one output layer (which corresponds to the computation of the network), and at least one hidden layer. Also, notice that an MLP is a kind of feedforward neural network, in which there are no cycles in the connections between neurons. Another characteristic of an MLP is that it is fully connected, meaning that an artificial neuron is connected to all neurons in the previous layer.

Let $\mathbf{a}^{(k-1)}$ be the vector of activations computed by the artificial neurons on the $(k-1)$-th hidden layer. Then, the computation of neurons in the k-th layer can be executed in a vectorized way, according to Eq. 4.3. In this equation, $W^{(k-1)}$ is a matrix which holds the weight vectors for layer $k-1$ in its columns.

$$\mathbf{a}^{(k)} = \phi(W^{(k-1)}\mathbf{a}^{(k-1)}) \tag{4.3}$$

As a whole, given an input vector $\mathbf{x}$, an MLP computes a function $f(\mathbf{x})$ such that $f(\mathbf{x}) : \mathbb{R}^n \rightarrow \mathbb{R}^C$, according to Eq. 4.4.

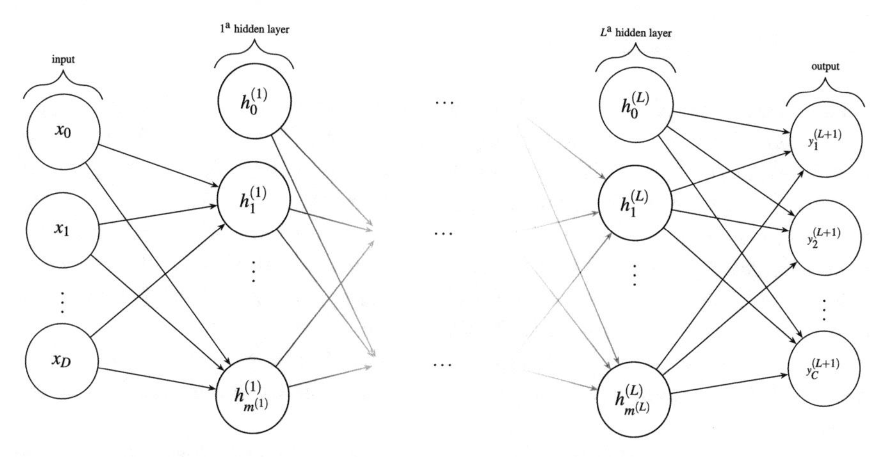

Fig. 4.2 Schema of an MLP with $(L+1)$ layers. There are n units on input layer and C units on the output layer. The l-th hidden layer has $m^{(l)}$ units

$$f(\mathbf{x}) = \mathbf{h}^{(L+1)}(\mathbf{x}) = \phi_{\mathbf{o}}(\mathbf{a}^{(L+1)}(\mathbf{x})) \tag{4.4}$$

In Eq. 4.4, $\phi_{\mathbf{o}}(\cdot)$ represents the activation function used by the units in the output layer. When the purpose of an MLP is to perform the classification task (as is the case in the identification of essential proteins), it is common to select for $\phi_{\mathbf{o}}(\cdot)$ the function called `softmax`, which allows to interpret the values of the output layer as posterior probabilities. The value produced by the `softmax` function for the i-th unit of the output layer is given by Eq. 4.5. The outputs of the units $a_i^{(L+1)}$, $1 \leq i \leq C$, can actually be interpreted as probability values, since they belong to the range [0, 1] and their sum is equal to 1.

$$\phi_{\mathbf{o}}\left(a_i^{(L+1)}\right) = \frac{e^{a_i^{(L+1)}}}{\sum_{j=1}^{C} e^{a_j^{(L+1)}}} \tag{4.5}$$

An MLP with just one hidden layer is proven to be a universal approximator [9]. However, the amount of neurons needed to produce a good approximation for a function might be way too large when compared to an ANN with more hidden layers. In general, an ANN is said to be deep if it has a large number of layers. It has been shown empirically that deep ANNs may be handy to solve complex problems in an efficient way. Indeed, increasing the amount of layers in neural network architectures has been a trend in the last decade and is one of the reasons deep neural networks are the current state of the art in several pattern recognition applications [25].

4.2.2.2 Convolutional Neural Networks

Convolutional Neural Networks (CNNs) are a type of architecture originally designed to process data with grid-like topology (such as images, speech, and time-series tasks) LeCun and Bengio [24]. CNNs are an efficient method for capturing spatial context and have recently attained state-of-the-art results for image classification using a 2D kernel [23]. In recent years, researchers expanded CNN actuation field to natural language processing, such as machine translation [15]. This novel architecture is built on a CNN with a 1D kernel, useful to capture temporal patterns in a sequence of the words to be translated. A CNN with 3D kernel is used to predict the future in visual data, which can be applied to action recognition [41]. In this domain, CNN performs 3D convolution operations over both time and space dimensions of the video.

Inspired by the visual cortex, some artificial neurons in a CNN learn to apply the convolution operation to scan the input data and detect different kinds of patterns located in a local region, called receptive field. A convolution is a set of transformations that apply the same linear transformation of a small local region across the entire input [17]. Formally, a convolution operation of two given functions f and g in terms of a variable v at point t is typically denoted with an asterisk and defined in the discrete domain according to Eq. 4.6 [17]. Here t can be seen as a time offset which allows $g(t-v)$ to slide along the v-axis.

$$s(t) = (f * g)(t) = \sum_{v=-\infty}^{\infty} f(v)g(t-v) \tag{4.6}$$

In CNN jargon, function f is referred to as the input and function g as kernel. Similarly, in Eq. 4.7, the convolution operation applied to more than one axis at a time, for instance, 2-dimensional image I as input, and a 2-dimensional kernel K, is denoted by

$$S(i,j) = (I * K)(i,j) = \sum_{m}\sum_{n} I(m,n)K(i-m, j-n) \tag{4.7}$$

where $I(m,n)$ is the pixel at location (m,n) and $K(i-m, j-n)$ gives the kernel's values (the weights of the artificial neurons responsible for applying the convolution) based on the pixel offsets i and j. In computer vision applications, convolution can be used for image processing to detect vertical and horizontal edges if weights are defined similar to

$$\begin{bmatrix} -1 & 0 & 1 \\ -2 & 0 & 2 \\ -1 & 0 & 1 \end{bmatrix} \text{ and } \begin{bmatrix} 1 & 2 & 1 \\ 0 & 0 & 0 \\ -1 & -2 & -1 \end{bmatrix}$$

respectively ([38]). One of the main advantages of applying CNN is that the kernel weights are learned by the network itself during the training phase. Figure 4.3 shows an example of a convolution operation [5]. The output value of the first convolution,

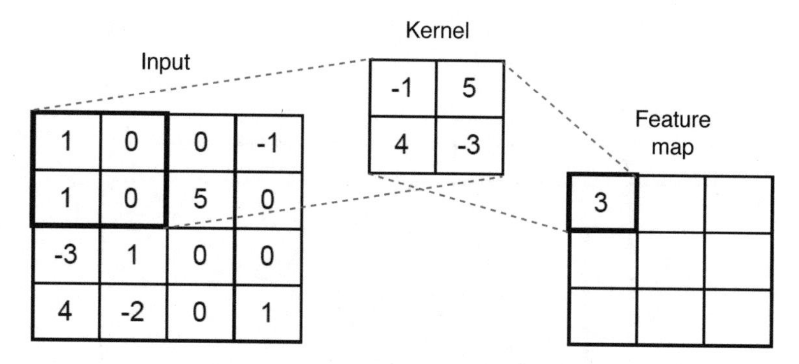

Fig. 4.3 A kernel convolving across an input. The size of the receptive field in the input is equal to the kernel size (2 × 2), the stride = 1, and padding = 0

3, was obtained as $(1 \times -1) + (0 \times 5) + (1 \times 4) + (0 \times -3)$. In the example, the input size is 4 × 4, the kernel size is 2 × 2, and the output will consist of a 3 × 3 matrix, result of sliding the kernel over the input. Thus, the output of each neuron after the convolution forms the feature map, and the kernel size defines the receptive field in the input.

The dimension of the feature map depends on two convolution properties: stride and padding. Stride controls how the kernel slides through the input, i.e., the position at which convolution operation must begin for each element. The padding technique can be applied to ensure that the output feature map has the same dimension as the input data. This technique surrounds each slice of the input volume with p cells containing zeros. Thus, the size of the output can be defined in terms of input size r_{in}, stride s, kernel size k, and padding p as $(r_{in} - k + 2p)/s + 1$.

Another important operation used on a CNN is *subsampling* (also known as *pooling*). In image processing, subsampling an image involves reducing its resolution without significantly altering its appearance. In the context of a CNN, a pooling layer reduces the dimensionality of a feature map provided as entry and produces another feature map, a kind of summary of the first. There are various pooling forms applicable to a feature map: select the maximum value (max pooling), the average (average pooling), and the norm of the set (L2-pooling), among others.

In the most common way of building a CNN, the network is organized in stages. Each stage consists of one or more convolution layers in sequence, followed by a pooling layer. CNN can contain several stages stacked after the input layer. Following the final stage of the network, one or more fully connected layers are stacked.

4.2.2.3 Recurrent Neural Networks

Recurrent Neural Networks (RNN) constitute a wide range class of neural networks whose evolution of the state depends on both the current input and the current states. This property (having dynamic state) equips RNNs with the possibility to perform

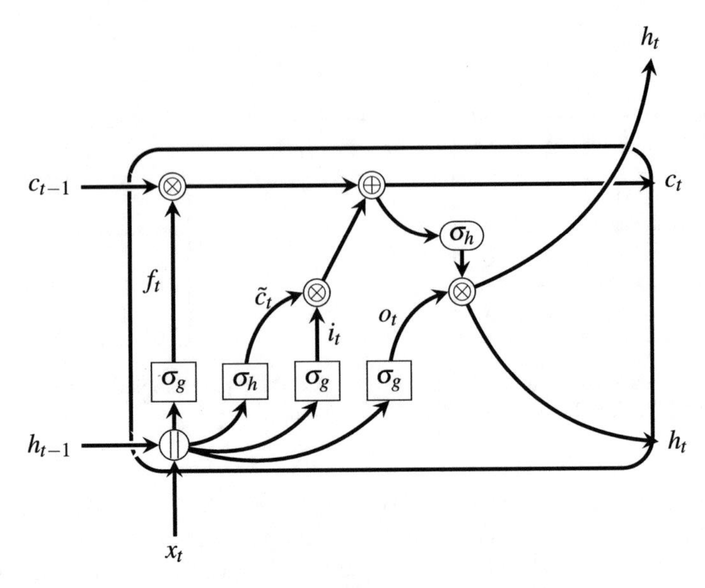

Fig. 4.4 An LSTM unit. Figure adapted from Veličković et al. [42]. The LSTM unit operates using three gates: input, forget, and output. The current item x_t from the input sequence and the previous hidden and cell states h_{t-1} and c_{t-1} are used to compute the next hidden and cell states h_t and c_t

context-dependent computing and learning long-term dependencies in sequential data: a signal that is supplied to a recurrent neural network at timestep t can change the behavior of this network at time step $t + k, k > 0$.

Long Short-Term Memory (LSTM)] is a type of RNN originally proposed in 1997 [19] and extended in 1999 [14] with the goal to be a remedy for the vanishing/exploding gradient problem in RNNs. An LSTM unit is composed of four elements (see Fig. 4.4): a memory block containing one or more cells and three logistic gates called read, write, and forget gate.

An LSTM unit is depicted in Fig. 4.4, where c_t and c_{t-1} are the current and previous cell states, σ_g and σ_h are the logistic sigmoid (Eq. 4.2a) and hyperbolic tangent (Eq. 4.2b) functions, respectively, and t is the time step. An LSTM unit also presents several gates, which are subnetworks (each with their learnable weights) meant to protect the memory cell from noisy and irrelevant data. Each step in the network consists of a forward pass, where a new cell state is predicted based on input data and the previous cell state, and a backward pass, where a variant of the backpropagation algorithm is used to correct the weights for each gate subnetwork and memory cell.

The input gate protects the cell in the forward pass, the output gate protects the cell in the backward pass, and the forget gate resets the cell once its content is outdated. These gates are mathematically defined in Eqs. 4.8, 4.9 and 4.10, where i_t, o_t, and f_t are the input, output, and forget gates activations, respectively. The matrices $W_.$, $U_.$, and $b_.$ hold the weight vectors of their corresponding subnetworks. In particular, the bias weight vectors $b_.$ are initialized with positive values for the forget gate and with negative values for the input and output gate. This way, the network only forgets when it has actually learned something. Notice how the gates are dependent on both the previous hidden state h_{t-1} and the current input x_t.

$$f_t = \sigma_g(W_f x_t + U_f h_{t-1} + b_f) \tag{4.8}$$

$$i_t = \sigma_g(W_i x_t + U_i h_{t-1} + b_i) \tag{4.9}$$

$$o_t = \sigma_g(W_o x_t + U_o h_{t-1} + b_o) \tag{4.10}$$

Equations 4.11, 4.12, and 4.13 show how the cell and hidden states c_t and h_t are updated. In these equations, $\circ$ is the Hadamard product. By definition $c_0 = 0$. In particular, Eq. 4.12 is responsible (1) for making the cell to forget something from the previous cell state (c_{t-1}) and (2) for holding something of the current input (x_t and h_{t-1}).

$$\tilde{c}_t = \sigma_h(W_c x_t + U_c h_{t-1} + b_c) \tag{4.11}$$

$$c_t = f_t \circ c_{t-1} + i_t \circ \tilde{c}_t \tag{4.12}$$

$$h_t = o_t \circ \sigma_h(c_t) \tag{4.13}$$

4.2.2.4 Graph Embedding Neural Networks

Protein interactions can be modeled as a graph. Therefore, a natural way of capturing their underlying patterns is through the analysis of the structure and information in the corresponding graph. The concept of learning representations on graphs has to do with learning an embedding that can be used to map nodes or subgraphs of a graph G to low-dimensional vectors. After being learned, these vectors can be used in several graph analysis tasks, such as link prediction, node classification, and regression.

A Graph Convolutional Network (GCN) is a model that can be used to extract representations for vertices of a graph. A GCN exploits the inherent structure of data that is organized into a network, analogously to how traditional convolutional neural networks exploit the spatial adjacency of a grid of Euclidean data. Let $G(V, E)$ be an attributed graph, that is, each $v \in V$ is associated to a d-dimensional feature vector. Let $x^{(i)}$ be the feature vector for the i-th vertex in G. Also, consider that $n = |V|$. The process of training a GCN involves two inputs:

- A $n \times d$ feature matrix X, which holds a vector of d feature vectors for each of the n nodes in G. In the above setting, X is be used to represent attributes associated to each node.
- G in matrix form, typically in the form of an adjacency matrix A.

A GCN can be trained to produce two kinds of outputs, namely, node-level and graph-level.

- A node-level output is an $n \times f$ feature matrix $\hat{Y}$, f being the number of output features, $f < d$.
- A graph-level is a scalar $\hat{y}$ for the entire graph. Graph-level outputs are usually produced through a pooling operation (e.g., Duvenaud et al. [12]).

Most GCNs follow a general architectural framework. Say we want to build a L-layer GCN. Let $H(i)$ be the i-th layer in a GCN. The computation done by the $l+1$ layer in this neural network can be written as presented in Eq. 4.14. In this equation, ϕ is the non-linear activation function chosen for layer $l+1$ (e.g., ReLU, sigmoid, etc). Notice that Eq. 4.14 is a recursive expression, for which there are two base cases. First $H(0) = X$. Secondly, $H(L) = \hat{Y}$ ($\hat{y}$ for the graph-level case).

$$H(l+1) = \phi(H(l), A) \tag{4.14}$$

Given the above general architecture framework, specific GCNs differ in how ϕ is chosen and parameterized.

One relevant example of GCN is node2vec [18], which can learn low-dimensional representations for vertices in a graph by optimizing a neighborhood preserving objective function. The algorithm supports many definitions of network neighborhoods by simulating biased random walks.

4.2.2.5 Probabilistic Neural Networks

A Probabilistic Neural Network (PNN) is a kind of feedforward neural network suited for the classification task [39]. A PNN is comprised of four layers. The input layer corresponds to the input example $\mathbf{x}$. The remaining layers are summarized below.

The second layer in a PNN is responsible for computing the distance of $\mathbf{x}$ to all the training examples. In order to do this, each training example is used as the mean (center) μ of a Gaussian distribution with covariance matrix Σ. Hence, if n is the amount of features (i.e., the dimensionality) of each training example, then $\mu \in \mathbb{R}^n$ and $\Sigma \in \mathbb{R}^{n \times n}$. If there are m training examples, then m Gaussian distributions will be fitted, one for each training example. These distributions are used to compute how close an unlabeled example $\mathbf{x}$ if from each training example, by applying Eq. 4.15.

$$p(\mathbf{x}; \mu, \Sigma) = \frac{1}{\sqrt{(2\pi)^n |\boldsymbol{\Sigma}|}} \exp\left(-\frac{1}{2}(x - \mu)^T \boldsymbol{\Sigma}^{-1}(\mathbf{x} - \mu)\right) \tag{4.15}$$

The result of computing Eq. 4.15 for each pair $(\mathbf{x}, \mathbf{x}^{(i)})$, $1 \leq i \leq m$, is a vector of m components, which is the result of the computation in the second layer of a PNN.

The third layer computes the average of the outputs from the second layer for each label. This way, the result of this layer is a vector of C units (one for each label).

Finally, the fourth layer performs a vote, selecting the largest value. The class label associated to the input pattern is then determined. A softmax layer (see Eq. 4.5) layer can be added in order to produce a distribution over the labels.

4.3 Research Scenario

Aiming to identify neural networks applied to predict essential proteins, we conducted a systematic mapping study using a query string involving the keywords "neural network", "essential gene", "essential protein", and "gene essentiality". We were interested in investigating if neural networks have been used in the context of predict essential proteins and which architecture of neural networks has been applied. The mapping was performed by the following questions:

- Q1: Which research works used neural networks to predict essential proteins?
- Q2: Which neural networks architecture was applied in this context of predicting essential proteins?
- Q3: Which input data were used in this context of predicting essential proteins?

In response to question 1, the query string was executed on the Scopus and PubMed databases in February 2020 and returned 20 references, considering both databases. We have also performed an additional snowballing search in Google Scholar. This resulted in the addition of 1 paper to base our presented. All of the 21 papers were analyzed. Nine papers remained, which were considered to answer questions 2 and 3 of the systematic mapping. The selected nine papers are presented in Table 4.1.

The papers are presented in Table 4.1 in a chronological way, to provide better visualization of the evolution of the use of neural networks for predicting essential proteins. Although the number of papers is low, we can notice that four of nine papers were published recently.

Question 2 is concerned to answer which neural network architecture was applied in those research works. The works of Xu et al. [44], Dai et al. [10], Campos et al. [6], Yang et al. [45], Palaniappan and Mukherjee [32], and Kondrashov et al. [22] used a classical multilayer perceptron neural network. On the other hand, Zeng et al. [46] adopted the multi-scale convolutional neural network. The works of Bakar et al. [4] and Bakar et al. [3] used a probabilistic neural network.

Answering question 3, sequence-based features, and topological network features in protein–protein interaction are the most common input data used for predicting essential proteins. The works of Xu et al. [44], Campos et al. [6], Palaniappan and Mukherjee [32], and Kondrashov et al. [22] adopted sequence-based features, while the works of Dai et al. [10], Bakar et al. [4], and Bakar et al. [3] used topological features. In addition, combinations of input data such as topological and gene expression features [46], and topological features and biological properties [45] are also considered.

4.3.1 Research Works

Xu et al. [44] proposed the Multilayer Perceptron (MLP) artificial neural network model (Sect. 4.2.2.1) to predict essential genes from 31 prokaryotic genomes using

Table 4.1 Publications presenting neural network algorithm to predicting essential genes

Year	Title	Reference
2020	Prediction of essential genes in prokaryote based on artificial neural network	[44]
2020	Network embedding the protein–protein interaction network for human essential genes identification	[10]
2019	DeepEP: a deep learning framework for identifying essential proteins	[46]
2019	An evaluation of machine learning approaches for the prediction of essential genes in eukaryotes using protein sequence-derived features	[6]
2014	Characterization of essential proteins based on network topology in protein interaction networks	[4]
2014	Analysis and identification of essential genes in humans using topological properties and biological information	[45]
2011	Predicting "essential" genes across microbial genomes: a machine learning approach	[32]
2011	Identifying hub proteins and their essentiality from protein–protein interaction network	[3]
2004	Bioinformatical assay of human gene morbidity	[22]

57 prokaryotic genomes features. The authors choose an MLP neural network model based on the test dataset, with the architecture of 57 input data, 12 neurons in a hidden layer, and 1 neuron in the output layer. This model was implemented in MATLAB R2014a using the feedforward net function.

The MLP prediction model's performance was evaluated with four methods: (i) Self-Predictions of Each Organism (SPEO): the test and training datasets were from a single organism; (ii) Leave One Genome Out (LOGO): neural network models were trained with genes of 30 organisms and tested in a selected organism; (iii) Predicting All by One Organism (PAOO): neural network models were trained using a single organism and the genes of the other 30 organisms were used as a test set; and (iv) Self-Predictions of All Organisms (SPAO): self-predictions were made in three datasets, which consisted of 10 Gram-positive organisms, 21 Gram-negative organisms, and 31 organisms of both types.

The 57 genome features used in the MLP prediction model are sequence-based features such as G + C content of the genes, silent base compositions, the frequency of rare amino acids, the frequency of aromatic amino acids, the frequency of each amino acid, number of the mole of amino acids, etc. To identify the features of the most strong genome associated with the gene essentiality, the authors applied the Weighted Principal Component Analysis method (WPCA), reducing the number of features from 57 to 29.

To evaluate the predictive capacity of the MLP models, the authors used the k-fold cross-validation method, with k=10, and the average scores were used as results of the evaluation. To obtain more reliable results from the models, the cross-validation was repeated 10 times, and so the mean values of the 100 models' scores were used for the final evaluation. The measure used in the evaluation was the Area Under the ROC Curve (AUC) score [36].

The presented results indicate that, for all 31 organisms, the average AUC score of SPEO was 0.7282 and for LOGO it was 0.694, respectively. Besides, SPAO's AUC scores for 31 organisms, 10 Gram-positive organisms, and 21 Gram-negative organisms were 0.7838, 0.8103, and 0.7794, respectively. For the PAOO method, as expected, due to the neural network training made with only one organism, the average AUC score was 0.6057. All of these results were obtained considering the 57 features based on the sequences. Considering the 29 features after the dimensionality reduction, the reduction in the average AUC score was less than 2%.

The authors describe that all prediction results indicated that these sequence-based features could be used to predict essential genes and confirm the effectiveness of the models. Also, they indicate that the results showed the effectiveness of the neural network model, but that the predictive performance still needs to be improved.

Another recent work that uses MLP neural network models was proposed by Dai et al. [10]. The authors proposed a new supervised method to predict essential human genes using network embedding the PPI network [18]. For input data, the authors considered that one of the most important features related to the gene essentiality is their topology properties in biological networks.

In this work, a network embedding method is used to extract useful information from biological networks. This method learns the vector representation of the nodes in the human PPI network by using a random walk-based network embedding method. A word2vec method inserts the pairs of nodes in an artificial neural network and learns the representation vectors for the nodes in the PPI network. The proposed approach performs a random biased walk on the human PPI network to generate the node context, which means that a node is likely to walk to the connecting nodes that share similar neighbors with it. After that, the pairs of nodes are inserted in a neural network to represent the features of the genes in the PPI network. The goal of the neural network is to learn the representation feature of each node (gene). These features can then be placed in a variety of classifiers to label whether genes are essential or not. In this work, to perform the prediction of essential human genes, the features were inserted in a Support Vector Machine (SVM) classifier.

The fivefold cross-validation technique was applied to test the prediction performance of the proposed method. Precision, recall, Specificity (Sp), Negative Predictive value (NP), F-measure, Accuracy (ACC), and Matthews Correlation Coefficient (MCC) were used to measure the method performance. The proposed method was compared with another four methods that used the SVM classifier to make predictions. The executions considered the prediction performance in two different human network datasets. The results showed that the method proposed by Dai et al. [10] has the best performance among all the methods compared in the two datasets, regardless of the ratio between essential and non-essential genes in the course of validation.

Besides the SVM classifier, six other classifiers were used to predict essential genes, to assess the effectiveness of the features of the genes that were extracted from the PPI network. The classifiers were Deep Neural Network (DNN), Decision Tree (DT), Naive Bayes, (NB), K-Nearest Neighbor (KNN), Logistic Regression (LR), Random Forest (RF), and Extra Tree (ET). Among all compared classifiers, the ET algorithm based on RF achieved the highest prediction performance.

Analogous to the previous work, Zeng et al. [46] proposed a new method to identify essential proteins, with the protein–protein interaction networks as the main input. Gene expression profiles and essential protein data complement the input data for this work. The method for identifying essential proteins, called DeepEP, is based on a deep learning structure that uses the node2vec technique, convolutional neural networks (Sect. 4.2.2.2) at various scales, and a sampling technique. The node2vec technique is applied to automatically learn the topological and semantic features of each protein in the Protein–Protein Interaction network (PPI). Convolutional neural networks at various scales are used to extract patterns in images of gene expression profiles. After this layer, the pooling layer is used to reduce the dimensionality. In this way, the outputs of each component (node2vec technique, convolutional neural networks at various scales and pooling layer) are concatenated and passed as input to a classification model. The classification model has a first fully connected layer that uses the Rectified Linear Unit (ReLU) activation function. After this layer, comes the output layer, which is fully connected and uses the softmax activation function to predict the final label of a protein.

DeepEP also uses a sampling method to alleviate the unbalanced data problem. The sampling method collects the same number of majority samples (non-essential proteins) and minority samples (essential proteins) in each training epoch as a new subset for training DeepEp. An advantage of this sampling method is to ensure that the results are not biased by any class in the training process.

With the propose to demonstrate the effectiveness of the method, the authors performed several experiments using the *Saccharomyces cerevisiae* dataset. To evaluate the performance of DeepEP with other methods, six Performance Measures were used: accuracy, precision, recall, F-measure, Area Under the Curve (AUC), and Average Precision (AP) score. The methods used for comparison were SVM, ensemble learning-based model, Decision Tree, and Naïve Bayes. The results were compared in two different ways. In the first way, raw data were used as training and testing datasets. In the second way, the random sampling technique was applied to collect samples of non-essential proteins. Then, non-essential proteins were added to the set of essential proteins to serve as input data for training machine learning models. The F-measure and AUC measures of DeepEP were 0.55 and 0.82, respectively. Such measures were superior to the measures of the other analyzed methods. Although DeepEP had not shown excellent results, the experimental results showed that DeepEP achieves a better result when compared to not deep machine learning methods.

The results obtained when using the node2vec technique were compared with those of six commonly used centrality methods: degree centrality, closeness centrality, eigenvector centrality, betweenness centrality, edge clustering coefficient centrality, and local average connectivity centrality. F-measure and AUC of DeepEP using the node2vec technique were 0.552 and 0.816, respectively. These metrics were superior to the measures of other evaluated methods. The analyzed results showed that the dense vectors generated by the node2vec technique contribute a lot to better performance. Also, the sampling method adopted proved to be a way to improve the performance of the identification of essential proteins.

The work of Campos et al. [6] trains and evaluates the prediction performance of five machine learning models. The focus is on the classification of essentiality within and between eukaryotic species using intrinsic features of protein sequences. The species include *S. cerevisiae*, *Schizosaccharomyces pombe*, *Caenorhabditis elegans*, *Drosophila melanogaster*, *Mus musculus*, and a dataset representing *Homo sapiens* cancer cell lines. From the species data, intrinsic features based on the protein sequences were extracted. The features are based on amino acid composition, protein autocorrelation, and chemical properties of individual protein sequences. For each species dataset, a design matrix was created, containing the various features extracted from the sequences of individual proteins, with labels for differential essential and non-essential proteins. Then, a standardized feature selection approach was performed. The most relevant features were selected for training in machine learning models. Random subsets of data were generated to estimate the fluctuation of the prediction performance when different sizes of datasets are used. The subsets varied from 10 to 90% of the sequences of essential and non-essential proteins in each dataset used for training the models. The rest of the data was directed to the set of tests.

In this work, the following machine learning models were trained: Generalized Linear Model (GLM), Artificial Neural Network (ANN), Gradient Boosting Method (GBM), Support Vector Machine (SVM), and Random Forest (RF). For comparison, the authors created a standard classifier (DF), which randomly classified the test sets using the essentiality probability calculated from the training sets, defined as the ratio between the number of essential proteins and the total number of reported proteins for each dataset.

The metrics ROC-AUC and area under the precision–recall curve (PR-AUC) were used for performance evaluation. All machine learning methods outperformed the standard classifier. RF achieved ROC-AUC and PR-AUC of approximately 1 for all tested datasets. Regarding ANN, performance decreased for most datasets, but increased for *S. cerevisiae*, *C. elegans*, and *Mus musculus* datasets. However, ANN achieved ROC-AUC > 0.8 using small training sets (10%). Subsequently, the performance of the machine learning models was evaluated considering the prediction of essentiality in the test sets of a species. Again, the trained machine learning methods outperformed the random classification (DF) and the ROC-AUC and PR-AUC of all machine learning models increased as more data was added to the training sets. In most cases, the performance of ANN models improved more slowly compared to other ML models. The authors conclude that this work showed that, using selected features of protein sequences linked to functional genomic datasets, machine learning methods can predict essential proteins in eukaryotes. However, they indicate that the approach can be extended to include other intrinsic or extrinsic features and to evaluate other machine learning methods, such as deep learning.

Bakar et al. [4] used a probabilistic neural network algorithm (Sect. 4.2.2.5), called EP3NN (Essential Proteins Prediction using Probabilistic Neural Network) as a classifier to identify essential proteins of studied organisms. The input data features were based on the network topology of the protein–protein interaction networks. The classifier uses the features of degree centrality, closeness centrality, local assortativity, and local clustering coefficient of each protein in the network. The experiments were carried out considering the *Saccharomyces cerevisiae* organism, for which protein interaction data and information on essential proteins were obtained. The EP3NN classifier was trained by a backward propagation algorithm with a decreasing gradient and adaptive learning rate and tested using tenfold cross-validation for each organism. EP3NN successfully has determined the essential and non-essential proteins for *S. cerevisiae* based on their network topology properties.

The EP3NN results were compared with the results of three other works available in the literature that also focused on *S. cerevisiae* as a study organism. The results showed that EP3NN was able to successfully perform the predictions of essential proteins, obtaining an accuracy of 95%, perfect sensitivity of 1, and specificity of 0.92 for the study organism.

The same authors proposed a method based on a topological network to identify hub and non-hub proteins in *Saccharomyces cerevisiae* [3]. The method, called HP3NN, employed a probabilistic neural network and a set of topological properties such as degree centrality, proximity centrality, local assortativity, and local clustering coefficient. The identification of hub proteins can provide more information about essential proteins and thus support methods for predicting essential proteins. The authors made a comparison with the results of three other methods that used essential proteins already known from *S. cerevisiae*. The hub proteins predicted by the HP3NN method consisted of 68.4% of essential proteins, a result superior to the three other compared methods (which consisted of 50.9%, 41.5%, and 52% of essential proteins, respectively).

Similar to those previously mentioned, the work of Yang et al. [45] was based on the features of the network topology of the protein–protein interaction networks and biological features of humans to build a machine learning classifier to predict essential proteins. To discriminate essential and non-essential genes with statistical significance were used 28 topological features and 22 biological features. Biological feature examples include GO scores, number of motifs, subcellular compartments, expression stage, and phyletic age. All features have been extensively analyzed by the authors. F-score was used to estimate the essentiality of each property. The functional enrichment analysis by GO was performed to investigate the functions of essential and non-essential genes. Good performances were obtained in the proposed machine learning approach using the jackknife test and tenfold cross-validation test, the Sensitivity (Sn), Specificity (Sp), overall prediction Accuracy (Acc), and Area Under the Curve (AUC).

The SVM approach, with radial basis function kernel, was initially used to build a classifier to identify essential and non-essential genes. The combination of both topological and biological properties significantly improved the classifier's performance concerning its execution using only one set of properties. For the jackknife test, the Sn, Sp, and Acc results were 68.81%, 75.02%, and 72.92%, respectively, and were higher than those obtained by network or biological features. For the tenfold cross-validation test, the results of Sn, Sp, and Acc were 68.65%, 74.98%, and 72.87%, respectively, and were higher than those obtained by other parameters. The corresponding AUC value was 0.8035 during the jackknife test and 0.8029 during the tenfold cross-validation test, which confirms the accuracy of the proposed model. The SVM results were compared with the logistic algorithms, IB1, J48, and the MLP neural networks. Based on the 50 traits, 1,292 essential genes and the same number of non-essential genes were predicted. The predictive results of the classifiers showed that the measures Sp and Acc of SVM (RBF) were 74.98% and 72.87%, respectively, which are higher than those of the other classifiers. The MLP algorithm was superior only to the IB1 and J48 algorithms.

Palaniappan and Mukherjee [32] evaluated machine learning approaches to the predicting problem of essential genes in microbial genomes using only sequence-based input features. For the task of predicting essential genes, three classification methods were investigated: Support Vector Machine (SVM), Artificial Neural Network (ANN), and Decision Tree (DT). The classifiers were trained and evaluated using data from 14 different microbial genomes whose essential genes were known. The classifiers used a set of 52 sequence-based features as inputs. The features refer to similar genes, properties of the genomic sequence, amino acid composition, codon frequency, physical–chemical characteristics, and subcellular location characteristics.

Two strategies were applied to reduce the problem of unbalanced data between the majority (non-essential genes) and minority (essential genes) classes. The first strategy reduces sequence redundancy in non-essential genes by grouping them by homology (40% similarity between sequences). In the new dataset of the majority class, the representative sequences of each cluster were included. The second strategy does a random sampling of 50% of the majority class.

The classification models were evaluated using the Leave One Genome Out (LOGO) and Leave One Taxon Out group (LOTO) test schemes and the tenfold cross-validation strategy performed 10 times. SVM was run with a radial base function kernel. A three-layer neural network was trained with the backpropagation algorithm. The J48 algorithm, which is a variation of the C4.5 algorithm, was used for the decision tree.

The experimental results indicated that the SVM and ANN models perform better than DT. The area under the Receiver Operating Characteristics (AU-ROC) presented scores of 0.80, 0.79, and 0.68, respectively.

In previous work in the area, Kondrashov et al. [22] performed an *in silico* analysis of the morbidity of human genes, using the known morbid genes and the other human (non-morbid) genes. The analysis was based on 18 features of the known morbid genes. Several of these features are quite informative and, together, produce a useful morbidity prediction for each human gene unequivocally known. Essential genes form a subset of morbid genes. By definition, an essential gene is morbid and a morbid gene responsible for a Mendelian disease is essential. Even those morbid genes that generally act only as risk factors for complex diseases are often essential [22].

The dataset was subdivided into 1273 known and 16580 generic morbid human genes. Examples of the 18 gene features included the length of the coding sequence, the length of the intron, the number of introns and the similarity with the orthologs of model organisms such as *D. melanogaster*, *C. elegans*, and *Arabidopsis thaliana*.

To recognize non-morbid genes, the classification of the genes was performed by an MLP neural network. All genes were divided into three subsets: the training set (50%), the test set (25%), and the validation set (25%). The validation set was used to assess the generalization capacity of the neural network. The neural network produced a value of the classification variable X for each gene, morbid, or generic. The results presented considered that if the threshold value of X were chosen so that 1, 5, or 10% of the known morbid genes were incorrectly classified as non-morbid, the proportions of generic genes classified as non-morbid would be 8, 34, and 44%, respectively.

4.3.2 Considerations

The objective of this research is to present works that adopted neural network algorithms for predicting essential proteins. Taking into account just this class of algorithms, the number of works available in the literature is low. However, when we consider all machine learning algorithms (methods), this number increases. Modifying the original query string for the keywords “machine learning”, “essential gene”, “essential protein”, and “gene essentiality”, it is returned 84 papers considering the execution on the Scopus database. Among these 84 articles, seven are included in our review and 28 were published in the last 3 years (2018, 2019, and 2020). Figure 4.5 shows the evolution of publications.

The most common methods used for predicting essential proteins are random forest and SVM, though decision trees, logistic regression, Bayesian network, and KNN are also cited. Similar to the reviewed papers, topological networks in protein–protein interactions, gene expressions, and sequence-based features are the main features used as input data in these machine learning works. In general, the methods used in these machine learning papers present good results for the performance evaluation. Even so, we can remark that there exists still a gap in new studies that use neural networks as a machine learning method to predict essential proteins. In particular, future works that use neural network architectures, such as convolutional or recurrent neural networks which are applied with success in other areas.

4.4 Conclusion

Studies on essential proteins are important to promote our understanding of biology and support further investigations, such as the development of new drugs. A large amount of data publicly available on gene essentiality greatly contributes to the

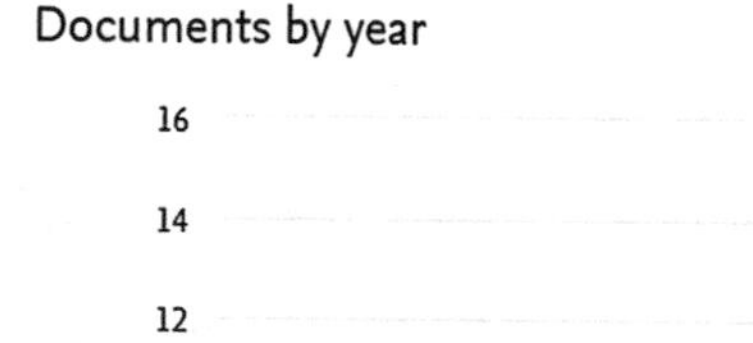

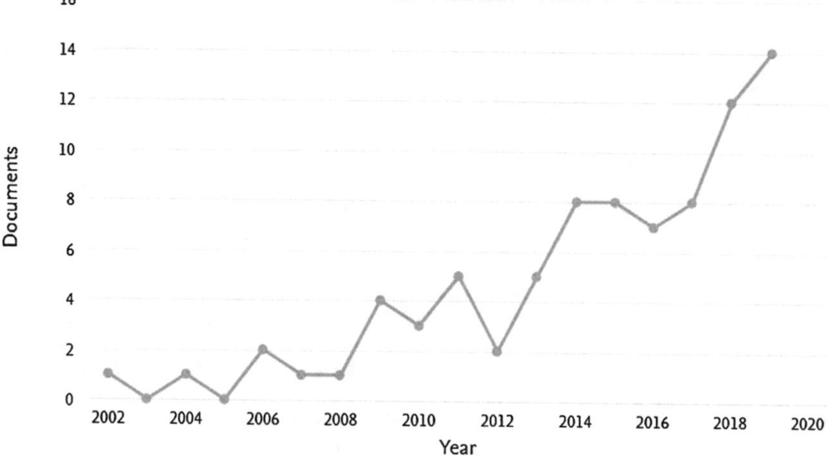

Fig. 4.5 Evolution of publications about machine learning and essential genes in scopus database

application of computational approaches. The use of computational machine learning methods to predict essential proteins reduces the costs and labor time associated with biological experiments. Advances and improvements in the use of these methods can be seen in the works published over almost 20 years. Support vector machines, random forests, decision trees, and artificial neural networks are examples of machine learning methods that stand out in these publications.

This review presented a background related to artificial neural networks, which encompasses different neural network architectures, and a specific survey of research papers that used artificial neural networks to identify essential proteins. We can observe that some of the reviewed works do not use the artificial neural network as the classification model in the prediction of essential proteins, but to perform the preprocessing of input data features.

Although most of the results presented in these research papers are satisfactory, it is possible to observe that most works adopt the classic multilayer perceptron neural network architecture (Sect. 4.2.2.1). Neural network architectures, such as convolutional neural networks (Sect. 4.2.2.2) and recurrent neural networks, like LSTM (Sect. 4.2.2.3), have been used successfully in other contexts such as computer vision and natural language processing. Due to the recent success in the use of these neural network architectures in the literature, for future works, it is plausible to investigate their applications in the prediction of essential proteins.

References

1. Acencio ML, Lemke N (2009) Towards the prediction of essential genes by integration of network topology, cellular localization and biological process information. BMC Bioinform 10:290. https://doi.org/10.1186/1471-2105-10-290
2. Ashtiani M, Salehzadeh-Yazdi A, Razaghi-Moghadam Z, Hennig H, Wolkenhauer O, Mirzaie M, Jafari M (2018) A systematic survey of centrality measures for protein-protein interaction networks. BMC Syst Biol 12(1). https://doi.org/10.1186/s12918-018-0598-2
3. Bakar S, Taheri J, Zomaya A (2011) Identifying hub proteins and their essentiality from protein-protein interaction network. In: Proceedings—2011 11th IEEE international conference on bioinformatics and bioengineering, BIBE 2011, pp 266–269. https://doi.org/10.1109/BIBE.2011.67
4. Bakar S, Taheri J, Zomaya A (2014) Characterization of essential proteins based on network topology in proteins interaction networks. AIP Conf Proc 1602:36–42. https://doi.org/10.1063/1.4882463
5. Burkov A (2019) The hundred-page machine learning book. Andriy Burkov, http://themlbook.com/wiki/doku.php
6. Campos T, Korhonen P, Gasser R, Young N (2019) An evaluation of machine learning approaches for the prediction of essential genes in eukaryotes using protein sequence-derived features. Comput Struct Biotechnol J 17:785–796. https://doi.org/10.1016/j.csbj.2019.05.008
7. Chen WH, Minguez P, Lercher MJ, Bork P (2011) Ogee: an online gene essentiality database. NuclC Acids Res 40(D1):D901–D906
8. Cullen LM, Arndt GM (2005) Genome-wide screening for gene function using RNAi in mammalian cells. Immunol Cell Biol 83(3):217–223. https://doi.org/10.1111/j.1440-1711.2005.01332.x
9. Cybenko G (1989) Approximation by superpositions of a sigmoidal function. Math Control, Signals, Syst (MCSS) 2(4):303–314. https://doi.org/10.1007/BF02551274
10. Dai W, Chang Q, Peng W, Zhong J, Li Y (2020) Network embedding the protein–protein interaction network for human essential genes identification. Genes 11(2). https://doi.org/10.3390/genes11020153
11. Deng J, Deng L, Su S, Zhang M, Lin X, Wei L, Minai AA, Hassett DJ, Lu LJ (2011) Investigating the predictability of essential genes across distantly related organisms using an integrative approach. Nucl Acids Res 39(3):795–807. https://doi.org/10.1093/nar/gkq784
12. Duvenaud DK, Maclaurin D, Iparraguirre J, Bombarell R, Hirzel T, Aspuru-Guzik A, Adams RP (2015) Convolutional networks on graphs for learning molecular fingerprints. In: Cortes C, Lawrence ND, Lee DD, Sugiyama M, Garnett R (eds) Advances in neural information processing systems 28, Curran associates, Inc., pp 2224–2232. http://papers.nips.cc/paper/5954-convolutional-networks-on-graphs-for-learning-molecular-fingerprints.pdf
13. Evers B, Jastrzebski K, Heijmans J, Grernrum W, Beijersbergen R, Bernards R (2016) CRISPR knockout screening outperforms shRNA and CRISPRi in identifying essential genes. Nat Biotechnol 34(6):631–633. https://doi.org/10.1038/nbt.3536
14. Gers FA, Schmidhuber J, Cummins F (1999) Learning to forget: continual prediction with LSTM. In: IET conference proceedings, vol 5, pp 850–855
15. Gehring J, Auli M, Grangier D, Yarats D, Dauphin YN (2017) Convolutional sequence to sequence learning. Int Conf Mach Learn 3:2029–2042
16. Gong X, Fan S, Bilderbeck A, Li M, Pang H, Tao S (2008) Comparative analysis of essential genes and nonessential genes in Escherichia coli K12. Mol Genet Genomics 279(1):87–94. https://doi.org/10.1007/s00438-007-0298-x
17. Goodfellow I, Bengio Y, Courville A (2016) Deep learning. MIT Press. http://www.deeplearningbook.org
18. Grover A, Leskovec J (2016) node2vec: scalable feature learning for networks. In: Proceedings of the 22nd ACM SIGKDD international conference on knowledge discovery and data mining—KDD '16. ACM Press, San Francisco, California, USA, pp 855–864. https://doi.org/10.1145/2939672.2939754

19. Hochreiter S, Schmidhuber J (1997) Long short-term memory. Neural Comput 9(8):1735–1780
20. Jansen R, Greenbaum D, Gerstein M (2002) Relating whole-genome expression data with protein-protein interactions. Genome Res 12(1):37–46. https://doi.org/10.1101/gr.205602
21. King Jordan I, Rogozin IB, Wolf YI, Koonin EV (2002) Essential genes are more evolutionarily conserved than are nonessential genes in bacteria. Genome Res 12(6):962–968. https://doi.org/10.1101/gr.87702. Article published online before print in May 2002
22. Kondrashov F, Ogurtsov A, Kondrashov A (2004) Bioinformatical assay of human gene morbidity. Nucl Acids Res 32(5):1731–1737. https://doi.org/10.1093/nar/gkh330
23. Krizhevsky A, Sutskever I, Hinton GE (2012) ImageNet classification with deep convolutional neural networks. In: Proceedings of the 25th international conference on neural information processing systems (NeurIPS). Curran Associates Inc., pp 1097–1105
24. LeCun Y, Bengio Y (1995) Convolutional networks for images, speech, and time series. In: Arbib MA (ed) The handbook of brain theory and neural networks. MIT Press, Cambridge, MA, USA, pp 1–14
25. LeCun Y, Bengio Y, Hinton G (2015) Deep learning. Nature 521(7553):436–444. https://doi.org/10.1038/nature14539
26. Leevy JL, Khoshgoftaar TM, Bauder RA, Seliya N (2018) A survey on addressing high-class imbalance in big data. J Big Data 5(1):42. https://doi.org/10.1186/s40537-018-0151-6
27. Lei X, Yang X (2018) A new method for predicting essential proteins based on participation degree in protein complex and subgraph density. PLOS ONE 13(6):e0198,998. https://doi.org/10.1371/journal.pone.0198998
28. Lu Y, Deng J, Carson M, Lu H, Lu L (2014) Computational methods for the prediction of microbial essential genes. Curr Bioinform 9(2):89–101. https://doi.org/10.2174/157489360899140109113434
29. Mobegi F, Zomer A, de Jonge M, van Hijum S (2017) Advances and perspectives in computational prediction of microbial gene essentiality. Brief Funct Genomics 16(2):70–79. https://doi.org/10.1093/bfgp/elv063
30. Mori H, Baba T, Yokoyama K, Takeuchi R, Nomura W, Makishi K, Otsuka Y, Dose H, Wanner BL (2015) Identification of essential genes and synthetic lethal gene combinations in Escherichia coli K-12. Methods Mol Biol 1279:45–65. https://doi.org/10.1007/978-1-4939-2398-4_4
31. Nigatu D, Sobetzko P, Yousef M, Henkel W (2017) Sequence-based information-theoretic features for gene essentiality prediction. BMC Bioinform 18(1):1–11. https://doi.org/10.1186/s12859-017-1884-5
32. Palaniappan K, Mukherjee S (2011) Predicting "essential" genes across microbial genomes: a machine learning approach. In: Proceedings—10th international conference on machine learning and applications, ICMLA 2011, vol 2, pp 189–194. https://doi.org/10.1109/ICMLA.2011.114
33. Peng C, Gao F (2014) Protein localization analysis of essential genes in prokaryotes. Scientific reports 4. https://doi.org/10.1038/srep06001
34. Peng C, Lin Y, Luo H, Gao F (2017) A comprehensive overview of online resources to identify and predict bacterial essential genes. Front Microbiol 8:1–13. https://doi.org/10.3389/fmicb.2017.02331
35. Rumelhart DE, Hinton GE, Williams RJ (1986) Learning representations by back-propagating errors. Nature 323(6088):533–536. https://doi.org/10.1038/323533a0
36. Sahoo G, Kumar Y (2012) Analysis of parametric & non parametric classifiers for classification technique using weka. Int J Inf Technol Comput Sci (IJITCS) 4(7):43
37. Seringhaus M, Paccanaro A, Borneman A, Snyder M, Gerstein M (2006) Predicting essential genes in fungal genomes. Genome Res 16(9):1126–1135. https://doi.org/10.1101/gr.5144106
38. Sobel I, Feldman G (1968) A 3 × 3 isotropic gradient operator for image processing. In: Pattern classification and scene analysis. Wiley, pp 271–272
39. Specht DF (1990) Probabilistic neural networks. Neural Netw 3(1):109–118. https://doi.org/10.1016/0893-6080(90)90049-Q

40. Sun S, Cao Z, Zhu H, Zhao J (2019) A survey of optimization methods from a machine learning perspective. IEEE Trans Cybern 1–14
41. Tran D, Wang H, Torresani L, Ray J, LeCun Y, Paluri M (2018) A closer look at spatiotemporal convolutions for action recognition. In: Proceedings of the IEEE computer society conference on computer vision and pattern recognition, IEEE computer society, pp 6450–6459. https://doi.org/10.1109/CVPR.2018.00675
42. Veličković P, Karazija L, Lane ND, Bhattacharya S, Liberis E, Liò P, Chieh A, Bellahsen O, Vegreville M (2018) Cross-modal recurrent models for weight objective prediction from multimodal time-series data. In: Proceedings of the 12th EAI international conference on pervasive computing technologies for healthcare, association for computing machinery. New York, NY, USA, PervasiveHealth '18, pp 178–186. https://doi.org/10.1145/3240925.3240937
43. Wang J, Peng W, Wu FX (2013) Computational approaches to predicting essential proteins: a survey. Proteomics-Clin Appl 7(1–2):181–192. https://doi.org/10.1002/prca.201200068
44. Xu L, Guo Z, Liu X (2020) Prediction of essential genes in prokaryote based on artificial neural network. Genes Genomics 42(1):97–106. https://doi.org/10.1007/s13258-019-00884-w
45. Yang L, Wang J, Wang H, Lv Y, Zuo Y, Li X, Jiang W (2014) Analysis and identification of essential genes in humans using topological properties and biological information. Gene 551(2):138–151. https://doi.org/10.1016/j.gene.2014.08.046
46. Zeng M, Li M, Wu FX, Li Y, Pan Y (2019) DeepEP: a deep learning framework for identifying essential proteins. BMC Bioinform 20. https://doi.org/10.1186/s12859-019-3076-y
47. Zhang R, Ou HY, Zhang CT (2004) Deg: a database of essential genes. Nucl Acids Res 32(suppl_1):D271–D272
48. Zhu J, Gong R, Zhu Q, He Q, Xu N, Xu Y, Cai M, Zhou X, Zhang Y, Zhou M (2018) Genome-wide determination of gene essentiality by transposon insertion sequencing in yeast pichia pastoris. Sci Rep 8(1). https://doi.org/10.1038/s41598-018-28217-z

Chapter 5
Transcriptograms: A Genome-Wide Gene Expression Analysis Method

Rita M. C. de Almeida, Lars L. S. de Souza, Diego Morais, and Rodrigo J. S. Dalmolin

Abstract In this chapter, we discuss the Transcriptogram method for statistically analyzing differential gene expression in a genome-wide profile. This technique suggests a method to hierarchically interrogate the data and, subsequently, narrow down to gene level. We present the method, discuss its reproducibility and enhanced signal-to-noise ratio, and discuss its application in investigating time series data as in cell cycle, therapy gene target identification, lineage and tissue classification and as a powerful test to identify error and assess the quality of normalization procedures. We finally present the software ready for download and discuss the R-plugin for BioConductor.

Keywords Transcriptograms · Wide genome gene expression · Differentially expressed gene sets

---

R. M. C. de Almeida (✉) · L. L. S. de Souza
Instituto de Física, Universidade Federal do Rio Grande do Sul, Porto Alegre, RS, Brazil
e-mail: rita@if.ufrgs.br

L. L. S. de Souza
e-mail: lars.sanhudo@gmail.com

R. M. C. de Almeida
Instituto Nacional de Ciência e Tecnologia: Sistemas Complexos, Universidade Federal do Rio Grande do Sul, Porto Alegre, RS, Brazil

R. M. C. de Almeida · D. Morais · R. J. S. Dalmolin
Programa de Pós-Graduação em Bioinformática, Universidade Federal do Rio Grande do Norte, Natal, RN, Brazil
e-mail: arthur.vinx@gmail.com

R. J. S. Dalmolin
e-mail: dalmolin_r@yahoo.com.br

R. J. S. Dalmolin
Departamento de Bioquímica, Universidade Federal do Rio Grande do Norte, Natal, RN, Brazil

F. A. B. da Silva et al. (eds.), *Networks in Systems Biology*, Computational Biology 32,
https://doi.org/10.1007/978-3-030-51862-2_5

5.1 Introduction

The possibility of genome-wide measures of gene expression opened many roads for the understanding of cell metabolism and the subsequent proposition of new therapies and treatments for pathologies. It also brought along many problems inherent to the utilization of big data.

Cell metabolism results from a very complex biochemical reaction that involves tens of thousands of components, produced by the cell itself or by its environment. The reaction, besides involving many agents, is highly nonlinear where the components that are produced by the cell are controlled by the current metabolism and present loops at many different levels. Although single-gene analyses produced important results, in general, we should not expect that a single-gene expression could explain the whole set of responses to stimuli, treatments, different conditions, mutations, etc., presented by cells.

A plethora of different methods and models have been proposed to analyze cell metabolism by changes in gene expression. The most direct—and naïve—method is represented by Volcano plots, where the fold change in gene expression is crossed with its *P*-value in the comparison among two classes of transcriptomic data. Soon the scientific community realized that the analyses should contemplate also gene sets linked to biological functions and/or metabolic pathways and tools as Gene Set Enrichment Analysis (GSEA) [1] and Significance Analysis of Microarray-Gene Set (SAM-GS) [2] were successfully proposed to consider differential expression of gene sets and then identify the altered genes.

However, cells are very changeable, due to cell cycle, response to environment, stress stimuli, or even aging, and this variability imposes confounding variables to the studies. On top of that, there is also the noise inherent of the techniques (microarrays or RNA-Seq), and the result is that the statistical tests power is greatly reduced due to the increase of standard deviations in the measures.

A further complication, inherent to both most used technologies (microarrays and RNA-Seq) lays in the preprocess procedures, known as normalization, required to compare expression across different genes and samples: there is not a clear criterion to decide whether the expression is adequately normalized, mainly because the effects that are necessary to fix by normalization may change from experiment to experiment. Not surprisingly, there are also many different normalization procedures, depending on the issue they fix. Evans and collaborators present a nice overview and compare different normalization methods, focusing on the underlying assumptions of each method, yielding a basis for adequate choices for each experiment [3]. However important, in what follows we will not focus on the normalization procedure, but rather at the steps after normalization, i.e., we assume the normalization procedure is adequately performed.

Here we review the Transcriptogram tool [4], which aims at two targets. The first is to controllably reduce noise such that the signal-to-noise ratio is increased and so is the power of statistical tests. In this way, both differential gene expression analyses and their reproducibility across laboratories are improved [5]. The second target is to provide a visualization of gene expression from a genome-wide point-of-view. As we will discuss in what follows, a consequence of paying attention to these two targets leads to a hierarchically interrogation of data and to suggestions of explanations based on systems biology. Transcriptograms do not substitute other techniques, it comes as a complementary step to help visualizing what is going on. Additionally, transcriptograms are useful in deciding about normalization procedures, as well as to prevent unwanted and probably disastrous accidental mixing of samples among classes.

In the next section, we present the transcriptograms and discuss the reasons why this technique works the way it does. We then present two case studies of previous, successful applications of the technique, and on Sect. 5.4 we discuss how transcriptograms may help in laboratory errors and normalization choices. In Sect. 5.5, we present the link for the Transcriptogramer tool and its facilities and the Transcriptogram R-Bioconductor plugin.

5.2 The Method

5.2.1 The Gene Ordering

Transcriptograms are projections of expression data on an ordered gene list, such that it is possible to run a window average over the expression data. The point is, when an average is performed on a set of expression data, the random noise is necessarily reduced in relation to single-gene comparison. On the other hand, if the set is randomly gathered, the average procedure will also reduce the signal. However, when the gene set is gathered by the gene products' biological function, one may expect that in most cases the signal will be preserved. This is the idea behind the transcriptogram technique. In this section, we explain how we cluster genes by the biological function exerted by their products.

We begin by retrieving information on Protein–Protein interaction (PPI) from STRING database [6] (https://string-db.org/). PPI is classified in STRING by organisms and consists of a list of proteins pairs that are inferred as associated. There are seven association inference methods, ranging from physical interactions discovered in real experiments to text mining in scientific publications, passing through retrieving information from reliable databases (as HUGO, or Flybase, for example) or neighborhood as measured in chromosomes. The important point for us is that STRING gives a confidence score for each protein–protein interaction [7], given by the probability that the methods used to infer the association predicts that these gene products participate in the same metabolic pathway as defined in KEGG database (Kyoto Encyclopedia of Genes and Genomes) (https://www.genome.jp/kegg/) [8]. We remark, however, that the source for gene–gene association could be other databases, provided the criterion to assign an association is based on the biological function the gene product participates.

The retrieved list from STRING comprises all proteins that have at least one association with some other protein. This acts as a filter and the resulting number of proteins/genes is less than the number usually reported for the organism. For example, *Mus musculus* have the order of 30,000 known genes and the list after the filter presents 13,384 genes.

After retrieving the information on the protein–protein associations we build an interaction matrix M_{ij} such that

$$M_{ij} = \begin{cases} 1, & \text{if the proteins at positions } i \text{ and } j \text{ are associated} \\ 0, & \text{otherwise} \end{cases}, \tag{5.1}$$

from there we calculate cost F for a given configuration of the matrix as

$$F = \sum_{i=1}^{N-1} \sum_{j=i+1}^{N} |i-j|^{\alpha} \big[|M_{ij} - M_{ij+1}| + |M_{ij} - M_{ij-1}| + |M_{ij} - M_{i+1j}| + |M_{ij} - M_{i-1j}| \big], \tag{5.2}$$

where N is the number of genes/proteins on the list and α is a parameter that regulates how diagonal the final gene ordering list is [8]. In what follows, we will keep $\alpha = 1$. Figure 5.1 illustrates how the different configurations of the association matrix may present different values for the cost function F: for the hypothetical site (i, j) in the figure, the contribution would be given by $|i - j|[0 + 0 + 1 + 1] = 2|i - j|$. Observe that $|i - j|$ is proportional to the distance from the site (i, j) to the matrix diagonal, the line for which $i = j$. F decreases when the number of interfaces between dark and white sites decreases and when these interfaces lay near the diagonal. In other words, the more diagonal M is, the lower F gets. One important

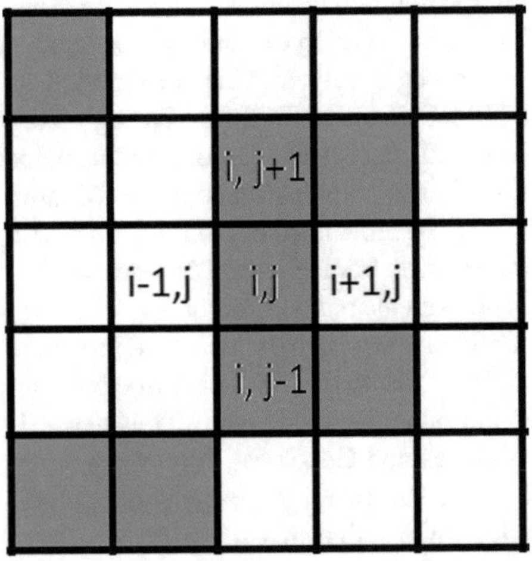

i, j+1
i-1,j
i,j
i+1,j
i, j-1

Fig. 5.1 Example of a portion of the association matrix M. Dark sites represent $M_{ij} = 1$, while white sites correspond to $M_{ij} = 1$. F increases when dark sites are neighbors to white sites

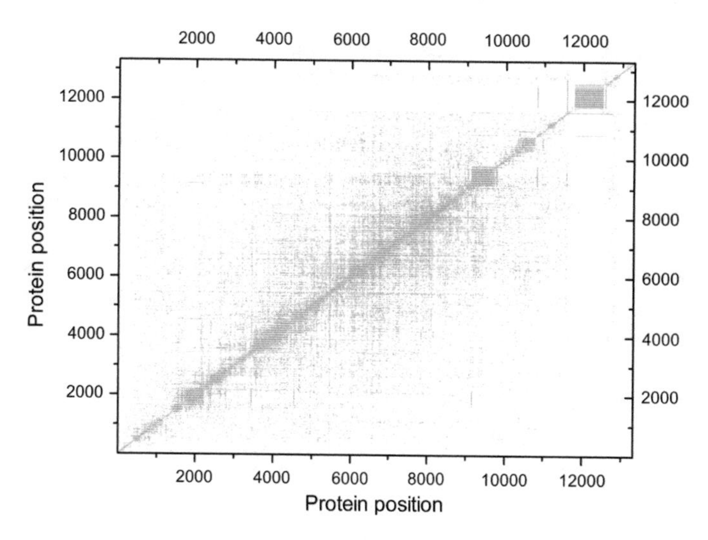

Fig. 5.2 Association matrix for *Mus musculus*. The gene list was retrieved from STRING with confidence score 0800, as described by Sanhudo et al. [5]. Observe that in this configuration, the matrix presents clusters of dark points near the diagonal. These clusters correspond to clusters of associated genes/proteins on the ordered list

consequence of M being as diagonal as possible is that associated genes/proteins are located in nearby positions on the ordered list.

The gene ordering of the list is implemented by a Monte Carlo code that minimizes F. The algorithm starts from the list of genes that have been retrieved from STRING that has been randomly arranged. It proceeds by randomly picking up a pair of genes and swapping their positions on the list. If this change reduces F, the swap is accepted and another pair is chosen. In case the value of F increases, the swap is accepted with probability given by a Boltzmann-like probability. Then another pair is chosen. This procedure is repeated until F stabilizes [4]. One example of the final configuration for an association matrix is given in Fig. 5.2. The square blocks on the diagonal corresponds to clusters of gene/proteins on the axes that are highly associated to one another, that is, they collaborate in the same biological process.

The clustering of genes/proteins collaborating in the same biological process opens the possibility of running window averages as follows. We can define a window of size $2r+1$ positions centered at a given position on the list. Due to the ordering, the window will cover a region of gene/proteins with a high probability of collaborating. Hence, for any given quantity, an average over that window represents an average of that quantity over a gene set where the components collaborate in the same biological function. The window center may run over the whole list resulting in a smooth profile of averages. A transcriptogram is produced when the averaged quantity is gene expression.

To endow the ordered list with biological meaning, we projected selected KEGG pathways and Gene Ontology: Biological Process terms on the ordered list, as shown in Fig. 5.3 for *Mus musculus*. The term profiles consist of the fraction of genes belonging to the term in running windows of radius 30. Observe that the succession of the terms has a functional logic: at the left end the list begins with energy metabolism terms, then terms related to energy metabolism, followed by genes involved in RNA translation and metabolism. After that, we find genes linked to DNA metabolism and cell cycle, and then pathways associated with cell fate and cell differentiation. Still in the center part, we find the terms associated with cytoskeleton dynamics and involved in the interaction with extracellular matrix. Finally, at the right end, we have pathways related to drug metabolism and the large gray peak, that corresponds to olfactory receptors.

Transcriptograms are expression profiles obtained by running a window average on the expression of the genes on the ordered list. The average is hence over gene sets that are functionally related. The average procedure is taken to improve the signal-to-noise ratio, as we discuss in the next section.

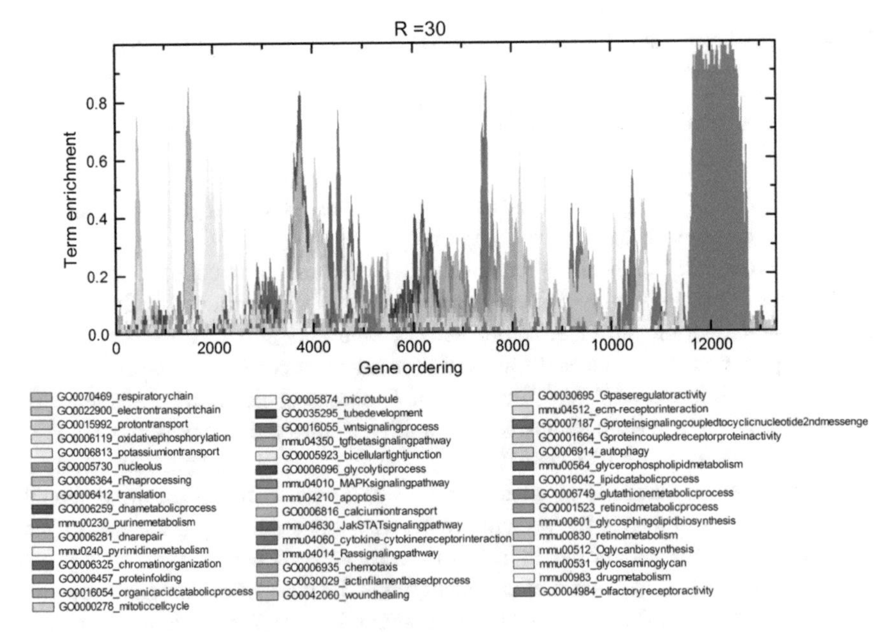

Fig. 5.3 Biological meaning of the genes/proteins ordered gene list for *Mus musculus*. Colored profiles represent the term enrichment of windows of radius 30 associated with selected Gene Ontology terms and KEGG pathways. Observe that at the left end of the ordering, we find terms related to energy metabolism (in olive hue), followed by genes involved in RNA translation and metabolism. After that, we find genes linked to DNA metabolism and cell cycle, and then pathways associated with cell fate and cell differentiation. Still in the center part, we find the terms associated with cytoskeleton dynamics and involved in the interaction with extracellular matrix. Finally, at the right end, we have pathways related to drug metabolism and the large gray peak, that corresponds to olfactory receptors

5.2.2 Transcriptograms

To demonstrate the advantages of the transcriptograms let us consider some gene expression samples and plot the result projected on the ordering before and after the running window average, as shown in Fig. 5.4. The profiles in this figure were obtained as follows. First, the PPI matrix for *Homo sapiens* is retrieved from STRING, version 9 with confidence score 800 and used to order the gene products list as explained above. Consider the expression data for a given sample and consider a position on the list as the center of an interval (window) of $2r + 1$ genes, with r being the radius of the window. It is possible to calculate the expression average for this interval and assign this value to the central gene. The window can be slipped one position and the process is repeated. We produce a smooth expression profile (the transcriptogram) when for each position on the ordered list the average of expression of this running window is calculated.

After we obtain the transcriptograms for all samples, they may be organized in classes. Figure 5.4 considers 3 replicates for cell lines obtained from human cystic tissue from a patient with Autosomal Dominant Kidney Disease and 3 replicates

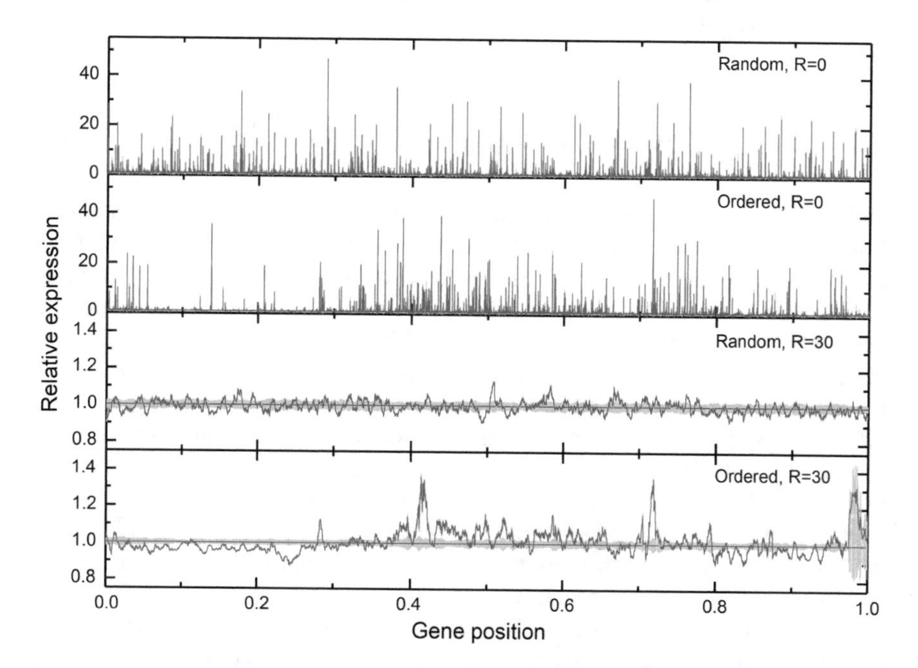

Fig. 5.4 Relative transcriptograms for cystic against normal (control) tissues from microarray samples from an Autosomal Dominant Polycystic Kidney Disease (ADPKD) [9] in *Homo sapiens*. The ordered gene list comprises 9684 genes. Red lines represent samples from a cystic tissue and gray horizontal profiles correspond to samples from normal tissue. Transcriptograms produced with (top) zero radius projected on a randomly ordered list, (second row) zero radius, projected on an ordered gene list, (third row) radius 30 projected on a random list and (bottom) radius 30 projected on an ordered list

for cell lines obtained from normal donor tissue samples [9]. We then calculate the average and standard deviation over the replicates in each class. The result is an average transcriptogram for each class for the transcriptogram value associated with each gene of the ordered list. We choose one class of samples as control (in this example the control represents samples from the normal tissue), and the values for each gene position is divided by the average transcriptogram value for the same gene in the control class, in such a way that the control class will be represented by a horizontal line at value 1. Figure 5.4 top presents the result of this process for a random gene list and running window with zero radius, hence representing the relative expression of a single gene. Figure 5.4 second row presents the same results when considering an ordered gene list. Figure 5.4 third row considers radius 30 (average over windows of 61 genes), projected on a random list, and finally Fig. 5.4 bottom considers radius 30 projected on an ordered list. Together, the panels in Fig. 5.4 show that both running window averages as well as the gene list ordering are required to enhance the difference between average transcriptograms for cystic and normal tissue samples. The biological meaning for the peaks presented by the red profile with respect to the horizontal gray profile, associated with the control sample (normal class), is obtained from analyzing the genes enriching the list at the locations of the peaks, as we discuss in what follows.

5.2.3 The Biological Meaning of Transcriptograms

Valleys or peaks in transcriptogram profiles as those shown in Fig. 5.4 bottom around gene positions 0.24 or 0.41 indicate that the transcriptogram values for the class of interest (here the cystic class) associated with genes at positions lying in these intervals are, respectively, lower or higher than the respective values in the control class. Both intervals are wider than 10% (0.1) of the whole gene list, which comprises 9684 genes. It means that each interval comprises more than 900 genes that share a neighborhood on the list with genes, whose average expression is significantly different from the respective intervals in the control class. The significance may be assessed by statistical tests as we discuss below but, for now, we do it visually: the thin light red and light gray shadows that follow, respectively, the red and black lines correspond to the standard error of each class. As the vertical distance between the shadows in those intervals are a few times larger than the standard errors, the significance of the difference is visual.

The question is, what those valleys and peaks stand for? The answer lays in finding the genes that enrich those intervals in the gene list that are associated with the valleys and peaks. This a direct task, by retrieving the gene or gene product names that populate those intervals. After that an inverse task is required: to discover which biological functions these gene products perform. At this point, the transcriptograms have done their work and other bioinformatics tools are required. There are plenty of possibilities and that is up to the user to choose, and the choice may depend on the analysis purpose. We can name David Tools: Functional Annotation [10, 11] (https://david.ncifcrf.gov/) as one possible choice.

The following steps depend on the purpose of the study. Below we discuss some possible uses, but before we want to address statistical tests for estimating the significance in differential expression for transcriptograms.

5.2.4 *Statistical Tests*

The simplest test is to consider the difference between average transcriptogram values for the two sample classes being compared. Then, by calculating the standard errors for each class at each gene position, we can perform a Welch's-t test, yielding a *P*-value for each gene position. Figure 5.5B completes the figure shown in Fig. 5.5 bottom with a panel for the *P*-values profile. Additionally, the differentially expressed peaks have been identified as enriched with the indicated biological functions. This has been accomplished by feeding David tools with the gene names corresponding to each peak or valley. The idea of such an analysis is to provide a genome-wide view of what is going on. From the scenario presented by Fig. 5.5, we can understand that cystic tissue presents a depressed cell cycle expression. As polycystic kidneys grow over years, this growth must have happened before the disease enters its active phase. Other relevant information that one can gather from this picture is that there is a drastic transformation of the metabolism, involving cell differentiation and cell fate pathways, where G-protein signaling overexpression is an important player. Furthermore, tight junction functioning is altered, which can suggest that kidney morphology alterations may stem from the pathways that characterize tissue morphology. Also, mitochondrion function is depressed, which could suggest that the cells are not producing the same amount of energy as the normal kidney. These are not conclusive explanations for the disease, but rather an intermediary step in the analysis, to indicate further steps to be taken.

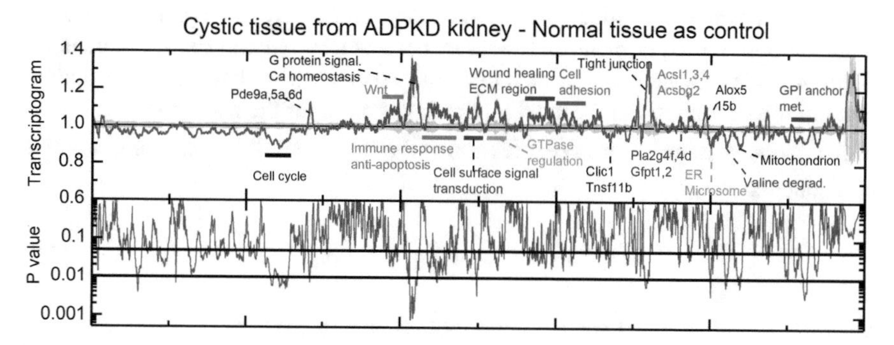

Fig. 5.5 (Top) Transcriptograms for the comparison between cystic tissue for ADPKD patients and normal tissue (control). (Bottom) The corresponding *P* value profile

P-value profile provided by Welch's t-test is the simplest test and correlates with the alterations demonstrated by the transcriptograms. They are not, however, conclusive. Another statistical test that is rather direct is a False Discovery Rate (FDR). The test is implemented as follows. For each test between two classes, scramble the samples and randomly divide into two groups with the same size as the two original classes. The original transcriptograms for each sample is still the same, but the average and standard errors will change. On average, the new classes should not differ, since each one should contain samples from the two original classes. The test consists in building a table where it gives the fraction of gene positions that are identified as differentially expressed for a given P-value. For low enough *P*-values, that should be zero. If it is not, this means the data structure is such that, for that *P* values, there are some discoveries that are false. Naturally, this is repeated for many different random selections of samples, and the false discovery rate is the average of these repetitions.

There are some other, more sophisticated tests under development, that we will not discuss here. However, the main purpose of transcriptograms is to provide a genome-wide visualization of the alterations in the gene expression. The validation of these alterations and the eventually suggested explanations based on systems biology may be obtained from other tests. As we remarked above, transcriptograms are not a competing technique with existing tools, but rather a complementary step to be taken.

5.2.5 Noise Reduction

The noise reduction observed in transcriptograms as compared to raw expression data has many advantages. In this section, we assume that the noise observed in transcriptomic data comes from two sources: a noise that is inherent to the technique and a biological variation that comes from biological differences between samples of the same class.

To demonstrate the effect of the transcriptogram on experimental data analyses, suppose an experiment that measures the whole-genome transcriptomes for 2 different classes, such that in each class the experiment is obtained for n_b biological different samples and repeated n_a times to access for the noise due to the technique. We may assume that the expression of the ith gene, $\left(t_i^{b,a}\right)_k$, has three additive components:

$$\left(t_i^{b,a}\right)_k = (s_i)_k + \left(v_i^b\right)_k + \left(\rho_i^{b,a}\right)_k, \tag{5.3}$$

where $(s_i)_k$ is the signal for the gene located at position i of the ordering under the experimental condition k. Different individuals under the same experimental conditions may transcribe differently, due to variation in DNA. The other two additive components are $\left(v_i^b\right)_k$, which responds for the biological variation and should depend only on the experimental condition and on the biological replicate and $\left(\rho_i^{b,a}\right)_k$ is a stochastic noise, which varies from a measurement to the other. The transcriptogram value $\left(T_i^{b,a}\right)_k$ for this data is the average of windows of radius r as follows:

$$\left(T_i^{b,a}\right)_k = \frac{1}{\sum_{j=i-r}^{j=i+r} \theta_j} \sum_{j=i-r}^{j=i+r} \left[\left(s_j\right)_k + \left(v_j^b\right)_k + \left(\rho_j^{b,a}\right)_k \right] \theta_j, \tag{5.4}$$

where $\theta_j = 1$ for RNA-seq experiments and for microarrays $\theta_j = 1$ only if gene at position j is a target for some probe set of the microarray platform used to generate the transcriptome. That is, the number $n(i)$ of genes measured by the technique (either microarray or RNA-seq) in the window centered at location i of the ordering may be written as

$$n(i) = \sum_{j=i-r}^{j=i+r} \theta_j. \tag{5.5}$$

For RNA-Seq, $n(i) = 2r + 1$, for all gene locations, except for the extremes of the ordering. By computing averages of the components of each class as well as the corresponding standard deviations, da Silva et al. [5] showed that when the difference between classes, $\left\langle \left(T_i^{b,a}\right)_2 - \left(T_i^{b,a}\right)_1 \right\rangle$, decreases slower than $\sqrt{n(i)}$ as the window radius increases, the transcriptogram method improves the signal-to-noise ratio and, hence, increases the power of the statistical test.

To quantify the signal-to-noise improvement due to the transcriptogram, it is useful to define $\Phi_i^k(r)$ as

$$\Phi_i^k(r) = \left| \left\langle \left(T_i^{b,a}\right)_k \right\rangle - \frac{1}{N} \sum_{j=1}^{N} \left\langle \left(T_j^{b,a}\right)_k \right\rangle \right|, \tag{5.6}$$

where $\langle \cdot \rangle$ stands for the average over the biological samples and technical replicates such that $\frac{1}{N} \sum_{j=1}^{N} \left\langle \left(T_j^{b,a}\right)_k \right\rangle$ represents the mean transcriptogram value for a given class k. In this way, for each class k and gene location i, $\Phi_i^k(r)$ measures the difference between the average over biological and technical replicates to the class mean value. These measures must be considered relative to the noise in the samples. For that, we define the contrast-to-noise ratio $\omega^k(r)$ as

$$\omega^k(r) = \frac{1}{N} \sum_{i=1}^{N} \frac{\Phi_i^k(r)}{\sqrt{\left\langle \left(T_i^{b,a}\right)_k^2 \right\rangle - \left\langle \left(T_i^{b,a}\right)_k^2 \right\rangle}}, \tag{5.}$$

that is, $\omega^k(r)$ measures the average deviations from the mean in the transcriptogram values of each gene from the mean of transcriptogram values in its class, measured in units of standard deviations.

Considering the transcriptomic data produced by Hwang et al. [12] that contemplates both technical and biological replicates for cytoplasmatic hybrid cells, da Silva et al. [4] showed that transcriptograms indeed enhances contrast-to-noise ratio due to the window averages over the gene ordering.

5.3 Case Studies

Transcriptograms have been applied to investigate cell cycle metabolic changes [4, 13], difference in metabolic response under different stress conditions [14–18], and difference in gene expression by different tissues or different cell lines [17], applying the software to the analysis of animals, plants, fungi, and bacteria genome-wide gene expression. In what follows we review two of these different applications.

5.3.1 Saccharomyces cerevisiae: *Cell Cycle*

Transcriptograms may be used to investigate expression variation during the cell cycle. Rybarczyk-Filho et al. [4] considered the experiment by Tu et al. [19] for microarray expression data obtained from yeast in a controlled culture. Cell cycle for the culture is synchronized by starvation and, after glycoses are added, the concentration levels of dissolved O_2 are constantly monitored. As yeast either respirates or ferments along the cell cycle, dissolved O_2 serves an indicator for cell cycle stage. These levels vary periodically in time and RNA content was measured for 12 different stages in three different oscillation periods, summing up 36 transcription profiles. Transcription data is available at the Gene Expression Omnibus (GEO) database under the accession GSE3431.

The gene ordering for yeast was performed using Protein—Protein interactions (PPI) from STRING database (8th version), with a 0.800 confidence score. It resulted in 4655 genes and 47,415 protein–protein associations. Figure 5.6 presents the projection of selected Gene Ontology (GO) terms, to inform how the biological functions are distributed on the ordering. Again, as shown in Fig. 5.3 for *Mus musculus,* there is a logic in the succession of the GO terms. Here window modularity is also shown. Window modularity assigns to the window central gene position (with a defined radius) the fraction of the number of interactions involving 2 genes in the window relative to the number of interactions involving 1 or 2 of these genes. When window modularity is near one, almost all interactions involving the genes in the window are with other genes also in the window.

Figure 5.7 (from Ref. [4]) presents a summary of the results for the transcriptogram

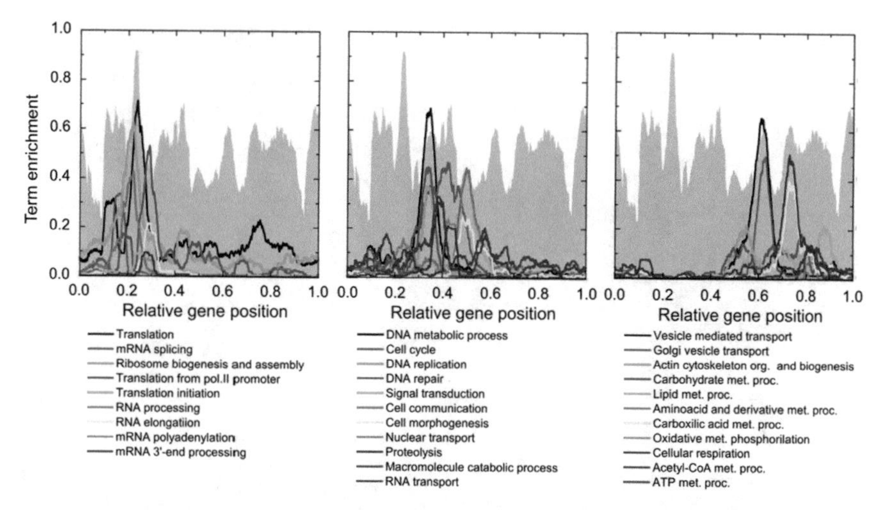

Fig. 5.6 Biological meaning of the genes/proteins ordered genes for Yeast, as presented in [4]. The colored profiles represent the term enrichment of windows of radius 250 associated to selected Gene Ontology terms and KEGG pathways. Observe that at the left end of the ordering (left panel) we find terms related translation and RNA processing, followed by genes involved in cell cycle and cell morphogenesis (center panel). After that, we find cytoskeleton and energy metabolism (right panel). The solid gray profile in three panels stands for the window modularity or radius 250, representing the fraction of interaction between 2 gene products in the window with respect to the number of interactions involving 1 or 2 genes in the window. A peak in the window modularity indicate the intervals with genes that interact preferentially within the interval

analysis for the yeast cell cycle, all results obtained considering radius 250. Panel (a) shows the transcriptograms for all 36 samples as colored lines, plotted over the window modularity profile, to help to identify the biological functions performed by the products of the gene localized where the expression alterations happen. Panel (b) presents the oscillations presented by the dissolved O_2, which is assumed to reflect the respiration/fermentation activity in yeast culture, associated with the phases of the cell cycle. The color code in all panels is the same as in panel (b), that is, orange, blue, and purple for transcriptomes obtained at time points at, respectively, the first, second, and third oscillation period. Panels (c–i) present transcriptograms for seven points (out of 12 time points) of the three cycles. Panel (c) represents the starting points (times 0, 300, and 600), relative to the mean of the three starting transcriptograms. At this starting point, the dissolved oxygen presents its lowest level, indicating that the fermentation phase begins. All other panels present transcriptograms relative to this mean, initial transcriptogram. Panel (c) shows how uniform the three relative transcriptograms are, evincing the noise reduction due to the transcriptograming process. From visual inspection, the panel (e) suggests that sample at time 675 min is not well normalized, probably due to a depth control problem (a multiplication by a constant puts the purple line on top of the other two), while panels (g) and (h) suggest the transcription peak is delayed for the first cycle (orange line).

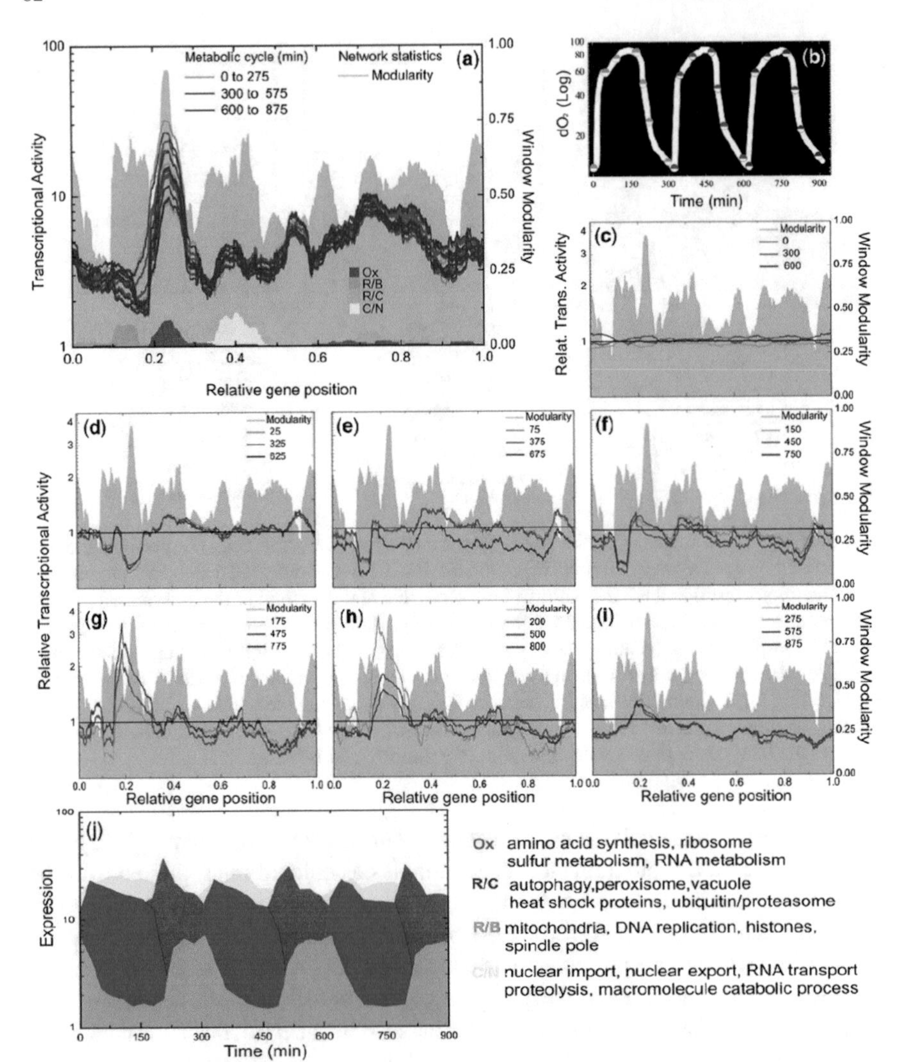

Fig. 5.7 Transcriptogram analysis for *Saccharomyces cerevisiae* from Ref. [4]. **a** Transcriptograms (radius 250) for all 36 samples. Different colors stand for different cycles. **b** Dissolved O_2 concentration versus experiment time. The time points at which the samples were taken are marked with colored dots. **c**–**i** Blackline fixed at 1 correspond to the average of the transcriptograms for the initial point of the three cycles (points at 0, 300, and 600 min). The colored lines are transcriptograms relative to the initial point average for time points as indicated in the panels. **j** Identified cycling gene sets. The localization of these gene sets on the ordering are presented in panel (**a**)

Tu et al. [14] found three gene sets that show maximum variation along the cycle and, from each gene set they gathered the top 40. Rybarczyk et al. produced enrichment profiles for these 40-gene-sets on the ordering, presented as solid gree, red, and blue profiles in Fig. 5.7a. From the transcriptogram inspection, the genes on the interval around position 0.4, linked to cell cycle genes, also show large time variation and are presented as a fourth varying gene set, presented in yellow in panels (a) and (j) in Fig. 5.7.

5.3.2 *ADPKD: Therapy Target Identification*

Genome-wide expression profiling is useful to indicate therapy targets from spotting expression alterations for genes or metabolic pathways. To exemplify how transcriptograms may help in this task, we will summarize the work by de Almeida et al. [9].

The experiment focuses on Autosomal Dominant Kidney Disease (ADPKD), caused by a mutation in a single gene, either polycystin-1 (PDK1) or polycystin-2 (PDK2). The patient lives asymptomatic until the fifth or sixth decade of life, when the accumulation of cysts in the kidney leads to eventual kidney failure, limiting treatment options to dialysis or kidney transplantation.

Reference [9] considers microarray data obtained for cell lines produced by immortalizing cells from cystic and non-cystic kidney tissue from an ADPKD donor and from normal tissue from a normal kidney. Each class comprises three replicates. The classes are, respectively, called C-ADPKD, NC-ADPKD, and NK.

The ordered gene list is produced as explained in Sect. 5.2, retrieving human PPI information from STRING database (version 9.1), confidence score 0800. The resulting ordered gene list for *Homo sapiens* comprises 9684 genes. Projection of selected GO terms and KEGG pathways on the ordered gene list shows that the succession of biological function terms follows the same logic as in Fig. 5.3, for *Mus musculus*, with some small variants (for example, energy metabolism in the right end for *Mus musculus* and in the far right for *Homo sapiens*) (Fig. 5.8).

Figure 5.9 considers the comparison of non-cystic tissue from an ADPKD patient against the normal kidney sample. It shows that there are few intervals that are represented as differentially expressed; the most clearly altered interval relates to cell cycle. ADPKD patients have their kidneys size strongly increased in the symptomatic phase and this increased expression of cell cycle related genes may explain that. To verify differential expression, de Almeida et al. considered all intervals showing P values lower than 0.05, as shown in the middle panel, and identified genes and gene sets responsible for the alteration, resulting in a list of gene sets and genes differentially expressed. Enrichment profiles for these gene sets are represented in the lower panel in Fig. 5.9.

The same analysis has been implemented to compare C-ADPKD and NK, as shown in Fig. 5.10. Now the scenario is very different. While NC-ADKD presented roughly the same transcriptogram levels as NK, with the exception for the interval

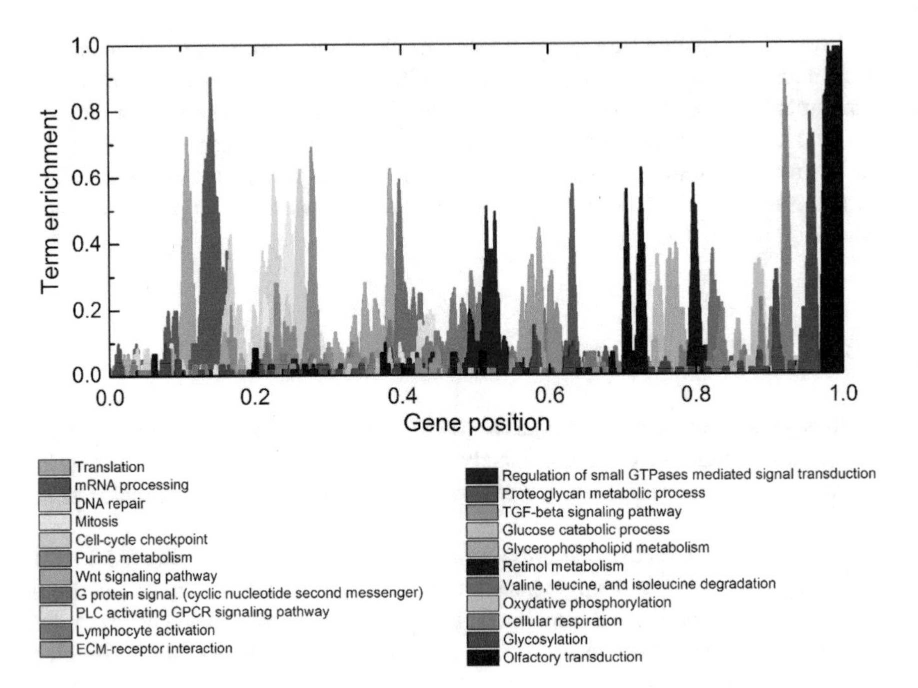

Fig. 5.8 (From Ref. [9]) Biological meaning of the genes/proteins ordered gene list for *Homo sapiens*. The colored profiles represent the term enrichment of windows of radius 30 associated with selected Gene Ontology terms and KEGG pathways. Observe that at the left end of the ordering, we find terms related to RNA translation and metabolism. Then followed by genes involved in DNA metabolism and cell cycle. After that, we find genes linked to cell fate and cell differentiation. Still in the center part, we find the terms associated to cytoskeleton dynamics and involved in the interaction with extracellular matrix. Finally, at the right end, we have pathways related to drug and energy metabolism and the large dark brown peak, that corresponds to olfactory transduction

enriched with cell cycle genes, C-ADPKD presents a whole-genome alteration. Differently from NC-ADPKD, in C-ADPKD genes involved in cell cycle have their expression depressed, suggesting that kidney size gain is due to non-cystic tissue, opening to the possibility of it happening even before the symptomatic phase. Additionally, marked peaks are related to G-protein signaling and processing, showing a cascade of expression alterations that ranges from Wnt signaling, processes involved in cell–cell juntions and cell–matrix interactions, suggesting a transformation in cell phenotype. In special, the sharp peak near 0.3 position is related to phosphodiesterases involved in control of c-GMP, which is probably related to the genome-wide expression alteration in the C-ADPKD expression profile opening to the possibility of therapies focusing on these molecules [15]. The transcriptome analysis for ADPKD cell lines proceeds from the transcriptogram presented in Fig. 5.10 with other statistical tests, by investigating possible biological explanations for such a scenario as presented in Ref. [9].

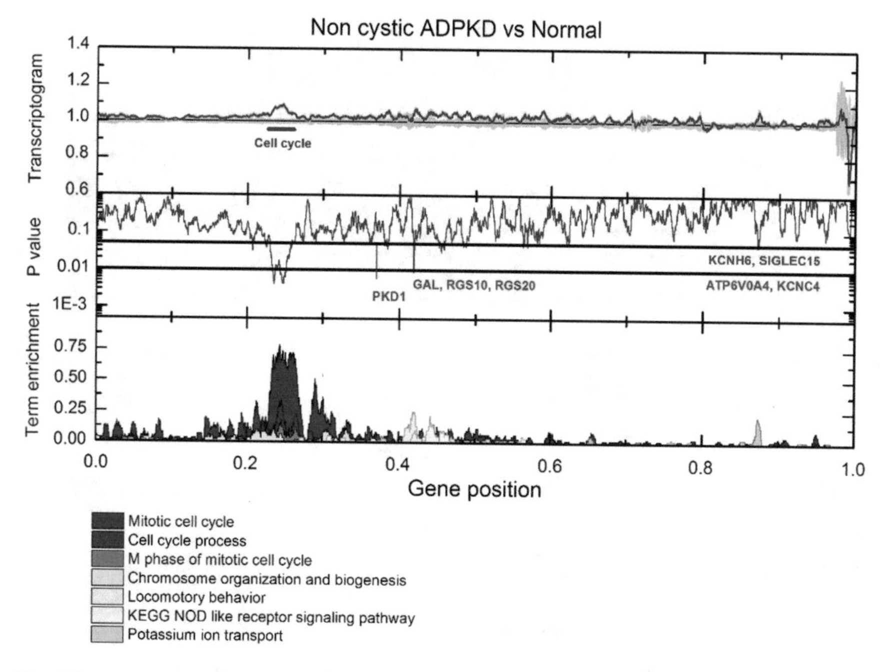

Fig. 5.9 Upper panel: Relative transcriptograms for the comparison NC-ADPKD (blue line) and NK as control (horizontal black line. The cyan and gray shadows following, respectively, the blue and back lines represent standard errors. Middle panel: *P*-value for the comparison ant each point. Lower panel: Projection of GO terms or KEGG pathways found differentially expressed

Whole-genome modifications suggest a change in cell phenotype and most commonly it is caused by a cascade of events. The root of the change, however, is not evinced in transcriptomes, that show the whole change. Previous knowledge of the disease and/or on the possible regulator genes are welcome. In this particular study, the sharp peak just before position 0.3 on the ordering presented in Fig. 5.10, linked to purine metabolism, could explain some of the disease development and suggested a therapy [9, 20].

5.4 Quality Control and Normalization Quality Assessment

Quality control protocols should always be welcome and we suggest transcriptograms may serve as a tool for quality control. To support our claim, we consider the analysis presented in [4] considering a publicly available dataset produced by Fry et al. [21]. The experiment was meant to investigate the role of SGS1 gene in yeast aging, to gather information on the role played by its human homologs WRN and BLM. Mutations in these two genes result in the human aging diseases known as Werner and Bloom syndromes, respectively. The experiment designed by Fry et al. consists

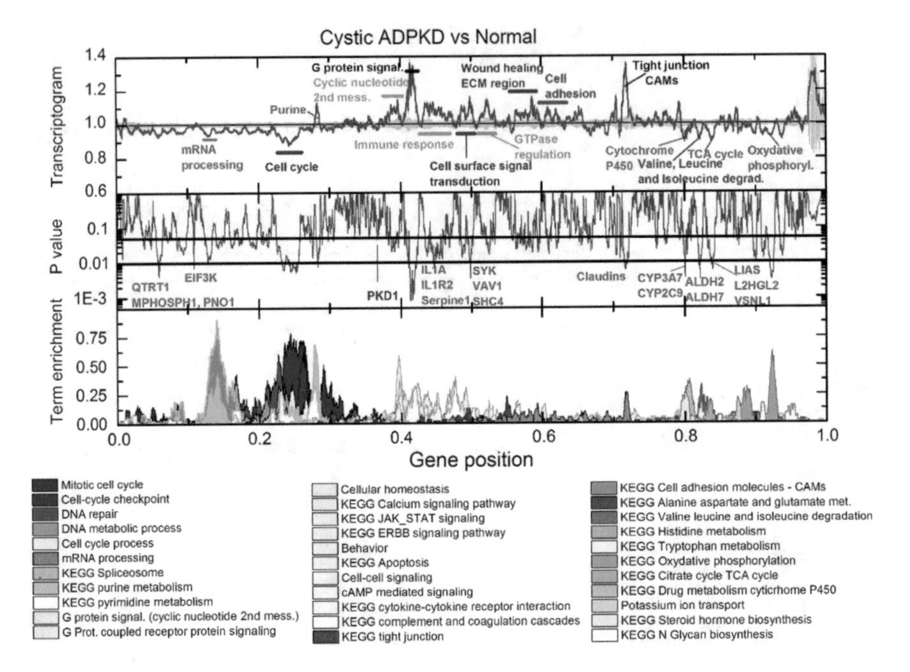

Fig. 5.10 Upper panel: Relative transcriptograms for the comparison C-ADPKD (blue line) and NK as control (horizontal black line). The light red and gray shadows following, respectively, the red and back lines represent standard errors. Middle panel: *P*-value for the comparison ant each point. Lower panel: Projection of GO terms or KEGG pathways found differentially expressed

of comparing the responses of SGS1 deleted and wild type strains when their culture medium is added with 0.1% methyl methane sulfonate (MMS), a DNA alkylating agent that produces DNA damage. Under such a stress, the lack of ability to protect or recover DNA may be linked to pathological aging. Fry et al. produced in duplicates the transcriptomes for wild type and SGS1 mutant strains cultured in a neutral medium and wild type and SGS1 mutant strains cultured in an MMS-altered medium. Rybarczyk et al. [4] produced transcriptograms, radius 250, for Fry et al.'s data and the results are presented in Fig. 5.11. Panel (a) shows that SGS1 deletion depresses gene transcription. Panels (b) and (d) suggest that on average no great changes happen due to SGS1 deletion, which is roughly one of the conclusions in Ref. [21]. However, panels (d) and (e) shows that the transcriptograms associated with one replicate for wild type in MMS medium and one replicate for the SGS1 mutant strain in MMS medium overlap almost perfectly. The explanation could be a cell arrest in a given phase of cell cycle, or that the samples have been wrongly labeled.

Normalization or preprocess protocols are intended to fix different issues, and may depend on the experiment design, purpose, or technology. To give an example of how transcriptograms may help in qualifying a normalization procedure, we refer to panel (e) in Fig. 5.7. The purple profile follows almost perfectly a parallel profile when compared to lines orange and blue. A multiplication by a constant of all expressions

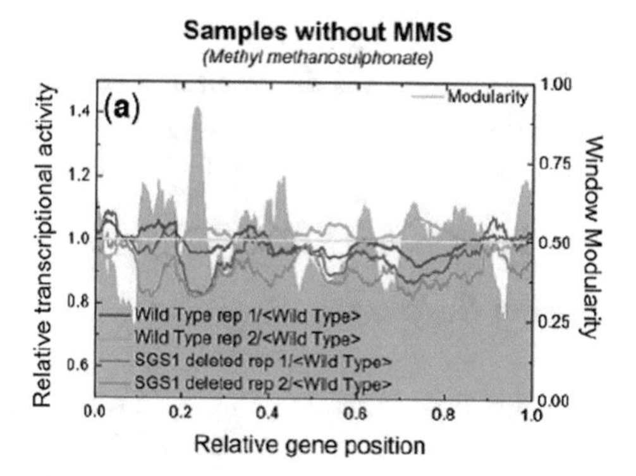

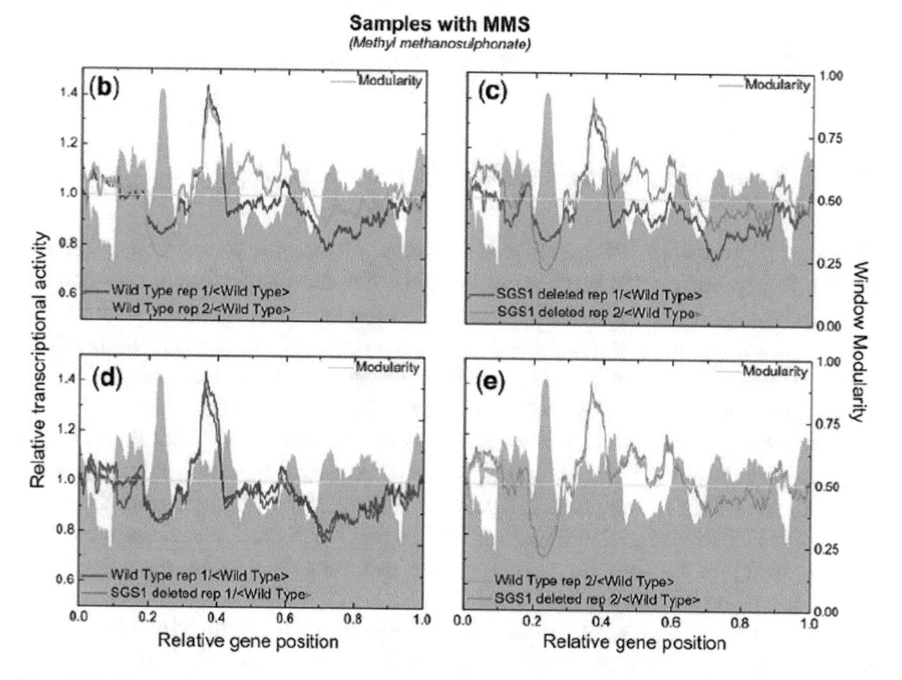

Fig. 5.11 Transcriptograms, radius 250, produced by Rybarczyk-Filho et al. [4] for Fry et al. [21]. For all panels, the gray profile stands for window modularity, radius 250. The panels present transcriptograms relative to the average transcriptogram in the neutral medium for the wild type (horizontal, yellow line). **a** Neutral medium: Orange and red lines describe the two replicates for the SGS1 mutant strain and cyan and blue lines stand for the wild type strains. **b** MMS medium: cyan and blue lines describe the two replicates for the wild type. **c** MMS medium: orange and red lines describe the two replicates for the SGS1 mutant strain. **d** and **e** MMS medium: each panel contains transcriptograms for one replicate for the wild type and one for SSG1 mutant strain (as labeled)

corresponding to the purple profile may overlay all three profiles, repeating what is seen in other panels. As transcription is the result of a very complex biochemical reaction, we then discard the possibility that a multiplicative constant may emerge from a biological origin, being rather due to a normalization fault.

5.5 Transcriptogram Softwares

There are two software to produce transcriptograms. One is a code available at https://lief.if.ufrgs.br/pub/biosoftwares/transcriptogramer/. This software is licensed under GPL open-source license that is completely free to use and download. In the software home page also presents a manual.

The second possibility is an R-Bioconductor package, as follows.

5.5.1 Transcriptogramer R/Bioconductor Package

The transcriptogramer R/Bioconductor package [22] is able to create a transcriptogram from preprocessed Microarray and RNA-Seq data, to perform a case-control differential expression, to assess the functional enrichment using Gene Ontology, as well to perform other features related to graph properties. The transcriptogramer R package does not compute an ordering from protein–protein interaction (PPI) data. Instead, the package contains prebuilt orderings for four organisms: *Homo sapiens*, *Mus musculus*, *Saccharomyces cerevisiae*, and *Rattus norvegicus*. Prebuilt orderings were computed for each organism using three combined score thresholds (700, 800, and 900) from STRING database PPI data. The PPI data were obtained from STRING database v11.0 for the package version 1.8.0, being the prebuilt orderings subject to updates in future versions. In addition to the data contained in the package, other orderings can be used as inputs.

The package requires four inputs to run the analysis: (i) the PPI data, (ii) an ordering compatible with the used PPI data, (iii) expression data from two different conditions, and (iv) a dictionary mapping the expression data identifiers to the PPI data. Initially, the expression data is mapped to the ordering using the dictionary. Then, a sliding window of size $2r + 1$ travels through each position in the ordering. The sliding window computes the average expression from groups of functionally associated proteins for each sample, as explained in Sect. 5.2.1. The R package performs this process taking into account periodic boundaries independently of data platform (Microarray or RNA-Seq).

The samples are then labeled according to their condition (i.e., case or control samples), and the differential expression is performed using functions from the limma R/Bioconductor package [23]. Thus, the differential expression is performed using the average expression of functionally associated proteins, previously stored in the ordering positions. The differential expression result is processed to adjust p-values, and to identify differentially expressed clusters of functionally associated genes. Being each position a group of genes, if a group is differentially expressed (DE), and there is another differentially expressed group reachable within the radius previously defined, these groups, as well as all groups between them, are clustered. As a group is composed by $2r + 1$ genes, a cluster with the first DE group at position x, and the last DE group at position y, involves genes from $r - x$ to $r + y$.

The transcriptogramer R package produces a plot to summarize the differential expression result (Fig. 5.12). Such plot depicts the difference of means, between the case and the control, from the average expression of groups at each position. The mean from the control samples is used as a reference, represented by a black line placed at 0 (y-axis), and the difference of means from the case samples to the control is represented by a gray line. Colored regions over the gray line represent the identified clusters.

The Gene Ontology (GO) functional enrichment analysis is performed after the differential expression. This analysis is done using functions from the topGO R/Bioconductor package [24] to statistically assess GO terms related to the clusters identified during the differential expression. As a preprocessing step, a dictionary is automatically built to map the proteins in the ordering to Gene Ontology identifiers from a given ontology (Biological Process, Molecular Function, or Cellular Component). The ontology used by default is the Biological Process, but all three ontologies are supported by the package. To statistically assess GO terms, the genes from each

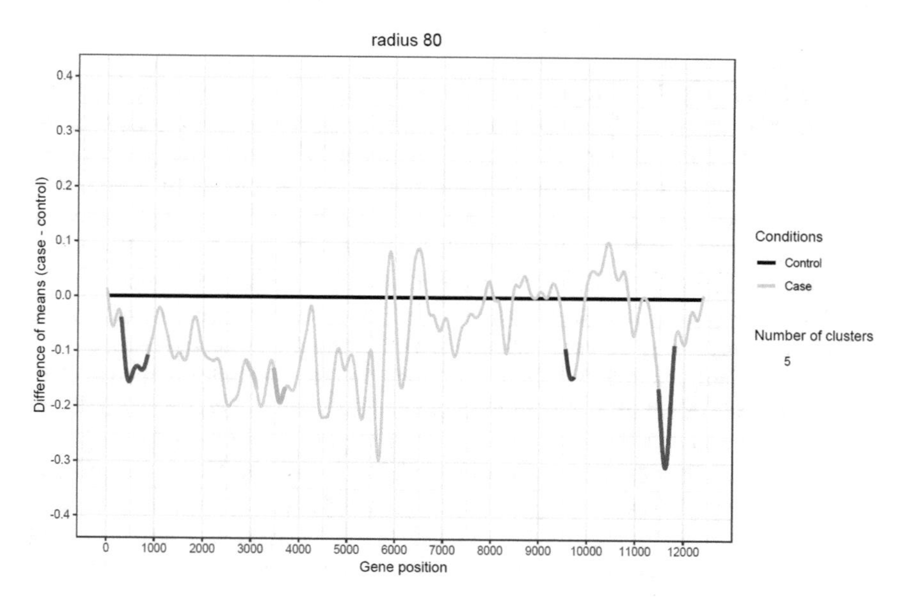

Fig. 5.12 Output figure of differential expression analysis performed by the transcriptogramer R/Bioconductor package. The black line represents control mean and the gray line represents the case mean. The color lines over the gray line represent the differentially expressed cluster of functional-related genes (Extracted from the transcriptogramer user's guide, https://doi.org/10.18129/B9.bioc.transcriptogramer.)

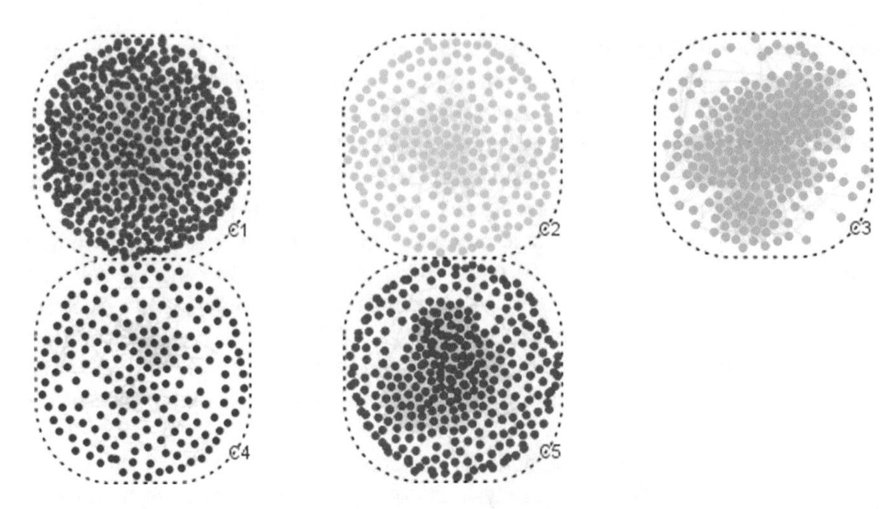

Fig. 5.13 PPI network graphs of the differentially expressed clusters of functionally associated genes. The colors are the same used to identify differentially expressed clusters in the transcriptogram in Fig. 5.12 (Extracted from the transcriptogramer user's guide, https://doi.org/10.18129/B9.bioc.transcriptogramer.)

cluster are considered as an independent list of genes of interest, and the genes in the ordering are considered as the universe of genes. The result is a list of GO terms statistically significant, given a 0.05 threshold for the adjusted *p*-values as default, related to each cluster.

After identifying the differentially expressed clusters and performing the Gene Ontology enrichment analysis of each differentially expressed cluster, the package also performs a PPI analysis. The transcriptogramer R/Bioconductor package has a function to plot PPI network of a differentially expressed cluster of functionally associated genes (Fig. 5.13). Here, each cluster is nested and the node colors are the same used to identify the clusters on the transcriptogramer plot (see Fig. 5.12). Therefore, by running the complete pipeline, the user can identify gene clusters differentially expressed, identify GO terms associated with each cluster, and plot the PPI networks involving the genes within each differentially expressed cluster.

The transcriptogramer R/Bioconductor package was successfully applied to identify the systemic effect of lead poisoning in human neural progenitor cells [18]. In the study, Reis and collaborators evaluated RNA-seq samples of human embryonic-derived neural progenitor cells treated with 30 μM lead acetate in 26-day time-course. In this paper, all the transcriptogramer R/Bioconductor packages were used. The methodology was able to identify the biological processes affected by lead poisoning as well as the transcriptional impact of lead poisoning in acute and chronic exposure [N4].

Currently, the transcriptogramer R/Bioconductor package (version: 1.8.0) is freely available at Bioconductor version: Release 3.10 (https://bioconductor.org/packages/release/bioc/html/transcriptogramer.html) for R version 3.6.

References

1. Subramanian A et al (2005) Gene set enrichment analysis: a knowledge-based approach for interpreting genome-wide expression profiles. PNAS 102:15545–15550
2. Dinu I et al (2007) Improving gene set analysis of microarray data by SAM-GS. BMC Inform 8:242
3. Evans C, Hardin J, Stoebel DM (2018) Brief Bioinform 19:776–792
4. Rybarczyk-Filho JL et al (2011) Towards a genome-wide transcriptogram: the Saccharomyces cerevisiae case. Nucleic Acids Res 39:3005–3016
5. da Silva SRM, Perrone GC, Dinis JM, de Almeida RMC (2014) Reproducibility enhancement and differential expression of non predefined functional gene sets in human genome. BMC Genomics 15:1181
6. Szklarczyk D et al (2017) The STRING database in 2017: quality-controlled protein-protein association networks, made broadly accessible. Nucleic Acids Res 45:D362–D368
7. Jensen LJ et al (2009) STRING 8—a global view on proteins and their functional interactions in 630 organisms. Nucleic Acids Res D412–D416
8. Kanehisa M, Goto S (2000) KEGG: kyoto encyclopedia of genes and genomes. Nucleic Acids Res 28:27–30
9. de Almeida RMC et al (2016) Transcriptome analysis reveals manifold mechanisms of cyst development in ADPKD. Human Genomics 10:37
10. Huang DW, Sherman BT, Lempicki RA (2009) Systematic and integrative analysis of large gene lists using DAVID bioinformatics resources. Nat Protocols 44–57
11. Huang DW, Sherman BT, Lempicki RA (2009) Bioinformatics enrichment tools: paths toward the comprehensive functional analysis of large gene lists. Nucleic Acids Res 37:1–13
12. Hwang S et al (2011) Gene expression pattern in transmitochondrial cytoplasmic hybrid cells harboring type 2 diabetes-associated mitochondrial DNA haplogroups. PLoS ONE 6:e22116
13. Sanhudo LS, Dinis JM, Lenz G, de Almeida RMC (2019) Cell cycle genome wide profiles in reveals varying temporal gene expression correlations (submitted)
14. de Oliveira-Busatto LA et al (2020) The soybean transcriptogram allos a wide genome to single gene analysis that evinces time dependent drought response (submitted)
15. Miotto YE et al (2019) Identification of root transcriptional responses to shoot illumination in Arabidopsis thaliana. Plant Mol Biol 101:487–498
16. Cadavid IC, Guzman F, de Oliveira-Busatto Luisa A, Margis R (2020) Transxriptomic ando post-transcriptional analyses of two soybean cultivars under salt stress. Mol Biol Rep (accepted for publication)
17. Ferrareze PAG et al (2017) Transcriptional analysis allows genome reannotation and reveals that Cryptococcus gatii CGII undergoes nutrient restriction during infection. Microorganisms 5:49
18. Reis CF et al (2019) Systems biology-based analysis indicates global transcriptional impairment in lead-treated human neural progenitor cells. Front Genet 10:791
19. Tu BP, Kudlicki A, Rowicka M, MacKnight SL (2005) Logic of the yeast metabolic cycle: temporal compartmentalization of cellular processes. Science 310:1152–1158
20. Bacallao R, Clendenon SG, de Almeida RMC, Glazier JA (2018) Targeting cGMP-related phosphodiesterases to reduce cyst formation in cystic kidney disease, and related materials and methods. US2018/0078559 A1, 22 March 2018
21. Fry RC, Sambandan TG, Rha C (2003) DNA damage and stress transcripts in Saccharomyces cerevisiae Mutant sgs1. Mech Ageing Dev 124:839–846
22. Morais DAA, e Almeida RMC, Dalmolin RJS (2019) Transcriptogramer: an R/bioconductor package for transcriptional analysis based on canonical protein-protein interaction data. Bioinformatics 35:2875–2876
23. Ritchie ME et al (2015) limma powers differential expression analyses for RNA-sequencing and microarray studies. Nucleic Acids Res 43(7):e47
24. Alexa A, Rahnenführer J, Lengauer T (2006) Improved scoring of functional groups from gene expression data by decorrelating GO graph structure. Bioinformatics 22(13):1600–1607

Chapter 6
A Tutorial on Sobol' Global Sensitivity Analysis Applied to Biological Models

Michel Tosin, Adriano M. A. Côrtes, and Americo Cunha

Abstract Nowadays, in addition to traditional qualitative methods, quantitative techniques are also a standard tool to describe biological systems behavior. An example is the broad class of mathematical models, based on differential equations, used in ecology, biochemical kinetics, epidemiology, gene regulatory networks, etc. Independent of their simplicity or complexity, all these models have in common (generally unknown a priori) parameters that need to be identified from observations (data) of the real system, usually available on the literature, obtained by specific assays or surveyed by public health offices. Before using this data to calibrate the models, a good practice is to judge the most influential parameters. That can be done with aid of the Sobol' indices, a variance-based statistical technique for global sensitivity analysis, which measures the individual importance of each parameter, as well as their joint-effect, on the model output (a.k.a. quantity of interest). These variance-based indexes may be computed using Monte Carlo simulation but, depending on the model, this task can be very costly. An alternative approach for this scenario is the use of surrogate models to speed-up the calculations. Using simple biological models, from different areas, we develop a tutorial that illustrates how practitioners can use Sobol' indices to quantify, in a probabilistic manner, the relevance of the parameters of their models. This tutorial describes a very robust framework to compute Sobol' indices employing a polynomial chaos surrogate model constructed with the UQLab package.

Keywords Mathematical biology · Global sensitivity analysis · Sobol' indices · Surrogate model · Polynomial chaos expansion · UQLab

M. Tosin · A. Cunha
Nucleus of Modeling and Experimentation with Computers—NUMERICO, Rio de Janeiro State University—UERJ, Rua S ão Francisco Xavier, 524, Rio de Janeiro 20550-900, Brazil
e-mail: michel.tosin@uerj.br

A. Cunha
e-mail: americo@ime.uerj.br

A. M. A. Côrtes (✉)
Multidisciplinary Computer Research Center—NUMPEX-COMP, Federal University of Rio de Janeiro—UFRJ, Rodovia Washington Luiz, Km 104.5, Duque de Caxias 25.265-970, Brazil
e-mail: adriano@nacad.ufrj.br

F. A. B. da Silva et al. (eds.), *Networks in Systems Biology*, Computational Biology 32,
https://doi.org/10.1007/978-3-030-51862-2_6

6.1 Introduction

The progress that computer models have made over the past few decades in various areas of biology is impressive, with an increasing demand for the use of these tools in several scenarios that require quantitative predictions, such as infectious diseases [1, 2], change in species dynamics due to climate change [3, 4], protein signaling in cells [5, 6], etc. For instance, the recent pandemic of the Coronavirus SARS-CoV-2, responsible for the disease COVID-19 [7, 8]. The rapid spread of the disease is problematic because of the risk of death, and also for most of the health systems are not prepared to receive so many sick people in hospitals [9]. This kind of emergency scenario illustrates the importance of cooperation between researchers from different areas to deal with these situations by creating some understanding about the disease and propose strategies to slow down its reach, as fast as possible. The use of mathematical models can be very useful in situations like this, as long as it has the necessary ingredients to describe the basic aspects of the biological system of interest, and it is well-calibrated with real data [10].

The utility of a model must precede a process of certification to guarantee its match with reality [11]. During this process, it is useful to understand how certain variations in a model input parameter can affect its outcome (response) [12]. This knowledge allows one to detect which phenomena are more important in the real system [13, 14]. For instance, in the Coronavirus example, the effect of the response time to the outbreak by local governments and the various mitigation strategies has been evaluated [9].

In this sense, the mathematical model parameters are the "fundamental blocks" of the predictive tool and understand how each of these "pieces" affects the outcomes related to the biological system of interest is a key point for proper use of this quantitative arsenal. Global sensitivity analysis can be very useful to clarify this understanding, once it can provide information about the dependence of the model outcome with respect to each one of its input parameters [15]. Following that idea, this chapter aims to present a tutorial that illustrates the process of global sensitivity analysis in biological systems via Sobol' indices [16].

In Sect. 6.2 we present the setting for the mathematical model representation. Sobol' indices method is described in the Sect. 6.3, where the sensitivity analysis ideas and goals are also characterized. After that, Sect. 6.4 brings a brief overview of surrogate models as a computationally efficient strategy to approximate the Sobol' indices. With the theory detailed, Sect. 6.5 is responsible for presenting the Sobol framework and developing three biological examples: the predator–prey model. an NF-κB signaling pathway model and the SIR epidemiological model. All of them reproducible by the reader applying the codes available in a public repository.

6.2 Mathematical Modeling

Modeling is the process of creating a (simplified) representation of the real process [17]. The objective is not being completely faithful to reality, but just being able to present its main characteristics or facilitate the identification and understanding of its mechanisms and processes [10]. To that goal, we can make several types of models for a single reality. With a good model, you can analyze many things about your problem to obtain more knowledge about it. Between the modeling objectives, we can highlight the understanding of the mechanisms that govern the phenomena of interest, prediction of the future or of some state that is currently unknown, and control by constraining a system to produce a desirable condition [18]. Because of this, mathematical modeling is often required to deal with biological processes.

The process of modeling involves some paradigms and steps but, more objectively, any model is composed by three elements: (i) the input, that comes from the previous useful information about the system state; (ii) the output, which is the information about the real problem; and (iii) the model, which is the system representation that maps input to the output [19]. We can make a parallel with a manufactured production defining the raw material as the input, the processed product as the output, and the machine responsible for transform one in another as the model. Intuitively, with a more sophisticated machine, we can create a more elaborate object as well as with a more detailed model we can obtain a better representation of the reality. However, including further details is necessary for a more complex representation. Greater complexity implies greater difficulty in analyzing all aspects of interest. Thereby, the hypotheses must balance the most important elements of the phenomena and a reasonable form of representing the relationship of these elements with what is interesting to quantify or evaluate. The model itself is this relation [20]. For example, the evolution of some animal species is related to the food resources available but also with the topographic characteristics of the environment. Assuming the species has lived in an environment for a long time, we can neglect then it and only take in count the first one. Sure, the effect of this approximation must be tested to guarantee your model is sufficiently compatible with reality. Additionally, the number of individuals of the species will increases (or decreases) according to some units of time. The change in a month is probably different than in a year. This time scale relation may be relevant or not, depending on the process treated.

Technically we represent the important input components of a real system in the model by parameters. In mathematical words, these parameters are quantities whose values are dependent on specific aspects of the phenomena and that controls how the model entails the desired predictions [10]. This prediction is called quantity (or quantities if are more than one) of interest (QoI). So, we can think of the mathematical model as a "machine" that uses the parameters as ingredients to obtain the QoIs. Formally, the model is defined by a mathematical operator $\mathcal{M}$, so that the QoIs are given by

$$\mathbf{y} = \mathcal{M}(\mathbf{x}), \tag{6.1}$$

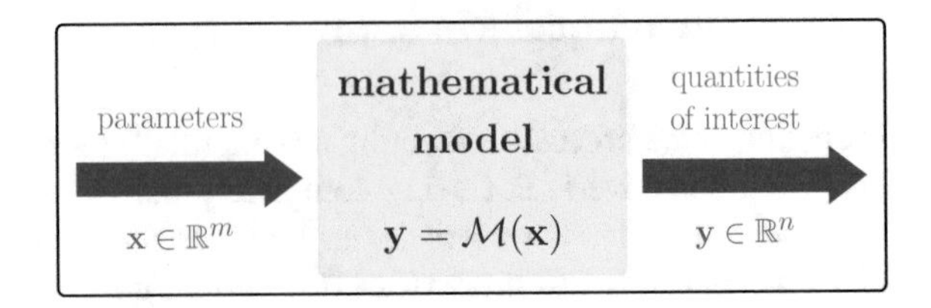

Fig. 6.1 Schematic representation of a general computational model

where the vector $\mathbf{x} \in \mathbb{R}^m$ contains all the model parameters and the n QoIs are reunited in a vector $\mathbf{y} \in \mathbb{R}^n$. The structure of a general computational model is illustrated in Fig. 6.1. If the model is time dependent, one can lump the QoI at the instants of interest into a vector. To simplify the notation, the time dependence will be omitted in the theoretical-based discussions.

6.3 Sensitivity Analysis

Sensitivity Analysis (SA) is a process in which the contribution of each input parameter of a mathematical model, to its response, is identified [21]. In particular, for nonlinear systems, not only the parameters individually affect the QoIs, but also the interactions between then. In this case, the joint-effect of parameter variation must also be quantified. This feature makes difficult the process of SA for high-dimensional systems because several orders of interactions must be computed, and their influences are not trivial in general. Another important issue in the modeling process is to take into account the underlying uncertainties. The lack of knowledge about the real system and the natural variability of some parameters create some difficulties in any attempt to understand a phenomenon, that can be compensated by the use of a stochastic model [14, 19]. This is the domain of investigation of the Uncertainty Quantification (UQ) theory.

UQ and SA areas gained greater notoriety in the last decades, where it was recognized that is extremely important not only to deal with multiple sources of uncertainties in a mathematical model but, in some contexts, also apply strategies to reduce their effects, especially if their presence can be catastrophic for the system [22]. An alternative approach is also possible, where one tries to take advantage of the uncertainties to improve the performance of a given system. Although UQ and SA literature are already considerable, it is common to find texts that confuse these two concepts [13]. The first one deals with the quantitative evaluation of how the uncertainties in the input variables are transported to the model response, which is done through the calculation of the so-called uncertainty propagation process. On the other side, the second concept is related to the quantitative evaluation of how uncertainty in the input parameters contribute (individually or jointly) to change the model response. In UQ it is mandatory to prescribe a characterization of the uncertainties so that after the calculation of their propagation, one has a general overview of the uncertainty in the model's response. In the SA perspective, the idea is to discover how a certain parameter changes the system response when changed [23].

The SA methods can be distinguished in two large groups: The Local SA (LSA) methods and the Global SA (GSA) methods [24]. From a general perspective, the difference between them is how the parameter space is explored in each case. Local methods are normally based in partial derivatives or gradients but are (totally) dependent on the point of the parameter space for which the model is evaluated. In this way, the LSA results do not reflect the general dependency of model outcome concerning the selected input parameter. Moreover, for nonlinear models, the interactions of the parameters are very important for the response and local analysis can not capture those effects. Differently, GSA methods explore screening or variance decomposition to cover the local analysis limitations [23]. In this chapter the focus of our interest in GSA, as a form to discover the most important parameters, that is, those that contribute the most to the model response (factor prioritization) and those which contribute very little and can potentially be fixed (factor fixing) [13].

6.3.1 *Sobol' Indices*

There are several methods to perform SA present in scientific texts [13, 14]. But as said in the previous section, here we have preference for global methods. For the global sensitivity analysis framework to be robust and general, it makes sense to select a method that is simple to implement and use. The Sobol' indices [16] is a variance-based method very popular in recent literature [12] after the dissemination of surrogate model ideas, and very soon will be clear why.

In this framework, the system is analyzed from a probabilistic perspective that considers the model input as a random vector $\mathbf{X}$, characterized by a joint Probability Density Function (PDF) $f_{\mathbf{X}}$ with support $I_{\mathbf{X}}$. The stochastic version of the model is represented as

$$\mathbf{Y} = \mathcal{M}(\mathbf{X}), \tag{6.2}$$

which has joint PDF $f_{\mathbf{Y}}$ that is unknown before the propagation of uncertainties [23]. Note that the notation with capital letters is chosen to describe random objects, and the one with small letters still make sense for the deterministic case. Assuming for simplicity that the QoI is a scalar value, with the random input parameters being composed by independent and identically distributed (iid) uniform parameters X_i, scaled to have support $[0, 1]^m$, the Hoeffding-Sobol decomposition is given by

$$Y = \mathcal{M}_0 + \sum_{i=1}^{n} \mathcal{M}_i(X_i) + \sum_{1 \leq i < j \leq n} \mathcal{M}_{ij}(X_i, X_j) + \cdots + \mathcal{M}_{1 \ldots m}(X_i, \ldots, X_n), \tag{6.3}$$

where

$$\begin{aligned}\mathcal{M}_0 &= \mathbb{E}[Y], \\ \mathcal{M}_i(X_i) &= \mathbb{E}[Y \mid X_i] - \mathcal{M}_0, \\ \mathcal{M}_{ij}(X_i, X_j) &= \mathbb{E}[Y \mid X_i, X_j] - \mathcal{M}_0 - \mathcal{M}_i - \mathcal{M}_j, \\ &\dots\end{aligned} \tag{6.4}$$

that is, $\mathcal{M}_0$ is the mean value, and the terms of increasing order are conditional expectations defined in a recursive way, that characterize an unique orthogonal decomposition of the model response [16, 21].

Following this idea, we can now decompose the total variance of the response as follows

$$\text{Var}(Y) = \sum_{\mathbf{u}} \text{Var}\left(\mathcal{M}_{\mathbf{u}}(\mathbf{X}_{\mathbf{u}})\right), \quad \text{for } \varnothing \neq \mathbf{u} \subset \{1, \dots, n\}, \tag{6.5}$$

where $\text{Var}\left(\mathcal{M}_{\mathbf{u}}(\mathbf{X}_{\mathbf{u}})\right)$ expresses the conditional variance for the subvector $\mathbf{X}_{\mathbf{u}}$, containing the variables which indices are indicated by the subset $\mathbf{u}$ [21]. Thus, the Sobol' index associated to the subset $\mathbf{u}$ is defined as the ratio between the contribution given by the interaction among the components of $\mathbf{u}$ for the model variance, and the total variance itself [14], i.e.,

$$S_{\mathbf{u}} = \frac{\text{Var}\left(\mathcal{M}_{\mathbf{u}}(X_{\mathbf{u}})\right)}{\text{Var}\,(Y)}. \tag{6.6}$$

As a result of this equation we can verify that, for $\mathbf{u} \subset \{1, \dots, n\}$, $\mathbf{u} \neq \emptyset$,

$$\sum_{\mathbf{u}} S_{\mathbf{u}} = \sum_{i=1}^{n} S_i + \sum_{1 \leq i < j \leq n} S_{ij} + \dots + S_{1\dots m} = 1, \tag{6.7}$$

that is, by construction the sum of all the Sobol' indices must be equal to the unit.

The terms

$$S_i = \frac{\text{Var}\left(\mathcal{M}_i(X_i)\right)}{\text{Var}(Y)}, \qquad i = 1, \dots, n \tag{6.8}$$

are called the first-order Sobol' indices for the single variable X_i and denote the individual effect of X_i for the total model variate. Similarly,

$$S_{ij} = \frac{\text{Var}\left(\mathcal{M}_{ij}(X_{ij})\right)}{\text{Var}(Y)}, \qquad 1 \leq i < j \leq n \tag{6.9}$$

are the second-order indices that contemplate the effect of the interaction between the variables X_i and X_j. Keep following we can construct the Sobol' indices of all orders until the mth order index, $S_{1,\dots,m}$, which represents the contribution of the interaction between all the variables in $\mathbf{X}$ [16].

To measure the full contribution of the ith random variable X_i for the total variance either by its single effect or by its interaction with others, we use the total Sobol' indices, which are defined by

$$S_i^T = \sum_{\substack{\mathbf{u}\subset\{1,\ldots,n\} \\ i\in\mathbf{u}}} S_{\mathbf{u}} \quad i = 1, \ldots, n. \tag{6.10}$$

6.4 Surrogate Models

To compute Sobol' indices, defined Sect. 6.3.1, it is necessary to calculate the underlying variances. Despite this task can be done with Monte Carlo (MC) simulation, the associated computational cost can be high (infeasible for high-dimensional systems) and subjected to numerical instabilities like cancelation errors. A very appealing alternative, which allows circumventing these two problems, is the use of surrogate models based on polynomial chaos expansions [15].

6.4.1 Polynomial Chaos Expansion

The Polynomial Chaos Expansion (PCE) of the computational model response is a sum of orthogonal polynomials weighted by coefficients to be determined [22, 23, 25, 26], which reads as

$$Y = \mathcal{M}(\mathbf{X}) = \sum_{\alpha=0}^{\infty} y_\alpha \Psi_\alpha(\mathbf{X}), \tag{6.11}$$

where $\Psi_\alpha(\mathbf{X})$ are multivariate orthonormal polynomials, associated with the density $f_{\mathbf{X}}$, and y_α are the deterministic coefficients to be determined in order to construct the expansion [22]. For computational implementation purposes, a truncated PCE is considered

$$Y \approx \mathcal{M}^{PC}(\mathbf{X}) = \sum_{\alpha=0}^{P} y_\alpha \Psi_\alpha(\mathbf{X}), \tag{6.12}$$

where P is the number of terms in the expansion, which depends on the number of input random variables m and the maximum degree allowed for the polynomial expansion p, according to the formula $P + 1 = (m + p)!/(m!p!)$. Note the quality of your PCE is directly dependent on the number of terms you have in the expansion [27].

The family of orthonormal polynomials to be used is chosen according to the model input distribution, in a sense that seeks to minimize the number of terms needed in the expansion to build a good computational representation of the model. Table 6.1 summarizes a list of the most classical polynomial families and underlying

Table 6.1 Correspondence between the random variable distribution and the optimal family of orthonormal polynomials [22]

Type of variable	Distribution	Support	Orthogonal polynomials
Uniform	$\mathbf{1}_{]-1,1[}(x)/2$	$[a, b]$	Legendre
Gaussian	$\frac{1}{\sqrt{2\pi}}e^{-x^2/2}$	$(-\infty, \infty)$	Hermite
Gamma	$x^a e^{-x}\mathbf{1}_{\mathbb{R}_+}(x)$	$[0, \infty)$	Laguerre
Beta	$\mathbf{1}_{]-1,1[}(x)\frac{(1-x)^a(1+x)^b}{B(a)B(b)}$	$[a, b]$	Jacobi

distributions. For further details about the construction of the polynomial basis see [22, 23, 27].

6.4.2 Calculation of the Coefficients

Several strategies can be adopted to calculate the PCE coefficients. In this section, we describe the basics of the calculation procedure based on a regression, employing the Ordinary Least-Squares (OLS) method because of its simplicity and generality [24]. The reader is encouraged to see further details about other methods for PCE coefficients calculation in [22, 27].

From the moment that we use the truncated PCE from Eq. (6.11) for computer simulations, there is an error that distances it from the "complete" PCE given by Eq. (6.12), which can be rewritten as follows:

$$Y = \mathcal{M}(\mathbf{X}) = \sum_{\alpha=0}^{P} y_\alpha \Psi_\alpha(\mathbf{X}) + \varepsilon_P = \mathbf{y}^T \mathbf{\Psi}(\mathbf{X}) + \varepsilon_P \,, \tag{6.13}$$

where ε_P is the truncation error, $\mathbf{y} = \left(y_0, \ldots, y_P\right)^T$ is the vector of coefficients and $\mathbf{\Psi}(\mathbf{x}) = \{\Psi_0(\mathbf{x}), \ldots, \Psi_P(\mathbf{x})\}$ is the matrix that assembles the values of all the orthonormal polynomials in $\mathbf{X}$ [27]. Therefore, coefficients of the truncated PCE are calculated in a way to reduce the truncation error [23]. For that we obtain via MC method a "small" set of N_s samples of each random variable X_i, called the experimental design

$$\chi = \left\{\mathbf{x}^{(1)}, \mathbf{x}^{(2)}, \ldots, \mathbf{x}^{(N_s)}\right\} \tag{6.14}$$

and compute the model response for that samples

$$y^{(1)} = \mathcal{M}(\mathbf{x}^{(1)}) \,, \tag{6.15}$$

$$y^{(2)} = \mathcal{M}(\mathbf{x}^{(1)}) \,, \tag{6.16}$$

$$\vdots \tag{6.17}$$

$$y^{(N_s)} = \mathcal{M}(\mathbf{x}^{(N_s)}) \,. \tag{6.18}$$

The key problem is calculate the PCE coefficients that force the PCE to better fit the responses obtained from the computational model. This is the classic least-squares regression problem

$$\mathbf{y}^T \Psi(\mathbf{x}) \approx \mathcal{M}(\mathbf{x}), \tag{6.19}$$

for which the general solution may be expressed as

$$\mathbf{y}^* = \underset{\mathbf{y}}{\text{argmin}}\, \mathbb{E}\left[\left(\mathbf{y}^T \Psi(\mathbf{x}) - \mathcal{M}(\mathbf{x})\right)^2\right], \tag{6.20}$$

where $\mathbf{x}$ values comes form the experimental design. Note that MC sampling can be costly, so the idea is to choose N_s small enough to ensure accuracy without increasing computational cost.

6.4.3 Surrogate Error Estimation

The construction of a good surrogate requires a rigorous process of validation of the response surface obtained. Use a good error metric is essential to characterize a good approximation [15].

Considering the previous section, where PCE coefficients are computed via OLS method, it is natural to evaluate the approximation error by the coefficient of determination [10]. In this case, we can calculate this quantity from the experimental design used in the regression as follows

$$R^2 = 1 - \frac{\frac{1}{Ns}\sum_{i=1}^{Ns}\left(\mathcal{M}(\mathbf{x}^{(i)}) - \mathcal{M}^{PC}(\mathbf{x}^{(i)})\right)^2}{\hat{V}(Y)}, \tag{6.21}$$

where $\hat{V}(Y)$ is the empirical variance of the model evaluations, given by

$$\hat{V}(Y) = \frac{1}{Ns-1}\sum_{i=1}^{Ns}\left(\mathcal{M}(\mathbf{x}^{(i)}) - \bar{y}\right)^2, \tag{6.22}$$

with

$$\bar{y} = \frac{1}{Ns}\sum_{i=1}^{Ns}\mathcal{M}(\mathbf{x}^{(i)}), \tag{6.23}$$

and $\mathbf{x}^{(i)}$ is the ith evaluation of the random input $\mathbf{X}$. Another option to measure the error is to apply the normalized empirical error [26]. Nevertheless, this measure can be problematic in cases of over-fitting. Thus, if you are not sure about the size of your experimental design, it is recommended to work with the Leave-One-Out (LOO) cross-validation error [27, 28], calculated by

$$\epsilon_{LOO} = \frac{\sum_{i=1}^{Ns} \left(\mathcal{M}(\mathbf{x}^{(i)}) - \mathcal{M}^{PC\backslash i}(\mathbf{x}^{(i)}) \right)^2}{\sum_{i=1}^{Ns} \left(\mathcal{M}(\mathbf{x}^{(i)}) - \bar{y} \right)^2}, \tag{6.24}$$

where $\mathcal{M}^{PC\backslash i}$ notation indicates the ith metamodel built using the reduced experimental design $\chi \backslash \mathbf{x}^{(i)} = \{\mathbf{x}^{(j)}, j = 1, \ldots, N_s, j \neq i\}$.

6.4.4 PCE-Based Sobol' Indices

Note that due the orthonormality of the surrogate PCE metamodel the model variances (partials and total) can be calculated only using the expansion coefficients. The sum of the squared of all the PCE coefficients provides the variance of the model's response and subtracting those coefficients associated with certain indices, the conditioned variances can be obtained [21, 29]. Therefore, an estimator for the Sobol' index associated with the subset $\mathbf{u}$ is given by

$$S_{\mathbf{u}} = \frac{\sum_{\alpha \in \mathbf{u}} y_\alpha^2}{\sum_{\alpha=1}^{P} y_\alpha^2}, \tag{6.25}$$

where $\mathbf{u}$ is a suitable subset of indices.

We observe that once the surrogate is already calculated, these indices can be obtained with negligible computational cost, since in general, a single evaluation of the model is much more expensive than the sums involved in Eq. (6.25) fraction. In addition to the computational cost issue, the use of a surrogate PCE to calculate Sobol' indices also eliminates the possibility of numerical cancelation errors in the variance calculation, since only sums of positive quantities are involved in the above algorithm [15].

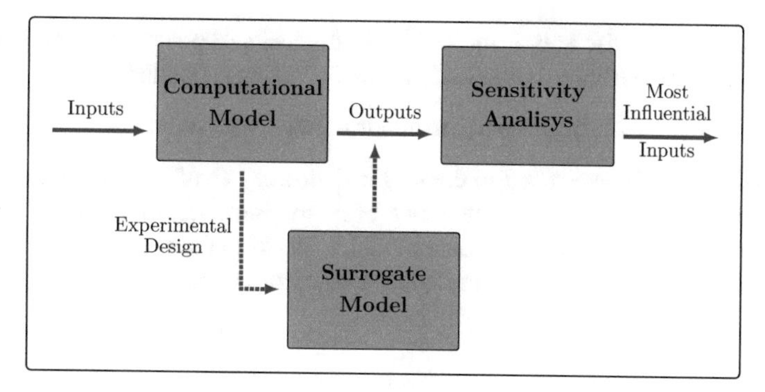

Fig. 6.2 Schematic representation of the process of obtaining the Sobol' indices by using a surrogate model

6.5 A Practical Tutorial

6.5.1 Tutorial Description

Finished the presentation of the theoretical background, we now move to the practice, after all, this is a tutorial goal. The computational methodology to compute Sobol' indices is composed of three steps: (i) Characterization of the random input; (ii) Construction of the PCE; and (iii) Calculation of the Sobol' indices. The framework of this process is illustrated in Fig. 6.2.

Here UQLab [30] library is explored to facilitate the implementation of a code to compute the Sobol' indices. We encourage the reader to look for more details in the UQLab manual [30]. To describe the random input, it is necessary to define the distribution for each random parameter as well as the respective hyperparameters and support. In the second step, you have to define a computational model to simulate your computational experiment, the maximum degree of your polynomials expansion, and the number of samples that will be used to generate the design set for the regression. The PCE coefficients are calculated for each time of observation defined, and you have everything you need to obtain the Sobol' indices upto the desired order. Note that you cannot select the maximum order for the Sobol' indices higher than the dimension of the random input vector. Further details about this construction can be seen in UQLab manual for Global Sensitivity Analysis [29].

6.5.2 SoBioS: Sobol' Indices for Biological Systems

To facilitate the manipulation of the UQLab packages, the authors developed a computational library called SoBioS—Sobol' indices for Biological Systems—focused on the simulation of the Sobol' indices for biological systems with the UQLab tool. This computational library is available in the following repository:

https://americocunhajr.github.io/SoBioS

The auxiliary routines used to create the following example results can be also found there, but, in resume, you need two basic routines: The main file to define the input details, to call the computational model, to perform the sensitivity analysis and plot results, and a QoI file, to define your quantity of interest.

6.5.3 Example 1: Predator–Prey Dynamics

The first example is the classical Lotka–Volterra model. Better known as predator–prey model, this is a dynamical system used to reproduce a simple predator–prey relationship [18]. It is assumed that preys are capable of reproducing spontaneously and the predators are only able to feed from the previous prey. The standard model considers preys and predators' reproduction, the natural death of the predator (independent of the prey), and death of preys by hunting. So, the relation between the two species is beneficial for the predator and harmful for the prey.

By assuming a simple representation of these mechanisms, we can construct the following pair of equations:

$$\frac{\mathrm{d}V}{\mathrm{d}t} = aV - bVP \,, \tag{6.26}$$

$$\frac{\mathrm{d}P}{\mathrm{d}t} = dVP - cP \,, \tag{6.27}$$

where V and P represent the populations of prey and predator, respectively, and the time is measured in years. In the first equation, we have the subtraction between the birth term and the hunt term. Before the hunt is assumed that the predator increases in a proportion of the energetic efficiency d times the number of preys and decreases with a constant ratio. The model parameters as well as the other simulation parameters are described in Table 6.2. The model parameters' intervals and initial conditions were extracted from the UQLab Bayesian Calibration manual [31] and the other simulation values were chosen by the authors. Numerical integration was performed using the Dormand–Prince adaptive Runge–Kutta method, implemented in `ode45`, and the response will be restricted to three times instants to simplify the visualization of the results. Our interest can be for the prey or predator population depending on the situation.

In this first example, we compare the results of sensitivity analysis by MC and Surrogate approaches. Before that, it is necessary to analyze how much samples to use for each one of those methods. Two different strategies are considered in order to evaluate the reliability of the computed Sobol' indices. For the MC method, the idea is to execute hierarchical simulations increasing the number of samples, using only

Table 6.2 Parameters description and values used to simulate the Sobol' indices for the predator–prey model

Name	Description	Value
a	Birth rate for preys	$[0.44, 0.68]$ year^{-1}
b	Search rate	$[0.02, 0.044]$ year^{-1}
c	Death rate for predators	$[0.71, 1.15]$ year^{-1}
d	Energetic efficiency	$[0.0226, 0.0354]$ year^{-1}
V_0	Initial quantity of preys	33 individuals
P_0	Initial quantity of predators	6.2 individuals
tspan	Time span	$[0, 10]$ years
tQoI	Time instants of interest	$\{1, 6, 10\}$ years
QoI	V or P	$V(t)$ or $P(t)$ individuals

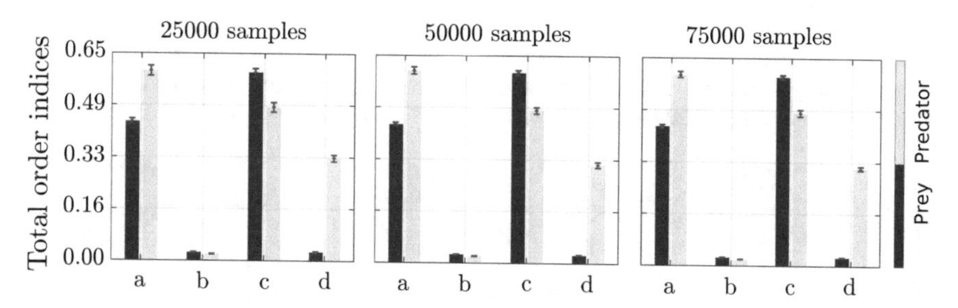

Fig. 6.3 Comparison between MC-Sobol' Total Order results using 25000 (left), 50000 (middle) and 75000 (right) samples with 95% confidential intervals (red) calculated with Bootstrap method considering both predators (yellow) and prey (blue) populations in the $10th$ year

one value of each QoI, and calculating 95% Confidence Intervals (CI) by Bootstrap method [23]. For PCE-Sobol the approach is to plot the validation graphs and compare the surrogate response with the full model one, to estimate also the ideal maximum polynomial degree. Some results for each strategy can be observed in Figs. 6.3, 6.4, 6.5, respectively. We can see the convergence as we increase the number of samples in MC because the CI amplitude decreases. On the validation plots of Figs. 6.4 and 6.5 we can observe the good match between the surrogate results and the response for the original computational model in addition to the low order for the ϵ_{LOO} cross-validation error. This comparison is done using 10000 samples. By these results, we define a size of 75000 samples for MC simulations and 1000 for PCE simulations. Additionally, the ideal maximum degree for the surrogate approximation was adopted as 6.

Finally, Figs. 6.6 and 6.7 present the comparison of the MC and PCE results for sensitivity analysis. In these graphs, we can observe that the results are pretty close. However, the total order results for MC are slightly smaller. This is due to the negative second-order indices obtained since it suffers from a cancelation error, something to which the calculation via PCE is immune once it does not involve subtractions.

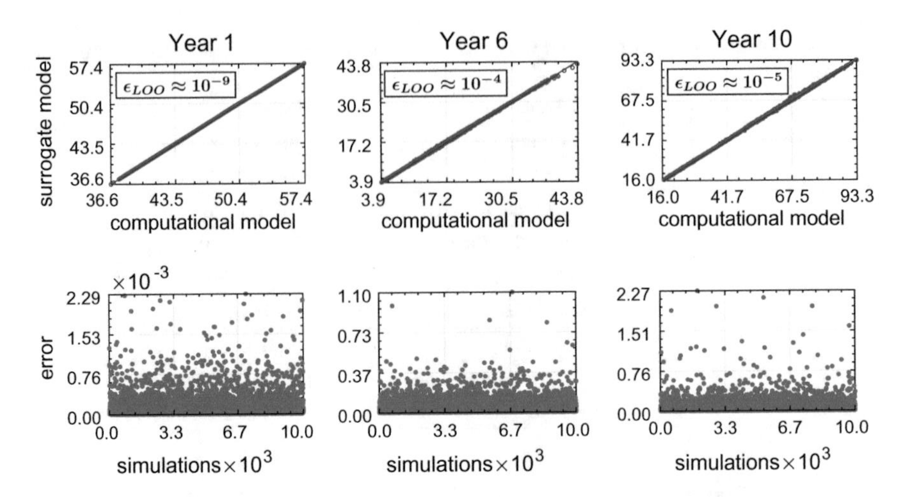

Fig. 6.4 Comparison PCE surrogate response and computational model response (top) for predators in years 1 (left), 6 (middle), and 10 (right), associated mismatch error (bottom), and leave-one-out error (top legend) using 1000 samples as experimental design and 10000 as validation set

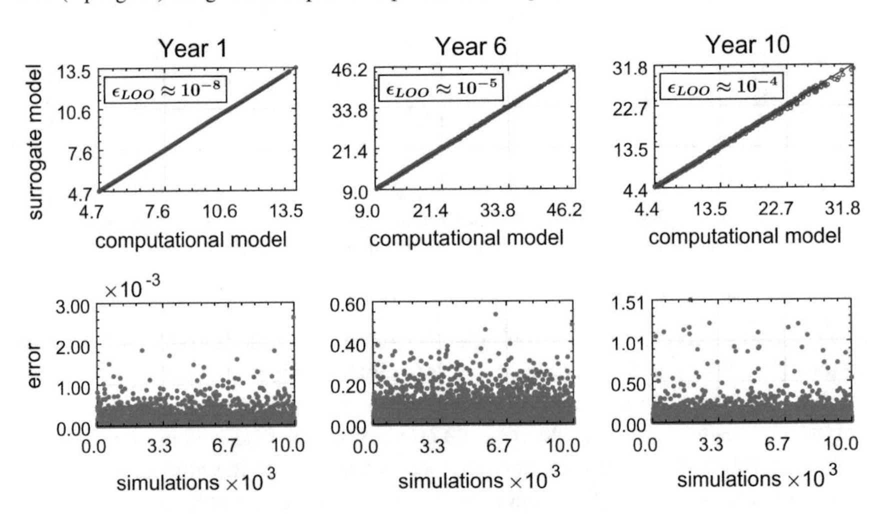

Fig. 6.5 Comparison PCE surrogate response and computational model response (top) for preys in years 1 (left), 6 (middle), and 10 (right), associated mismatch error (bottom), and leave-one-out error (top legend) using 1000 samples as experimental design and 10000 as validation set

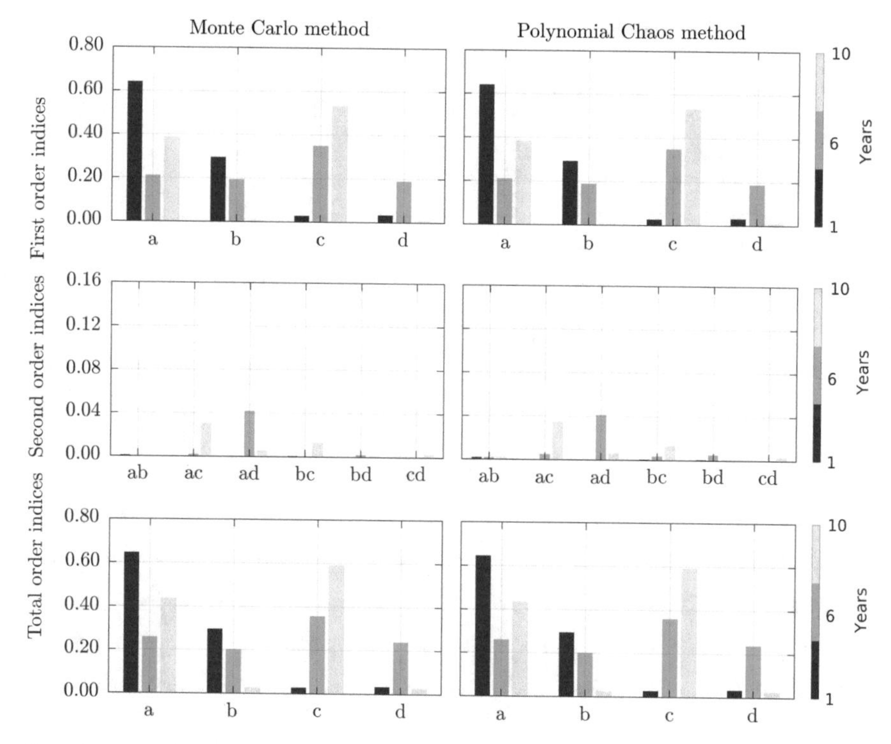

Fig. 6.6 Comparison between MC-Sobol' results (left) and PCE-Sobol results (right) for the prey population of the Lotka–Volterra model. The simulation was perform with 75000 samples for the MC and 1000 for the PCE

This fact has been omitted from the associated figures to avoid any confusion about negative Sobol' indices, which do not make sense.

Note that for this simple scenario the cancelation error is not that problematic but wondering a situation of high order indices, the error propagation can affect your results' conclusions. The second advantage is, of course, in the time of simulation. Even for this simple analysis, the difference is notable. Using an Intel Core i5-8250U 1.6Ghz × 8 (8GB of RAM), the MC method needed 2 min, while the PCE performed in 45s, even with the second one plotting more results due to the validation graphs.

Besides all that discussion about methods, the Sobol' results reveal that, for this parametric intervals and initial conditions, the parameter d is more important in the first year, b in the sixth and a in the final one when analyzing the predator population. For prey the results are completely different. The parameter a control the most part of the variance in the first week but the c parameters assume the control after that.

The great difference of scenarios for each population can sound weird as well as the change of the most important parameter during the passage of time. The truth is that is not only possible but common. This detail for systems with time dependency is essential for control measures. It will be necessary some adaptive strategy to affect

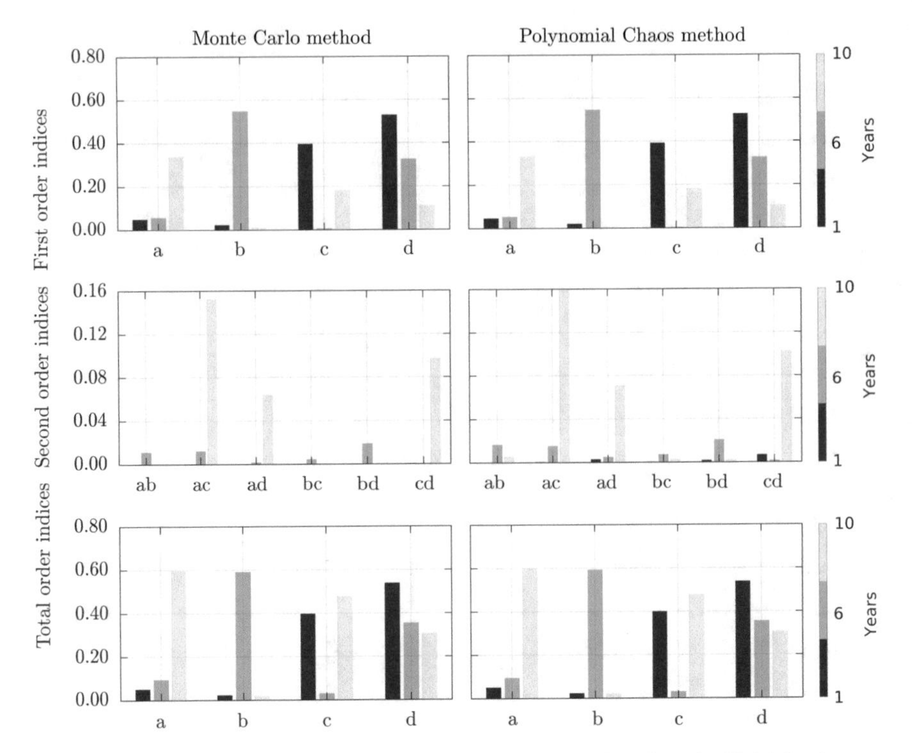

Fig. 6.7 Comparison between MC-Sobol' results (left) and PCE-Sobol results (right) for the prey population of the Lotka–Volterra model. The simulation was perform with 75000 samples for the MC and 1000 for the PCE

different parameters of phenomena at each step of implementing the measure. Also, note that in this case the second-order interactions were not so much important for the total order indices in case of prey but increase significantly the total indices for predators.

6.5.4 Example 2: NF-κB Signaling Pathway

NF-κB is a family of Transcription Factors (TF) that takes part in the regulation of several mammalian cellular processes, to mention some: cell division, apoptosis, inflammation, immune response, and cancer disease [32]. The family consists of five subunits (p50, p52, p65, c-Rel, and RelB), which associate to form functional dimers. The NF-κB complex when bound to the IκB inhibitor protein is inactive in the cytoplasm, being released after the phosphorylation of the IκB protein by the IκB kinase (IKK) complex. This dissociation of the NF-κB:IκB complex also promotes the IκB ubiquitination and than its degradation. At this moment, NF-κB is free to be translocated into the nucleus, where it activates gene transcription by binding to specific DNA κB sites. Among the activated ones are the genes for IκB, that is, IκB mRNA are transcripted and translated into new IκB proteins. Part of it binds to the NF-κB in the cytoplasm, and part enters the nucleus where it binds to nuclear NF-κB forming a NF-κB:IκB that is exported to the cytoplasm. In both cases, the complex is again the target of the IκB kinase (IKK), characterizing a negative feedback mechanism.

A mathematical model for the NF-κB pathway was introduced in 2002 by Hoffman et al. [5]. Their model consists of 26 molecular species (variables) and 64 reaction coefficients (parameters). Besides setting up the system of differential equations for such a complex biological system, they studied the influence of the negative feedback on the sustained and on the damped oscillation of the NF-κB concentration over time. An interesting follow up of their group's work was presented in [6]. It is clear from the references list how their mathematical model was influential and how the NF-κB pathway model is a well succeed system biology example of the interplay between mathematical modeling and experimental analyses.

Taking the bi-compartmental model of [5] as departure point, removing species that have no feedback from NF-κB and removing slow reactions at the expense of faster ones, Krishna et al. [33] were able to extract the core feedback loop of the model, coming up with a reduced model constituted of 7 species (variables) and 12 reaction coefficients (parameters). The variables of the reduced model are: N_n and N, the free nuclear and free cytoplasmic NF-κB concentration, respectively; I_n and I, the free nuclear and free cytoplasmic IκB concentration, respectively; I_m, the IκB mRNA concentration; $\{NI\}_n$ and $\{NI\}$, the nuclear and cytoplasmic NF-κB:IκB complex concentration, respectively; and IKK, the IκB kinase concentration. The dynamics of the seven species are depicted in Fig. 6.8, resulting in the system of equations

$$
\begin{aligned}
\frac{\mathrm{d}N_n}{\mathrm{d}t} &= k_{N_{\text{in}}} N - k_{f_n} N_n I_n + k_{b_n} \{NI\}_n \, , \\
\frac{\mathrm{d}N}{dt} &= -k_{N_{\text{in}}} N - k_f N I + (k_b + \alpha)\{NI\} \, , \\
\frac{\mathrm{d}I_n}{\mathrm{d}t} &= k_{I_{\text{in}}} I - k_{I_{\text{out}}} I_n - k_{f_n} N_n I_n + k_{b_n} \{NI\}_n \, , \\
\frac{\mathrm{d}I}{\mathrm{d}t} &= -k_{I_{\text{in}}} I + k_{I_{\text{out}}} I_n - k_f N I + k_b \{NI\} + k_{tl} I_m \, , \\
\frac{\mathrm{d}\{NI\}}{\mathrm{d}t} &= k_{NI_{\text{out}}} \{NI\}_n + k_f N I - (k_b + \alpha)\{NI\} \, , \\
\frac{\mathrm{d}\{NI\}_n}{\mathrm{d}t} &= -k_{NI_{\text{out}}} \{NI\}_n + k_{f_n} N_n I_n - k_{b_n} \{NI\}_n \, , \\
\frac{\mathrm{d}I_m}{\mathrm{d}t} &= k_t N_n^2 - k_m I_m \, .
\end{aligned}
\tag{6.28}
$$

Table 6.3 Parameters descriptions and nominal values of the seven species NF-κB model [33]

Parameter	Values (Units)	Description
$k_{N_{\text{in}}}$	5.4 min^{-1}	NF-κB nuclear import rate
$k_{I_{\text{in}}}$	0.018 min^{-1}	IκB nuclear import rate
$k_{I_{\text{out}}}$	0.012 min^{-1}	IκB cytoplasm export rate
$k_{NI_{\text{out}}}$	0.83 min^{-1}	NF-κB:IκB cytoplasm export rate
k_f	30 $\mu\text{M} \cdot \text{min}^{-1}$	NF-κB:IκB cytoplasm association rate
k_{f_n}	30 $\mu\text{M}^{-1} \cdot \text{min}^{-1}$	NF-κB:IκB nuclear association rate
k_b	0.03 min^{-1}	NF-κB:IκB cytoplasm dissociation rate
k_{b_n}	0.03 min^{-1}	NF-κB:IκB nuclear dissociation rate
k_{tl}	0.24 min^{-1}	IκB mRNA translation rate
k_t	1.03 $\mu\text{M}^{-1} \cdot \text{min}^{-1}$	IκB mRNA transcription rate
k_m	0.017 min^{-1}	mRNA degradation rate
α	$1.05 \times IKK$ min^{-1}	IκB degradation rate (by phosphorylation)

The description and the nominal values of the parameters are shown in Table 6.3, according to [33]. In the model, the NF-κB pathway is activated by an extracellular stimuli, considered by a IκB kinase (IKK) input, incorporated in the parameter α (top of Fig. 6.8).

Before proceeding we actually note that for IκB concentration is meant IκBα concentration, since as elucidated in [5] is the IκBα isoform of the IκB protein that enters in the negative feedback loop of the NF-κB pathway and generates a sustained oscillation on the NF-κB concentration. The solution of the system for the nominal values in Table 6.3 is shown in Fig. 6.9, with initial conditions $(N_n, N, I_n, I, \{NI\}, \{NI\}_n, I_m)_0 = (0, 1, 0, 0, 0, 0, 0)$ and $IKK = 0.7$, and it was obtained by numerically integrating the system (6.28) with the ODE solver, `ode15s`, for stiff equations, where it is observed the spiky oscillation of the NF-κB concentration in time. With further considerations on the NF-κB:IκB complex association and dissociation reactions, they reduced the model even more, coming up with an over reduced model constituted of 3 species: N_n, I, I_m (the bold boxes in Fig. 6.8), and 5 parameters (here they are not straight reaction coefficients as before, but combinations of those).

As pointed out in [33], the spiky type oscillations are robust with respect to the IKK input variation and that the down-regulated genes by NF-κB have different times of response, related to this oscillation, that is why the QoI for our sensitivity analysis was the peak duration, that was defined in [33] as the time the NF-κB concentration spends above its mean. For the solution in Fig. 6.9, the mean is represented by

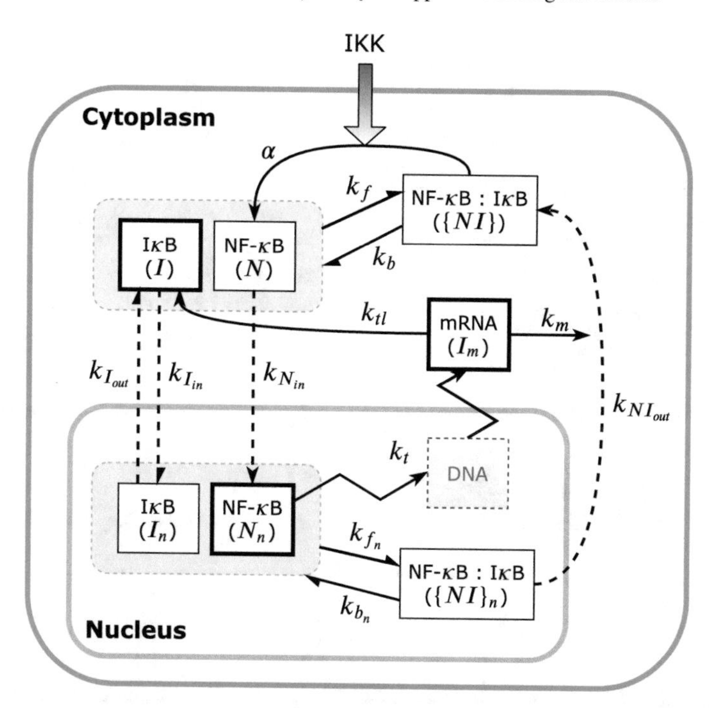

Fig. 6.8 Diagram representing the dynamics of the seven species NF-κB model [33]. The parameters are described in Table 6.3. The bold boxes are the species present on the reduced three species model

a constant dashed (red) line in the last (bottom) subplot. To build the PCE surrogate model 400 experimental design samples were taken with a maximum polynomial degree of 12. The probabilistic input model consisted of independent uniform random variables for each of the parameters, with lower and upper bounds given by 1.5% of dispersion around the mean, taken as the nominal values considered at Table 6.3. The quality of the surrogate approximation is appreciated in Fig. 6.10, showing to be reasonable to the purpose. According to the total indices, displayed in Fig. 6.11, the most five influential parameters, in decreasing order, are k_t, k_{tl}, $k_{I_{\text{in}}}$, k_m, $k_{N_{\text{in}}}$. It is interesting to note that these parameters are the ones that enter the over reduced model composed of three species; this is because these parameters appear in the equations of (6.28) related to the species N_n, I, I_m. The transcription and the translation parameters of the IκB mRNA are very influential as can be noted from the first-order indices, Fig. 6.12 and its degradation rate k_m does not have such a first order influence, but k_m appears with significant coupling with five other species in the second-order level, Fig. 6.13. Actually, the three highest second-order indices, namely, $k_{N_{\text{in}}} - k_m$, $k_t - k_m$ and $k_{I_{\text{in}}} - k_m$ are coupled with it. Another parameter that has no first-order influence, but appears coupled with five other species in the second-order level, is the parameter $k_{N_{\text{in}}}$. That is why both k_m and $k_{N_{\text{in}}}$ express themselves in the total order

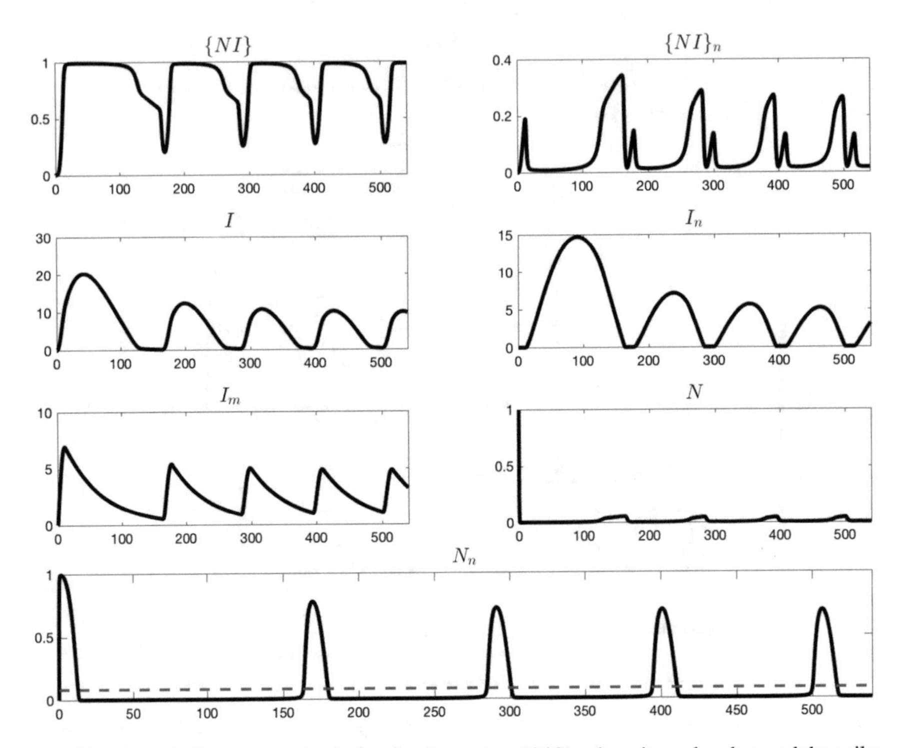

Fig. 6.9 Numerically integrated solution for the system (6.28), where it can be observed the spiky oscillation of the NF-κB concentration in time. The constant dashed (red) line is N_n = const. = $mean(N_n)$

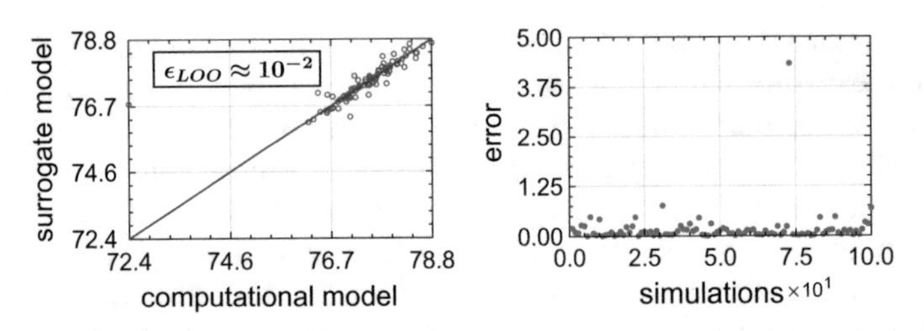

Fig. 6.10 Comparison PCE surrogate response and original computational model response for the peak duration QoI

indices but not in the first-order ones. Finally, we need to mention that in Fig. 6.13, we are showing only the second-order indices above the threshold value 10^{-4}.

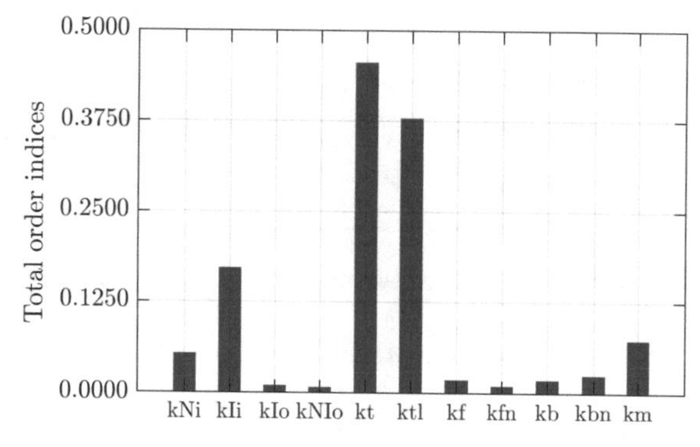

Fig. 6.11 Total order Sobol' indices for the peak duration QoI

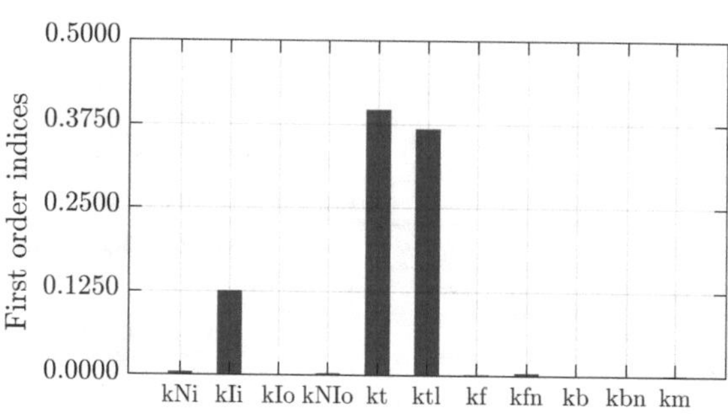

Fig. 6.12 First-order Sobol' indices for the peak duration QoI

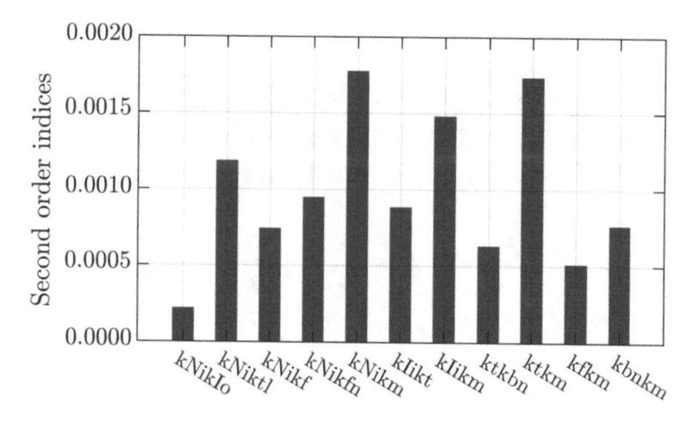

Fig. 6.13 Second-order Sobol' indices for the peak duration QoI

6.5.5 *Example 3: The SIR Model*

The famous SIR model is a classical tool to study epidemics [1]. Recently several research groups are using variations of the SIR model in the new coronavirus pandemic research [9]. Of course this model, as will show below, is not ideal to reproduces the Corona phenomena correctly, however, can be useful to test hypotheses and analyze scenarios. The model is based on separating the host population from a disease in three compartments: The Susceptible, S, is the population of individuals able to become infected by the disease upon contact with the pathogen; Infected, I, indicates those who carry the pathogen and has the ability to infect the susceptible;

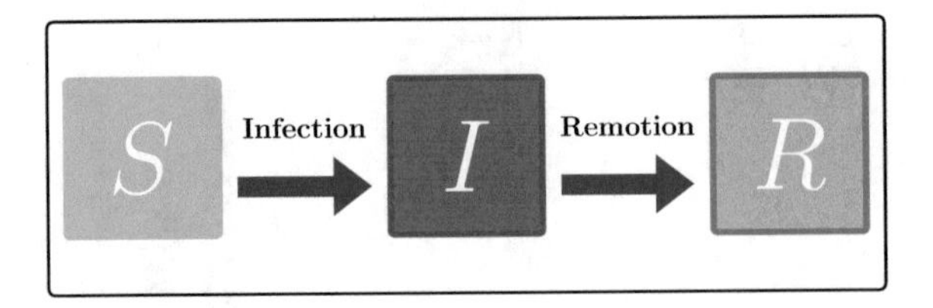

Fig. 6.14 Schematic representation of the SIR compartmental model's mechanisms

Table 6.4 Parameters description and values used to simulate the Sobol' indices for the SIR model

Name	Description	Value
b	Transmission rate	0.37 days^{-1}individuals^{-1}
a	Removal rate	0.15 days^{-1}
N	Total population	1000 individuals
R_0	Initial number of removed	0 individuals
I_0	Initial number of infected	200 individuals
S_0	Initial number of susceptible	$(N - I_0)$ individuals
tspan	Time span	[0,42] days
tQoI	Time instants of interest	$\{7, 14, \ldots, 42\}$ days
QoI	I	$I(t)$ individuals

The Removed, R, reunites the individuals who recovered from the disease (gaining immunity) or died from it [1]. The evolution between the compartments is represented in Fig. 6.14 and mathematically formulated by the following set of equations:

$$\frac{\mathrm{d}S}{\mathrm{d}t} = -\frac{bSI}{N}, \tag{6.29}$$

$$\frac{\mathrm{d}I}{\mathrm{d}t} = \frac{bSI}{N} - aI, \tag{6.30}$$

$$\frac{\mathrm{d}R}{\mathrm{d}t} = aI, \tag{6.31}$$

where bSI/N is the infections rate term and aI indicates the removal rate. The QoI in this model is the infected population $I(t)$. The description and the values of the simulation parameters are gathered in Table 6.4. As in the previous example, we have reference for the mean values of the random variables, and the dispersion factor around the nominal values will be of 0.4 for each one. The parameter a can be interpreted as the inverse of the infectious period. Thereby, we took the mean value of 6.5 days estimated by the researchers of the Imperial College group (in their 9th Report) [9]. Besides that, in the same report it is estimated the basic number of reproduction $\mathcal{R}_0 = b/a$ for several countries. Using this equation of $\mathcal{R}_0$ for the SIR model [34] and $a = 1$, we estimate a mean for the transmission rate.

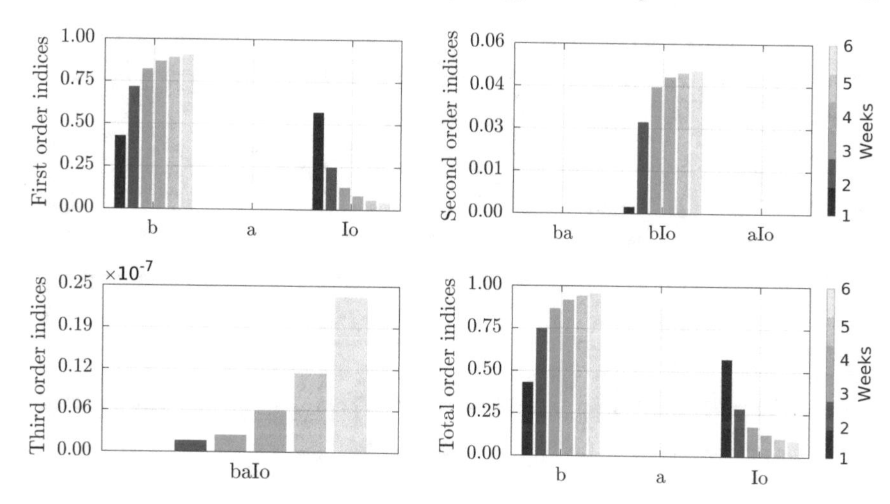

Fig. 6.15 Sobol' indices for the infected population of the SIR model in the weeks of interest simulated using 1000 samples

Different from the previous examples, now we will consider also an initial condition as a random variable, to account for the uncertainty in the initial number of infected individuals, that is not well known. If we assume that the host population is constant, the initial number of susceptible is also a random variable, depending on the initial number of infected. Finally, although our system is measured in days, we will display results in weeks to facilitate visualization.

In Figs. 6.15 and 6.16, the reader can find the Sobol' indices results and validation plots, respectively. The maximum degree for the PCE was 10 and the experimental design was composed of 1000 samples. It can be noted a very good match between the computational response and the surrogate. To differentiate the previous two examples, here we calculate the third-order Sobol' index. In this scenario, this index does not show so much information but reveals that the influence of this third-order term maybe increases over time. The same seems to occur with the b importance while the opposite can be observed for I_0. So, maybe if we analyze a previous time windows, the initial number of infected individuals can be more influenced for the sensitivity of the response. Similarly, for a posterior time window we could have a I_0 not important. Of course, new studies are needed to support these hypotheses but are quite interesting to imagine these possibilities. Curiously, the removal rate had practically zero importance. This is not intuitive because we expected some kind of influence over time. But, it is important to be clear that this not mean that the a parameter not change that response if it changes. This result showed that if the two parameters and the initial number of infected individuals changes, b and I_0 will domain how the response will change.

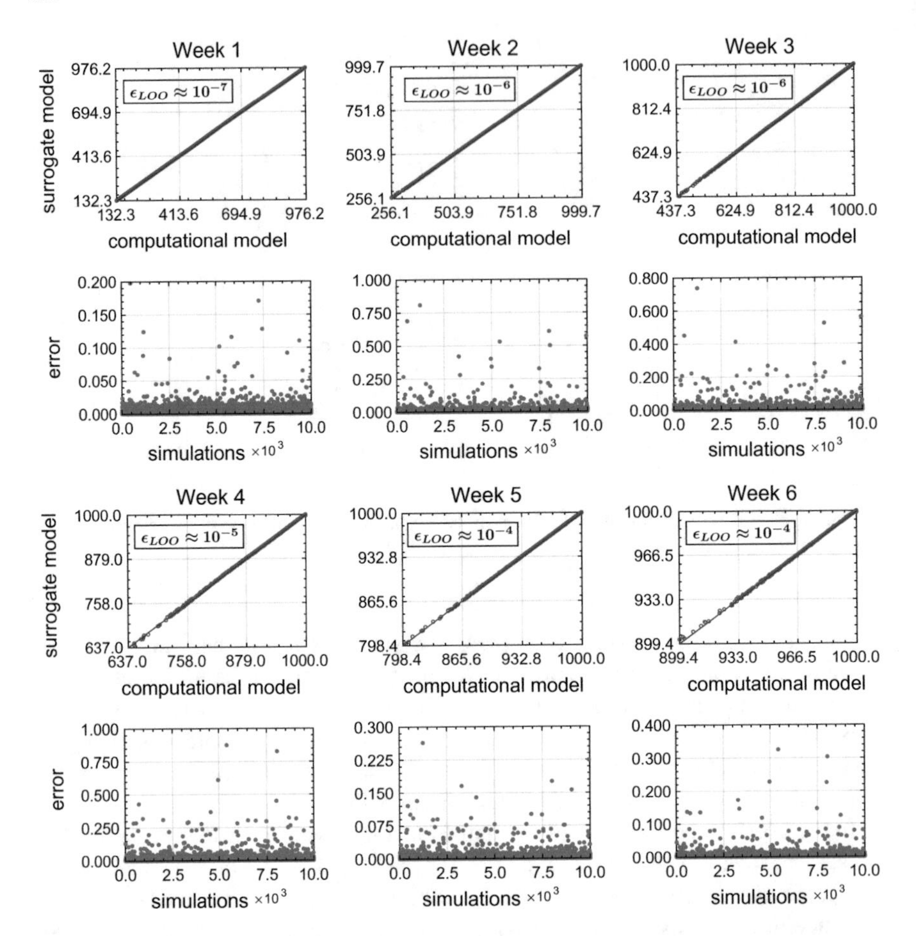

Fig. 6.16 Comparison PCE surrogate response and original computational model response for infected population in weeks of interest (red curves), and associated leave-one-out error (legends) using 1000 samples as experimental design and 10000 as validation set

6.6 Final Remarks

This chapter presented the use of Sobol' indices for global sensitivity analysis in biological systems. This type of formalism allows us to statistically measure how each of the system's parameters contributes to its response. This tool can be very useful to identify which factors have a greater contribution in a certain quantity of interest associated with the biological system, which can be interesting, for instance, in the calibration process of a computational model concerning a certain set of observations (data). This analysis tool is presented in the form of a tutorial, with a simplistic summary of the theory, without much mathematical formalism in favor of a better understanding of the fundamental ideas, and application in three biological systems with different levels of complexity. A code in MATLAB was developed and is available in a public repository to facilitate the use of this tool in the most diverse types of biological systems.

Acknowledgements The authors would like to thank Prof. Flávio Codeço Coelho (EMAp/FGV) by discussing several ideas at the beginning of the preparation of this manuscript. They are also grateful, for the financial support provided, to Carlos Chagas Filho Research Foundation of Rio de Janeiro State (FAPERJ) under grants 210.021/2018 and 211.037/2019, and Coordenação de Aperfeiçoamento de Pessoal de Nível Superior - Brasil (CAPES) - Finance Code 001.

References

1. Brauer F (2017) Infect Dis Model 2(2):113. https://doi.org/10.1016/j.idm.2017.02.001
2. Hethcote HW (2000) SIAM Rev 42(4):599. https://doi.org/10.1137/S0036144500371907
3. Boukal DS, Bideault A, Carreira BM, Sentis A (2019) Curr Opin Insect Sci 35:88. https://doi.org/10.1016/j.cois.2019.06.014
4. Sekerci Y, Petrovskii S (2015) Bull Math Biol 77:2325–2353. https://doi.org/10.1007/s11538-015-0126-0
5. Hoffmann A, Levchenko A, Scott ML, Baltimore D (2002) Science 298(5596):1241. https://doi.org/10.1126/science.1071914
6. Cheong R, Hoffmann A, Levchenko A (2008) Mol Syst Biol 4(1):192. https://doi.org/10.1038/msb.2008.30
7. Centers for Disease Control and Prevention (2020) Preventing the spread of coronavirus disease 2019 in homes and residential communities. https://www.cdc.gov/coronavirus/2019-ncov/hcp/guidance-prevent-spread.html, 20 March 2020
8. World Health Organization (2020) Coronavirus disease (COVID-19) outbreak. https://www.who.int/emergencies/diseases/novel-coronavirus-2019, 19 March 2020
9. I.C.C.R. Team (2020) Covid-19 reports. Technical Report 1–14, Imperial College. https://www.imperial.ac.uk/mrc-global-infectious-disease-analysis/covid-19/
10. Motulsky H, Christopoulos A (2004) Fitting models to biological data using linear and nonlinear regression: a practical guide to curve fitting, 1st edn. Oxford University Press, Oxford
11. Dantas E, Tosin M, Cunha A Jr (2018) Appl Math Comput 338:249. https://doi.org/10.1016/j.amc.2018.06.024
12. Dantas E, Tosin M, Cunha A (2019) In: Proceedings of the conference of computational interdisciplinary science
13. Saltelli A et al (2019) Environ Model Softw 114:29. https://doi.org/10.1016/j.envsoft.2019.01.012
14. Smith RC (2014) Uncertainty Quantification, 1st edn. SIAM, Philadelphia
15. Gratiet LL, Marelli S, Sudret B (2017) Metamodel-based sensitivity analysis: polynomial chaos expansions and gaussian processes. Springer, Berlin, pp 1289–1326
16. Sobol IM (2001) Math Model Civ Eng 55(1–3):407
17. Oden JT (2011) An introduction to mathematical modeling: a course in mechanics, 1st edn. Wiley, New Jersey
18. Haefner JW, Spanos P, Akin E (2015) An introduction to mathematical epidemiology, 1st edn. Springer, New York
19. Soize C (2017) Uncertainty quantification: an accelerated course with advanced applications in computational engineering, 1st edn. Springer, New York
20. Müller J, Kuttler C (2015) Methods and models in mathematical biology: deterministic and stochastic approaches, 1st edn. Springer, New York
21. Konakli K, Sudret B (2016) Comput Methods Appl Mech Eng 263:42. https://doi.org/10.1016/j.ress.2016.07.012

22. Xiu D (2010) Numerical methods for stochastic computations: a spectral method approach, 1st edn. Princeton University Press, New Jersey
23. Ghanem R, Higdon D, Owhadi H (eds) (2017) Handbook of uncertainty quantification, 1st edn. Springer, New York
24. Sudret B (2008) Reliab Eng Syst Saf 93(7):964. https://doi.org/10.1016/j.ress.2007.04.002
25. Ghanen RG, Spanos PD (2003) Stochastic finite element method: a spectral approach, Revised edn. Dover Publications
26. Sudret B, Blatman G, Berveiller M (2013) Response surfaces based on polynomial chaos expansions, pp 147–168. ISTE/Wiley. https://doi.org/10.1002/9781118601099.ch8
27. Marelli S, Sudret B (2018) UQLab user manual—polynomial chaos expansions. Technical report, Chair of risk, Safety & uncertainty quantification, ETH Zurich. Report # UQLab-V1.1-104
28. Blatman G, Sudret B (2010) Probabilistic Eng Mech 25(2):183. https://doi.org/10.1016/j.probengmech.2009.10.003
29. Marelli S, Lamas C, Sudret B, Konakli K, Mylonas C (2018) UQLab user manual—sensitivity analysis. Technical report, Chair of risk, Safety & uncertainty quantification, ETH Zurich. Report # UQLab-V1.1-106
30. Marelli S, Sudret B (2014) In: Proceedings of the 2nd international conference on vulnerability, risk analysis and management (ICVRAM2014), Liverpool, United Kingdom, pp. 2554–2563. https://doi.org/10.1061/9780784413609.257
31. Wagner PR, Nagel J, Marelli S, Sudret B (2019) UQLab user manual—bayesian inversion for model calibration and validation. Technical report, Chair of risk, Safety & uncertainty quantification, ETH Zurich. Report # UQLab-V1.2-113
32. Pires BRB, Mencalha AL, Ferreira GM, de Souza WF, Morgado-Díaz JA, Maia AM, Corrêa S, Abdelhay ESFW (2017) PLOS ONE 12(1):1. https://doi.org/10.1371/journal.pone.0169622
33. Krishna S, Jensen MH, Sneppen K (2006) Proc Natl Acad Sci 103(29):10840. https://doi.org/10.1073/pnas.0604085103
34. Martcheva M (2015) An introduction to mathematical epidemiology, 1st edn. Springer, New York

Chapter 7
Reaction Network Models as a Tool to Study Gene Regulation and Cell Signaling in Development and Diseases

Francisco José Pereira Lopes, Claudio Daniel Tenório de Barros, Josué Xavier de Carvalho, Fernando de Magalhães Coutinho Vieira, and Cristiano N. Costa

Abstract Advances in molecular biology and experimental techniques have produced a large amount of data with unprecedented accuracy. In addition to computational methods for handling these massive amounts of data, the development of theoretical models aimed at solving specific questions is essential. Many of these questions reside in the description of gene regulation and cell signaling networks which are key aspects during development and diseases. To deal with this problem, we present in this chapter a set of strategies to describe the dynamic behavior of reaction network models. We present some results selected from broad mathematical results, focused on what is closely related to gene regulation and cell signaling. We give special attention to the so-called Chemical Reaction Network Theory (CRNT) that has been developed since the first half of the past century. Some of the most important and practical results of this theory are the analysis of the conditions for a given network to exhibit emergent proprieties like bistability. The description of this behavior has been recurrent in recent studies of developmental biology or cellular signaling networks [1–5]. However, while referring to the same phenomena, the

F. J. P. Lopes (✉) · J. X. de Carvalho
Universidade Federal do Rio de Janeiro, Campus Duque de Caxias Professor Geraldo Cidade, Grupo de Biologia do Desenvolvimento em Sistemas Dinâmicos, Duque de Caxias, Brazil
e-mail: flopes@ufrj.br

J. X. de Carvalho
e-mail: jxcarvalho@gmail.com

C. D. T. de Barros
Laboratório Nacional de Computação Científica, Data Extreme Lab (DEXL), Petropolis, Brazil
e-mail: claudiodtbarros@gmail.com; cdtb@lncc.br

F. de Magalhães Coutinho Vieira
Divisão de Físico-Química, Instituto de Química, Universidade de Brasília, Brasília, Brazil
e-mail: fdmcv@unb.br

C. N. Costa
Universidade Federal do Rio de Janeiro Campus Duque de Caxias Grupo de Biologia do Desenvolvimento em Sistemas Dinâmicos Programa de Pós-Graduação em Nanobiosistemas Instituto Federal do Rio de Janeiro Campus, Rio de Janeiro, Brazil
e-mail: cristiano.costa@ifrj.edu.br

F. A. B. da Silva et al. (eds.), *Networks in Systems Biology*, Computational Biology 32,
https://doi.org/10.1007/978-3-030-51862-2_7

connection between bistability or oscillations in CRNT and biology is not obvious and has been a challenge, despite some well-succeeded examples. Contribution to bringing these two areas of knowledge together is the primary purpose of this chapter.

Keywords Gene regulation · Bistability · Reaction network · Developmental biology · *Drosophila melanogaster* · Dynamical systems theory · Genetic algorithm

7.1 Introduction

Although the availability of large amounts of data is an extraordinary advance, the solution of problems related to different diseases requires the understanding of specific molecular mechanisms. Some of these problems are related to how the interaction between biochemical components gives rise to specific functions in a normal or pathological cell in response to an external stimulus or a programmed step during development. Some of these functions are closely related. For example, the biochemical steps in the NFkB signaling network during embryonic development are similar to the NFkB response in the immune system or uncontrolled growth in a malignant tumor. The extreme complexity of this network is a great challenge nowadays in theoretical modeling of biological systems. The safer way to navigate through these difficult waters is by building theoretical models based on rigorous mathematical approaches. However, the common idea that resides in the back of many researchers' minds is: how to build a complex, mathematically consistent model that I am still able to solve? In this chapter, we aim to contribute to answering this question trying to establish a connection between experimental problems in biology and the theoretical tools nowadays available.

We organized this chapter as follows. In Sect. 7.2. *Gene Regulation and Cell Signaling*, we present the very basic biological knowledge necessary to understand the following sections; in this section, we present a practical strategy to simplify a huge signaling network like NFkB signaling. In Sect. 7.3. *Dynamic System Theory*, we present the bases of this theory with simple and practical examples. We believe that this section provides sufficient knowledge for readers without training in this subject. A middle-level experience in mathematics will be necessary to understand Sect. 7.4. *Parameter Estimation Strategies*, however, even for readers with no sufficient background to understand all the content of this section, the basic definitions for sure will be accessible there. In Sect. 7.5. *Bistability in developmental biology*, we present a detailed discussion about a specific *systems biology approach* followed by a practical example of application. We end this section with a discussion about how to verify if a theoretical model can be assumed as a good representation of a biological system in the real world and how to improve its capacity to propose solutions to biologically significant questions.

7.2 Gene Regulation and Cell Signaling

Living beings of different species keep their features across generations by the information stored in their deoxyribonucleic acid (DNA) molecules. Human beings (*Homo sapiens*), fruit fly (*Drosophila melanogaster*), mouse (*Mus musculus*), *Escherichia coli*, the bacteria of intestinal flora, and all different species of plants and microbes have different DNA molecules. All these molecules are made of the same unit blocks, the nucleotides. The difference between them lies in the sequence of these blocks. Specifically, in particular regions of the DNA molecule called genes, which contain the necessary information to synthesize the cell proteins. This term information here is just the specific sequence of nucleotides that determines the sequence of amino acids, which are the blocks that build a protein molecule. However, there is an intermediate molecule, the *messenger* RNA (mRNA, a ribonucleic acid), Fig. 7.1. In a simple way: the gene nucleotides sequence determines the mRNA nucleotide sequence that, in turn, determines the protein amino acid sequence, which, ultimately, is predominant to determine the protein structure and so its functionality. The mRNA and protein synthesis is called gene transcription and translations, respectively. This whole process is called *gene expression.*

Different cells in a multicellular organism express different sets of genes, even though they contain the same DNA. In eukaryotic cells, the housekeeping genes are constitutively expressed in all cells irrespective of tissue type, developmental stage, or cell cycle state. These genes are related to sustaining life and internal control of the essential cell functions. On the other hand, the inducible genes are transcribed in a tissue-specific way at different steps during development. Some of these genes are active as a cell response to an external event; it occurs by a sequential process called

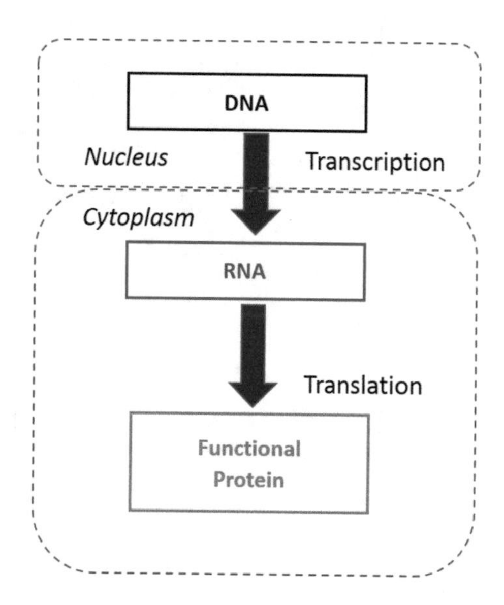

Fig. 7.1 Flowchart showing the central dogma of molecular biology

cell signaling. In this process, the external event generates intermediate messengers that can be hormones, proteins, peptides, chemicals as phosphate, and nitric oxide (NO) or even a pathogen like a virus can act as those extracellular messengers or signals, which are recognized by specific proteins (receptor proteins). After being recognized by receptors located in the cell membrane, these signals can trigger multiple responses in the recipient cell, such as metabolism regulation, alter or maintain the differentiated cell state, start or stop cell division, and in some circumstances, cell death. Some receptor proteins are located in the plasma membrane, while others in the cytoplasm or even in the nucleus. Signals transductions are complex multistep process; different signals can interfere positively or negatively in response to the primary stimuli.

7.2.1 NFkB Signaling Pathway

The NFkB (nuclear factor kappa B) is a transcription factor family involved in gene regulation and cell signaling during cell proliferation, cell death, inflammation, cancer, and others [4]. NFkB signaling was conserved during evolution, and the sequence of many proteins involved have conserved sequence in arthropods as *fruit-fly*, mouse, mice, and human (Fig. 7.2). In *Drosophila melanogaster*, the fruit-fly, the nuclear localization of Dorsal, the NFkB is controlled by the transmembrane

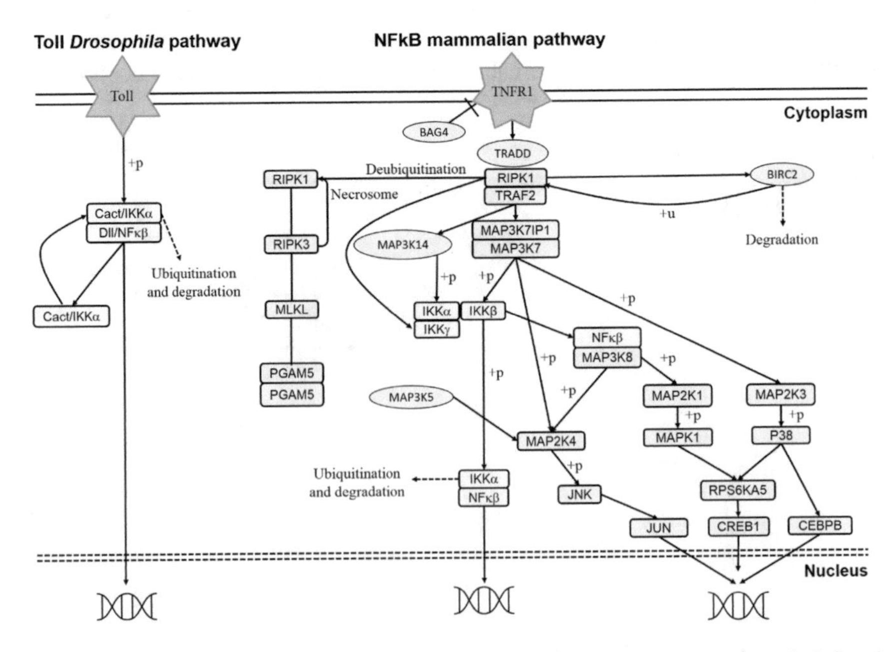

Fig. 7.2 Two conserved signaling networks. The toll *Drosophila* signaling network on the left and NFkB signaling on the right

receptor Toll [6]. Toll pathway is one of the most important examples of conserved pathways to date. It has a general and high common signaling scheme verified in vertebrates and invertebrates: Briefly, a transmembrane receptor undergoes conformational changes after receiving an external signal leading to the phosphorylation of cytoplasmic proteins. Consequently, several protein–protein interactions in the cytoplasm lead to the translocation of transcription factors (TF) to the cell nucleus, thus initiating gene regulation. In the case of Toll, this signaling is responsible for both the innate immune response and the embryonic axis formation. The Dorsal protein regulates the expression of developmental genes responsible for generating the three different tissues in the adult organism: mesoderm, endoderm, and ectoderm [7]. The homology between the signaling network in *Drosophila* and mammalians has the potential to generate valuable insights for the study of cancer, as this regulation family is also involved in oncogenesis in mammalians cells. In fact, *Drosophila* has been an important tool to map cancer pathways [8, 9]. For example, recent studies have found that the activation of NFkB (the mammalian homologs of *Drosophila* protein Dorsal) increases the expression level of its downstream-like genes like Snail, Slug, Sip1, and Twist1, playing a critical role in the epithelial-to-mesenchymal transition, a phenotypic transition associated with cancer metastasis [10]. All these proteins have *Drosophila* homologous with the same regulatory relationship. It indicates that not only the structural homology between these proteins is conserved, but also its regulatory relationship. For example, just as NFkB activates the expression of Snail, Dorsal also activates the expression of the Snail in *Drosophila* homologs. Several signaling and gene regulation processes were first described in *Drosophila melanogaster*, making this organism one of the most studied. Many reasons have contributed to this. Among them, the low maintenance cost and the relative ease of genetic manipulation compared to other organisms are undoubtedly among the most important. However, today, as the generation of quantitative data is no longer restricted to small organisms, why is studying a simple organism like the fruit fly still relevant? Why should we spend great efforts on construction, calibration, and analysis of theoretical models in *Drosophila*? To find a reasonable answer to this question, let us assume that our goal is to build a theoretical model that aims to contribute to the understanding of NFkB signaling in mammals.

A reasonable and perhaps indispensable step is to look for the best, or at least, a reasonable simplification for the signaling network in Fig. 7.2. At first glance, we can see that many sets of proteins follow the general scheme: an activated transmembrane cell receptor induces cytoplasmic proteins' phosphorylation resulting in nuclear translocation of transcriptional factors. We can also see that some receptors participate in more than one signaling pathway. It adds a layer of complexity and makes it even more challenging to look for a reasonable simplification for the NFkB signaling network. To circumvent these difficulties, we can propose a set of requirements that a simplified network should meet:

i. It must contain the general structure: cell receptor → cytoplasm membrane → nuclear localization of TF.
ii. It must be experimentally verifiable.

iii. It must be functional: represent a realist rather than only with academic importance.

Requirement (i) will be easily fulfilled if the network is an NFkB homologous. Condition (ii) is satisfied if it belongs to know a model organism. If we choose a specie less complex than mammalians and follow the opposite direction in the evolution process, the requirement (iii) will be accomplished because all signaling network selected during the evolution process is functional despite the different levels of complexity. Also, it is expected that the further away from mammals we are, the simpler the signaling network will be. Following this time reverse strategy, we find the Toll *Drosophila* pathway (Fig. 7.2). At the same time, we find a good network representing all the above requirements, we are also inspired to answer the above question: *Why modeling Drosophila melanogaster*? Our best answer: *Because this is a homologous signaling network evolutionary associated with the above complex mammalian network.*

7.3 Dynamic System Theory

A system can be broadly defined as a collection of interacting elements that forms an integrated whole. Examples include the mathematical model that describes the concentration of reagents in a chemical reaction, the number of individuals in a growing population problem, and so on. At any given time, the state of a system can be described by a set of mathematical variables that can assume continuous or discrete values. The system temporal evolution results from the interactions between system elements or between the system and its surroundings. The mathematical prescription for the temporal evolution of a system states is called a dynamical system. If time is continuous, the dynamical system is represented by a set of differential equations. In its turn, iterated maps represent a dynamical system when time is discrete. Since differential equations are much more used in sciences here, we will concentrate on them. Regarding the temporal evolution, it is worth mentioning that it is basically an update rule for the system states; it is fixed over time and, in the most general case, depends on the time and the state of the system. Although it was a branch of physics nowadays, the dynamical system theory is an interdisciplinary subject with applications in fields as diverse as physics, biology, economy, etc. Solving the system or integrating system corresponds to determine the state for all future times given an initial condition. This collection of states is called trajectory or orbit of the system.

7.3.1 Fixed Points and Stability

Consider the autocatalytic reaction below:

$$A + X \underset{k_2}{\overset{k_1}{\rightleftharpoons}} 2X \tag{7.1}$$

where one molecule of A combines with one molecule of X to form two molecules of X. According to the Law of Mass Action, the rate of a reaction is proportional to the concentration of the reactants. The temporal rate concentration for molecule X is given by the differential equation below:

$$\frac{\mathrm{d}x}{\mathrm{d}t} = k_1 a x - k_2 x^2 \tag{7.2}$$

where x and a represent the concentration of species A and X, respectively. Assuming that the concentration of specie A is kept constant, the state of the system depends only on the variable x, which is continuous and can assume values equal to or greater than zero. Equation (7.2) is a dynamical system that represents the reaction in (7.1).

In this particular case, it is possible to find an analytical solution

$$x(t) = \frac{x_0 a k_1 e^{a k_1 t}}{a k_1 - x_0 k_2 \left(1 - e^{a k_1 t}\right)} \tag{7.3}$$

For $t \to \infty$, the solution above converges to $x_S = a k_1 / k_2$. That means the concentration of a molecule X, for a sufficiently long interval of time, reaches a stead -state S where $\mathrm{d}x/\mathrm{d}t = 0$. However, there is a second fixed point. If we write (7.2) as

$$\frac{\mathrm{d}x}{\mathrm{d}t} = f(x) \tag{7.4}$$

where $f(x) = x(ak_1 - k_2 x)$, fixed points correspond to the solutions of the equation

$$f(x) = 0 = x(ak_1 - k_2 x) \tag{7.5}$$

that means a state where the rate chance of the dynamical variable x is zero. It is easy to see that there are two solutions: $x_1^* = 0$ and again $x_2^* = ak_1/k_2$. Each solution corresponds to a fixed point that has different behavior. If the system starts at $x_0 = x_1^*$, it stays at this point forever. However, even for an initial condition very close to zero, the system flows in the direction of the second fixed point x_2^*. The first fixed point is called *unstable* or repeller because small fluctuation drives the system away from this point. In contrast, the second fixed point is called *stable* or attractor because the dynamics of the system flow in its direction. When the system reaches this, it remains there forever. This state is called a steady state.

7.3.2 Steady-State Analysis

The fixed of a one-dimensional dynamical system

$$\frac{dx}{dt} = f(x) \tag{7.6}$$

can be found by solving $f(x) = 0$. To determine the stability of fixed points, let us define $(t) = x(t) - x^*$ that measures how close to the fixed point x^* the trajectory $x(t)$ is. Its temporal derivative is given by

$$\frac{d\in}{dt} = \frac{dx}{dt} = f(x) = f(\in + x^*) \tag{7.7}$$

Small values of $\in$ justify a Taylor expansion for $f(\in + x^*)$ around x^*, that is,

$$f(\in + x^*) = f(x^*) + f'(x^*) \in + \vartheta(\in^2) \tag{7.8}$$

where $\vartheta(\in^2)$ corresponds to quadratically small terms in ε. From (7.7) and (7.8), it is possible to find a linear relation between $d\in/dt$ and $\in$ if we observe that for a fixed point $f(x^*) = 0$, consider $f'(x^*) \neq 0$ and take sufficiently small values for $\in$. We then have

$$\frac{d\in}{dt} \approx f'(x^*) \in \tag{7.9}$$

and results

$$\in(t) = \in(t_0) e^{f'(x^*)(t-t_0)} \tag{7.10}$$

Thus, if $f'(x^*) > 0$ the trajectory $x(t)$ runs away from x^* since $\in(t)$ grows exponentially. In this case, the fixed point is unstable. If $f^{'(x^*)} < 0$, the trajectory gets closer and closer to the x^* since $\in(t)$ decays exponentially. This fixed point is stable.

Applying this result for autocatalytic reaction (7.2), we have

$$f(x) = x(ak_1 - k_2x) \Rightarrow f'(x) = ak_1 - 2k_2x \tag{7.11}$$

Fixed points are solution of $f(x) = 0$, that is, $x_1^* = 0$ and $x_2^* = ak_1/k_2$. Then

$$f'(x_1^*) = ak_1 \tag{7.12}$$

$$f'(x_2^*) = -ak_1 \tag{7.13}$$

Therefore $x_1 = 0$ is an unstable fixed point, whereas $x_2 = ak_1/k_2$ is a stable one. This result agrees with the previous analysis using the analytical solution.

7.3.3 *Bifurcation*

The qualitative structure of solutions of a dynamical system can change abruptly as some control parameter is varied. Fixed points can be created, destroyed, or the nature of stability

$$\frac{dx}{dt} = r - x^2 \tag{7.14}$$

The fixed point is $x^* = \pm\sqrt{r}$. Thus, there is a fixed point only for $r > 0$. The fixed point $+\sqrt{r}$. is stable, whereas $-\sqrt{r}$ is unstable. For $r = 0$, they merge into one. It is possible to show that this fixed point is half-stable. For r > 0, there is no more fixed point. This kind of bifurcation is called saddle-node bifurcation. Although very simple, the behavior of all saddle-node bifurcation looks like $dx/dt = r \pm x^2$.

7.3.4 *Bistability and Oscillation*

The concept of bistability refers to the existence of two stable steady states of a dynamical system. The simplest example is good enough to give a glimpse of every other multistable system, i.e., systems with more than two stable states. In Fig. 7.3,

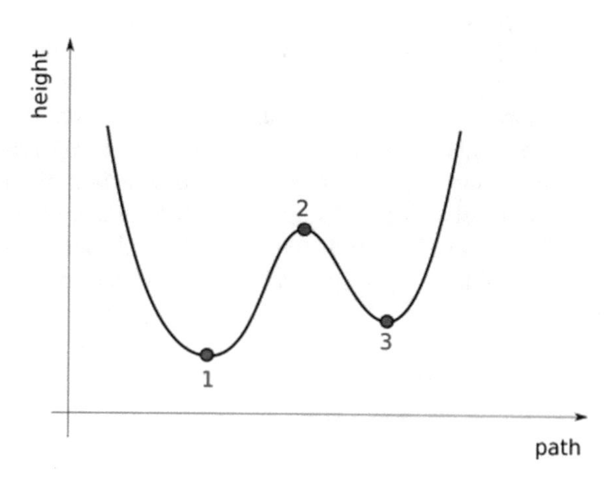

Fig. 7.3 A simple system with bistability

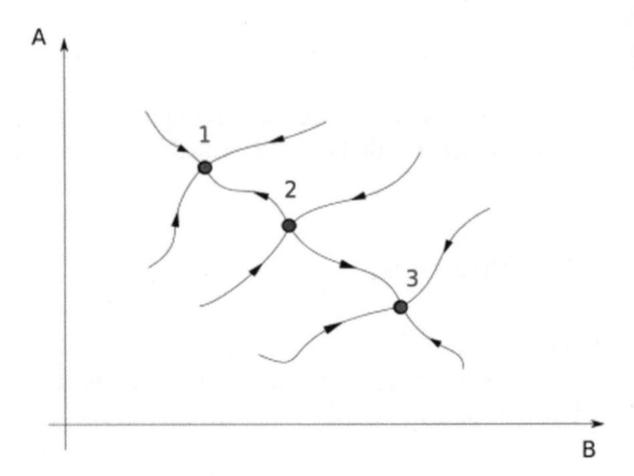

Fig. 7.4 Phase space of a simple bistable system

we could imagine the height of a given path. The points 1 and 3 are the local minimum of height, and if a ball is thrown close to them, it will stop at those points. The same ball could be left at rest at position 2; however, we know that any perturbation would remove the ball from this point, and it would fall either to point 1 or 3, which means that point 2 is an unstable steady state.

Remarkably, it is necessary to exist an unstable steady state in any system with multiple stable states. Also, it is essential to differentiate steady states from equilibrium states. Every equilibrium state, in the thermodynamic sense, is a steady-state; however, most steady states are nonequilibrium ones; this is very common in the case of chemical and biochemical systems. The bistable behavior can also be seen in the dynamical evolution of, say, two chemical species A and B. In Fig. 7.4, again, points 1 and 3 are stable, while point 2 is unstable. The curves shown represent some of the time evolutions of concentration of these species A and B from different initial conditions. The arrows indicate the concentrations chance in time.

Another especially important kind of time evolution behavior appears in the oscillatory systems. In this case, instead of going to a steady-state, the system continues to come back again and again to the same (nonequilibrium) states. For instance, we could imagine two species, say, lions and zebras. If the number of zebras grows, the lions have more available food, and its population may grow (Fig. 7.5). However, a huge number of lions would consume too many zebras, reducing their number. Fewer zebras will ultimately imply the reduction of the population of lions. If the number of lions is low enough again, the number of zebras restarts to grow, and so on. Strictly speaking, from the thermodynamic point of view, oscillatory systems are only possible for open systems, those for which energy is continuously provided.

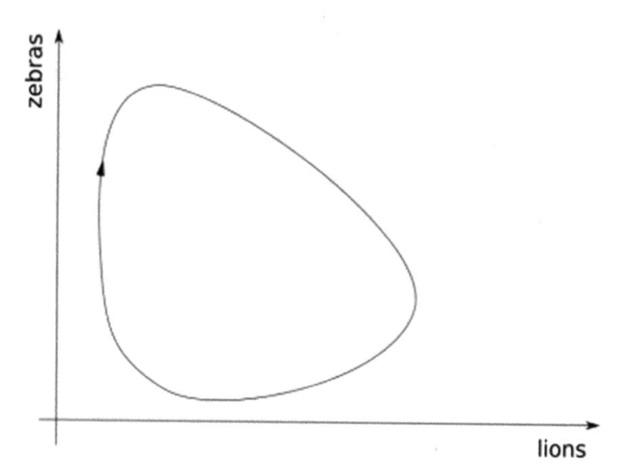

Fig. 7.5 A simple oscillatory system describing the temporal evolution of lions and zebras

The systems of ordinary differential equations (ODE) for chemical and biochemical reactions networks can be described by a special class of ODEs, namely:

$$\dot{x} = N.v$$

where x is the vector of the species present in the system, the dot means time derivative, N is a purely numerical matrix, more often than not rectangular, and represents the stoichiometric relations of the associated system of reactions. The vector v gives the velocities of those reactions and depends on time. This class has a lot of interesting and useful properties. Since this equation is a consequence of the intrinsic stoichiometric relations, we could even speculate that chemical and biochemical features are a consequence of stoichiometry. Some early results for these systems came from the work of Horn, Jackson, and Feinberg [11, 12]. At about the same time, Clarke [13] has also discovered a lot of basic features of this class of systems. Most of these results indicate whether or not a given reaction network may or may not present multiple steady states or oscillatory behavior. Some other important questions are also studied, like the possibility of extinction of any given species. A quite complete characterization of the possibility of multiple steady states has been studied by Willamowsky [14]. Among other things, it has been proven that no chemical system with two species will show oscillations. Moreover, these systems will show either only one stable or two stable steady states. The same kind of results is given for every other number of species. The main result from the work of Horn, Jackson, and Feinberg is the Zero Deficiency Theorem. It introduces the concept of deficiency of a network. This number is easily calculated from the number of combinations of reactants and products present in the network, the number of disconnected nets in the network, and the rank (a well-known concept from linear algebra) of the stoichiometric matrix N introduced above. If the deficiency of a network is zero, then no matter the value of the kinetic constants, the system will always have only one thermodynamically stable steady state. Later on, Feinberg [15] developed a number of results to understand the properties of systems with nonzero deficiencies. His theorems **allowed** the determination of many systems that would present multistability. It also allows us to discover which combinations of kinetic constants would allow the existence of multistability. Also, the use of these results to the understanding of real chemical and biochemical systems depends, among other things, on better ways to determine kinetic constants, which is not an easy task. The group of Prof. Martin Feinberg, now at Ohio State University, provides a computational tool, named Chemical Reaction Network Toolbox, to analyze any system of reactions using many of the known results in this field. More details are presented in Sect. 7.5.1 *A system biology approach.*

7.3.5 Bistability: A Practical Example

To explore the concept of bistability, we can consider the following dynamical system

$$\frac{dx}{dt} = x(1-x) - \frac{bx}{a+x} \tag{7.15}$$

a populational growth model capable of exhibiting all effects, that is, a reduced per capita growth rates at low population density. Strong allee effect results in negative growth rates below a critical population density leading to population extinction.

The solution of the equation

$$\frac{dx}{dt} = 0 = x\left[(1-x) - \frac{b}{a+x}\right] \tag{7.16}$$

are the fixed points of the system given by

$$x_1^* = 0 \tag{7.17}$$

$$x_2^* = \frac{(1-a) - \sqrt{(1+a)^2 - 4b}}{2} \tag{7.18}$$

and

$$x_3^* = \frac{(1-a) + \sqrt{(1+a)^2 - 4b}}{2} \tag{7.19}$$

Real values for x_2^* and x_3^* imply

$$(1+a)^2 - 4b \geq 0 \Rightarrow b \leq \frac{(1+a)^2}{4} \tag{7.20}$$

If $b = (1+a)^2/4$, we have

$$x_1^* = 0, x_2^* = x_3^* = \frac{(1-a)}{2} \tag{7.21}$$

On the other hand, if $a = b$

$$x_1^* = x_2^* = 0, x_3^* = 1 - a \tag{7.22}$$

Therefore, bistability is found in the region

$$a < b < \frac{(1+a)^2}{4} \tag{7.23}$$

and for $0 \leq a \leq 1$, since

$$\lim_{a \to 1} \frac{1}{4}(1+a)^2 = 1 \tag{7.24}$$

Figure 7.6 provides an overview of the present analyses. Figure 7.6a shows the change rate of variable x for a fixed $a = 0.1$ and three values of parameter b. The upper line corresponds to $b = 0.05$. There are two fixed points, that is, $(dx/dt = 0)$. The fixed point $x_1 = 0$ is unstable (you can see it directly from the graph observing the inclination on the tangent line at the fixed point). The other one $(x_3 = 0.9)$ is stable. The middle line corresponds to $b = 0.28$. In this case, there are three fixed points $(x_1 = 0; x_2 = 0.3 and x_3 = 0.6)$. In the b *versus* a diagram, Fig. 7.6b, this point $(a = 0.1, b = 0.28)$ is inside the bistability region indicating the existence of three fixed points. The boundary for bistability is given by $b(a) = (1+a)^2/4$ and

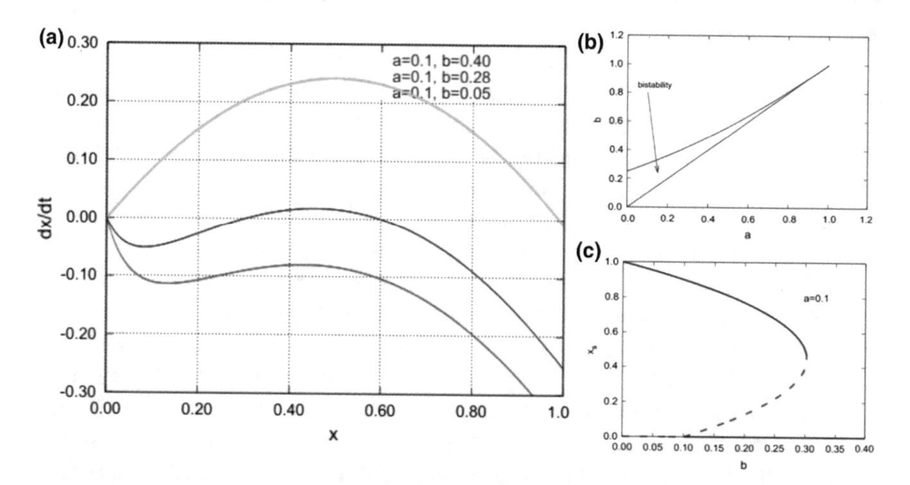

Fig. 7.6 Dynamical behavior of a differential Eq. (7.16). **a** Change rate of variable x of the dynamical system. Graphs correspond to a $= 0.1$, $b = 0.40$ (lower line), $b = 0.28$ (middle line), and $b = 0.05$ (upper line). **b** bistability in the parameter space. **c** Solid and dotted lines indicate stable and unstable steady states for the x variable, respectively

$b(a) = a$. Both curves collapse at $a = 1$. Finally, the lower line corresponds to $b = 0.40$, where the allee effect is strong. In this case, there is only one stable fixed point, x_1^*. It implies population extinction, Fig. 7.6c.

7.4 Parameter Estimation Strategies

Reaction networks, which encompass the pathway structure of biological processes, are usually modeled by Ordinary Differential Equations (ODEs) Systems, describing the evolution over time of the number of molecules or the concentration of chemical species, including proteins, mRNA, and DNA binding sites [16–23]. Also, when the network lies in a defined region in space, such as along an embryo or in biological tissues, these reactions are further represented by Partial Differential Equations (PDEs) Systems. The corresponding equations are written down using kinetic laws, such as the Law of Mass Action and the Michaelis–Menten Law, along with Fick's Law for diffusion. Therefore, the parameters are often kinetic constants and diffusion coefficients, which may be obtained in the literature if it is on hand, or they must be estimated from the measurement of other quantities in the system, i.e., the available data.

The parameter estimation in systems biology is thereby part of an iterative process to either develop data-driven models for biological systems or propose a set of solutions to physics-based models, both of them being explainable and having predictive value. A commonly used method for estimating parameters in reaction networks is to convert the estimation problem into an optimization problem, hence optimization algorithms well-established in the literature can be applied to find the best-fit parameters [24–27].

However, reaction network models have, in general, multiple sets of parameters lying in nonsmooth and nonconvex search spaces; thus, there are many local minima that do not have any biological meaning. Traditional deterministic algorithms are thus usually very time-consuming and highly computationally intensive, being inefficient when applied to bigger systems biology problems. Since the value of parameters and model variables may range over several orders of magnitude, the algorithm may get stuck in a local minimum or in a very flat part of the solution space.

Furthermore, since datasets are noisy and sparse, given a particular set of experimental data and one model parameterization obtained by a parameter estimation method, one should validate and test the model calibration in order to determine the relation between computational results and biological conclusions. Modern optimization approaches, such as evolutionary algorithms and other metaheuristics, can achieve near-global optima in a reasonable time for large size problems. Notably, evolution strategies are robust in noisy environments and exhibit an inherent parallel algorithmic structure that can be taken advantage of with modern computational hardware.

7.4.1 Formal Problem Definition

Reaction network models represent, in general, nonlinear systems, hence despising as a nonlinear programming problem with differential-algebraic constraints, i.e., given by a system of Differential-Algebraic Equations (DAEs) is of the following form:

$$\frac{\mathrm{d}C}{\mathrm{d}t} = f(C(\theta, t), \theta, u(t), t)t \in \left(t_0, t_f\right]$$

$$C(\theta, 0) = C_0(\theta)$$

$$y(\theta, t) = F(C)$$

The state vector C represents the number of molecules or the concentration of chemical species present in the reaction network, the parameters θ are the kinetic constants and other unknown parameters that one seeks to estimate, the input signal u denotes some external forcing of the process (such as temperature changes) which can depend on the time t, the initial state vector C_0 may depend on θ if there are some initial concentrations not known, and y is the output signal, usually being a vector of observables related to quantities measured experimentally. Moreover, there is the possibility of n_g equality constraints and n_h inequality constraints that the system must obey, as well as upper and lower bounds θ_U and θ_L, respectively, for parameter values:

$$g_i(C, \theta, u(t), t) = 0 \quad i = \left\{1, \ldots, n_g\right\}$$

$$h_j(C, \theta, u(t), t) \leq 0 \quad j = \{1, \ldots, n_h\}$$

$$\theta_L \leq \theta \leq \theta_U$$

Given $\hat{y}_t^{\exp}$ a vector of observed data over time, the parameter estimation problem is to find a set of parameters θ^*, such that, within the analyzed time range $\left(t_0, t_f\right)$, the following objective function is minimized:

$$J\left(y, \hat{y}_t^{\exp}, \theta\right) = \int_{t_0}^{t_f} \left(\hat{y}_t^{\exp} - y(\theta, t)\right)^T W(t)\left(\hat{y}_t^{\exp} - y(\theta, t)\right)\mathrm{d}t$$

where $W(t)$ is a scaling matrix. Since in practice the experimental measures are discrete, the objective function can be written as

$$\underline{J}\left(y, \hat{y}_t^{\exp}, \theta\right) = \sum_{i=1}^{N} \left(\hat{y}_{t_i}^{\exp} - y(\theta, t_i)\right)^T W(t_i)\left(\hat{y}_{t_i}^{\exp} - y(\theta, t_i)\right)$$

Despite the vast possibility of objective functions according to the choice of scaling matrix, the simplest form of network problems is the standard least-squares problem:

$$\underline{J}\left(y, \hat{y}_t^{\text{exp}}, \theta\right) = \sum_{i=1}^{N} \left(\hat{y}_{t_i}^{\text{exp}} - y(\theta, t_i)\right)^2$$

The extension to reaction–diffusion networks is straightforward, adding the spatial coordinates x to the equations constrained by a domain Ω, in addition, to insert the possibility of temporal evolution depending on the spatial derivatives (gradient, Laplacian, ...), thus leading to

$$\frac{\partial C}{\partial t} = f(C(\theta, x, t), \nabla C, \ldots, \theta, u(t), x, t) \quad t \in \left(t_0, t_f\right] \quad x \in \Omega$$

$$C(\theta, x, 0) = C_0(\theta, x)\ x \in \Omega$$

$$C(\theta, x, t) = C_{d\Omega}(\theta, t) \quad t \in \left(t_0, t_f\right] \quad x \in d\Omega$$

$$y(\theta, x, t) = F(C)$$

$$g_i(C, \theta, u(t), t) = 0 \quad i = \left\{1, \ldots, n_g\right\}$$

$$h_j(C, \theta, u(t), x, t) \leq 0 \quad j = \{1, \ldots, n_h\}$$

$$\theta_L \leq \theta \leq \theta_U$$

Thus, the objective function, considering that both time and spatial coordinates are discrete, is given by

$$\underline{J}\left(y, \hat{y}_t^{\text{exp}}, \theta\right) = \sum_{i=1}^{N} \sum_{j=1}^{M} \left(\hat{y}_{t_i,x_j}^{\text{exp}} - y\left(\theta, x_j, t_i\right)\right)^T W\left(x_j, t_i\right)\left(\hat{y}_{t_i,x_j}^{\text{exp}} - y\left(\theta, x_j, t_i\right)\right)$$

7.4.2 Parameter Estimation Methods

Biological models are often dynamic and highly nonlinear; thus, in order to find the parameter estimates, we must employ nonlinear optimization techniques, since no general analytic result can be determined. The traditional gradient-based methods, such as Levenberg–Marquardt or Gauss–Newton, may fail to identify the global solution and converge to a local minimum rather than the global minimum.

7.4.2.1 Local Optimization Methods

If the gradient of the objective function can be computed, it is possible to solve the minimization problem by finding the point where the gradient vanishes using gradient-based methods [16]. Along with them, direct-search methods try to find the minimizing point of the objective function without explicitly using derivatives, that is, by comparing the value of the function over different points in the parameter space.

Direct-Search Methods: These methods select a finite number of candidate solutions each step and check which one is the best in terms of minimizing the objective function. The **Hooke–Jeeves method** consists of two steps. Firstly, a series of exploratory changes of the current parameter vector are made, forming a basis with information about which direction the objective function decreases therein. Then, the information obtained is used to find the best direction for the minimization process. The **Nelder–Mead method** evaluates the objective function in all vertices of a simplex, and the vertices are ordered according to the value. A *simplex* $S \in \Re^n$ is defined as the convex hull of $n + 1$ vertices $x_0, \ldots, x_n \in \Re^n$. For instance, a simplex in two dimensions is a *triangle*, and simplex in three dimensions is a *tetrahedron* (the dimensions are defined by the number of unknown parameters). The next step tries to replace the vertex with the highest objective function value by performing a line search along the line through this vertex and the centroid of the remaining vertices. If the line search succeeds, the simplex is adapted by replacing the old vertex by the new one. Else, a shrinking procedure is performed: the best vertex stays in the simplex, while all other ones are replaced by a vertex half-way along the line from the best vertex.

Gradient-Based Methods: These are iterative methods that extensively use the gradient information of the objective function during iterations. For the minimization of $J(\theta)$, the general equation that searches an optimal parameter vector θ^* is such that, for each iteration i,

$$\theta^{(i+1)} = \theta^{(i)} + \alpha g\big(\nabla J, \theta^{(i)}\big)$$

where α is the step size which can vary during iterations, $g\big(\nabla J, \theta^{(i)}\big)$ is a function of the gradient ∇J and the current parameter vector $\theta^{(i)}$. The main difference between the methods lies on the chosen function. The **steepest descent method** uses $g\big(\nabla J, \theta^{(i)}\big) = \nabla J$ directly, which is shown to be a slow convergence. There are other methods, such as **Newton's method**, which also considers the Hessian of the objective function, i.e., $g\big(\nabla J, \theta^{(i)}\big) = -\nabla^{-2} J(\theta) \nabla J(\theta)$, thus needing more information about the behavior of the objective function in relation to the parameters. In **quasi-Newton methods**, including BFGS and Broyden's method, the Hessian is approximated by successive gradient vectors. When solving a least-squares problem, **Gauss–Newton algorithm** is usually applied instead, in which the Hessian is not used. The **Levenberg–Marquardt algorithm** also known as the **Damped least-squares (DLS)**, interpolates between the Gauss–Netwon method and the steepest descent by regulating an additional parameter such that, in the initial stage of the iterations, the steepest descent is desired in order to slowly explore the parameter space searching for global optimum, whereas in the final stages, it guarantees a faster convergence that targets the global minimum.

Although direct-search methods are generally applicable, they are less efficient when exploring high-dimensional parameter spaces.

7.4.2.2 Simulated Annealing

Simulated Annealing (SA) is a stochastic optimization algorithm based on the Metropolis algorithm, a Monte Carlo method to sample a thermodynamic system [28]. The name and inspiration come from annealing in physics and metallurgy, a technique involving heating and controlled cooling of a material to increase the size of its crystals and reduce their defects. Both are attributes of the material that depend on its thermodynamic free energy, affected by heating and cooling the material. In more detail, cooling must occur at a sufficiently slow rate, otherwise, the system will end up in an amorphous or polycrystalline state rather than at its minimum energy state. In optimization, the SA algorithm attempts to mathematically capture the process of controlled cooling associated with physical processes, hence the analogy to the minimum energy state is the minimum value for the objective function.

The basic components of SA include an objective function J, a definition of neighborhood, a cooling schedule, an acceptance criterion, and a solution perturbation scheme. Starting from a candidate solution θ, the SA perturbs θ in its defined neighborhood to create a new solution θ', which will be checked for whether it is accepted as the current solution depending on the values $J(\theta)$ and $J(\theta')$, according to the Boltzmann–Gibbs probability distribution:

$$P(\theta) = K \exp\left(-\frac{J(\theta) - J(\theta')}{k_B T}\right) J(\theta') < J(\theta)$$

Therefore, in addition to accepting solutions θ' such that $J(\theta') > J(\theta)$, the algorithm allows solutions whose objective function is bigger weighted by a probability with takes into account a parameter T(which is the temperature, in analogy to the annealing process), a normalization constant K and the Boltzmann constant k_B. That strategy prevents the algorithm from getting stuck in a local minimum, especially at the beginning of the iterations, where the parameter space exploration is important.

In this way, a Markov chain is obtained, which, if it is sufficiently long, describes the required probability distribution. The minimizing parameter vector is thereby the average over all states in the Markov chain. In SA, the Metropolis algorithm is applied with a slowly decreasing T regulated by a cooling schedule. Thus, SA starts with a high temperature, implying that all parameter vectors are equally probable. There are two variants of SA: the **homogeneous Markov chain method** SA computes, for a constant temperature, a complete Markov chain, then the temperature is slowly decreased, and the next distribution is sampled. On the other hand, the **inhomogeneous Markov chain method** decreases the temperature every time a new state has been found. Hence a cooling schedule, as in the physical system, is important so that, as the algorithm evolves, it is more likely that a local minimum found by it is the global minimum, and the solution, therefore, remains around its neighbor, similar to an exploitation strategy.

There are several attempts to find an efficient cooling schedule scheme. Among the most popular ones, the **logarithmic cooling schedule** decreases the temperature accordingly to the formula $k_B T_i = \gamma \log(i + d)$, where i is the iteration, d is an offset constant (usually set to 1), and γ is a maximum energy constant. Although it has been proven that this method can find the global minimum given some specific conditions, it is impractical because of its asymptotically slow temperature decrease. The **geometric cooling schedule** updates the temperature by $k_B T_i = \alpha k_B T_{i-1}$, where $\alpha < 1$ is a positive constant called a cooling factor. Finally, the **adaptive cooling scheme** controls the temperature depending on the state of the system, the number of previously accepted solutions, and so on.

7.4.2.3 Genetic Algorithms

Genetic Algorithms (GAs) are stochastic search algorithms inspired by biological evolution, processing potential solutions (parameter vectors composing a population) by transformations regarding mechanisms as reproduction, natural selection, mutation, recombination, and survival of the fittest [29].

Initially, several candidate solutions are randomly or heuristically chosen, thus generating the first population. Next, the corresponding objective functions are computed, defining the fitness of an individual. The selection process is employed by assigning probabilities to individuals related to their fitness, thus indicating the chance of being selected for the next generation. For the next generation, individuals are created by two operators: recombination and mutation.

Recombination consists of selecting parents and results in one or more children (new candidates). Mutation acts on one candidate by disturbing some of its characteristics, resulting in a slightly different candidate. These operators create the offspring (a set of new candidates), which will compete with old candidates in the following generations (survival of the fittest). This process can be repeated until a candidate with sufficient quality (a solution) is found or a predefined computational limit is reached.

Parameter Codification: Since each candidate solution is represented as an individual within a population, each unknown parameter can be understood as a genetic characteristic of the individuals. Thus, it is essential to codify these parameters in an efficient way for the subsequent application of the transformation operators. In terms of a numerical basis, two main codifications are used: *binary* and *floating-point* numbers.

- **Binary Codification**: Each parameter is converted to a binary sequence; therefore the individual is represented as a string of bits. When the parameters are originally real numbers, the **number of bits** representing each parameter must be kept in mind, as this indicates the **resolution** in the parameter space (the difference between two consecutive values). Given the lower and upper bounds θ_L and θ_U and the number of bits b, the following equation codifies the real parameter θ into the binary representation θ^{bin}:

$$\theta^{bin} = \frac{\theta - \theta_L}{\theta_U - \theta_L}(2^n - 1)$$

  and then apply a decimal-to-binary conversion. To decode the binary representation θ^{bin}, recovering the real parameter θ, one must use

$$\theta = \frac{\theta_2^{bin}}{2^n - 1}(\theta_U - \theta_L) + \theta_L$$

  where θ_2^{bin} is the binary-to-decimal conversion of θ^{bin}.
  Also, it is important to emphasize that when the parameters oscillate between several orders of magnitude, the logarithmic scale must be used instead to ensure scale homogeneity. In this case, the equations that hold are

$$\theta^{bin} = \frac{\log(\theta) - \log(\theta_L)}{\log(\theta_U) - \log(\theta_L)}(2^n - 1)$$

$$\log(\theta) = \frac{\theta_2^{bin}}{2^n - 1}(\log(\theta_U) - \log(\theta_L)) + \log(\theta_L)$$

- **Floating Point Representation**: The real number representation is more intuitive for the applications discussed here, as each individual is described by the actual values of its parameters. However, the recombination and mutation operators become more complicated, as they have a much more extensive range of possibilities, and less guaranteed to be efficiently searching the optimal solution in the parameter space.

Population Size: A Genetic Algorithm must be able to balance the efficient exploration of the parameter space and the computational cost. Thus, the importance of several individuals in each population that is large enough to guarantee the variability of the candidate solutions is remarkable, without exceeding the available computational cost. However, there is no fixed value for a good number of individuals, some numbers typically found in the literature are 30, 50, 100, depending on the complexity of the problem involved. There are also strategies in which the population size varies along with the iterations, thereby controlling the evolutionary pressure exerted on the population.

Fitness Function: The fitness function measures the quality of a solution, thereby defining the chances of a particular individual to survive over the generations and of passing on their characteristics to new candidates. The fitness function is, therefore, applied to the values obtained in the objective function. Conventionally, because parameter estimation problems are minimization problems, and the concept of fitness induces the idea of "the bigger, the better", some possible choices are

- The fitness value of $\max(0, J_U - J(\theta))$, where J_U is an upper-bound objective function value predefined by the user or defined by the program itself (for instance, the maximum value registered so far in the iterations).
- Other descending functions, such as $\frac{1}{J(\theta)}$, although some of them may have problems, especially when $J(\theta)$ is close to zero, which is undesirable.
- A rank-based fitness, where the population is sorted according to their objective values, thus the fitness assignment depends only on the position.

Genetic Operators: Perform transformations of candidate solutions according to the value of the fitness function and intrinsic parameters to each operator. The selection operator is responsible for convergence to the minimum, the recombination operator for exploring the parameter space and the mutation operator gives nearby solutions a chance to be candidates.

- **Selection**: This operator determines which individuals are chosen for the next generation and how many offspring each selected individual produces. A **truncated selection** picks m best individuals to be eligible for reproduction, thus discarding the remaining candidates. Afterward, there are several methods to choose the parents: a **roulette-wheel** selects individuals with probability proportional to their fitness values, while a **tournament** picks k random individuals as competitors, then the one with the highest fitness wins. Even further, an **elitism** strategy involves copying a small proportion of the fittest candidates, unchanged, into the next generation. This ensures that the algorithm does not waste time rediscovering previously discarded solutions since the highest fitness value of the population never goes down. Candidate solutions that are preserved unchanged through elitism remain eligible for selection as parents when breeding the remainder of the next generation.
  The selection process is an extremely important part of the convergence of the algorithm. If the selection pressure is high, such as with roulette-wheel, the convergence time is much faster, but the solution is less likely to be the global minimum. By the other hand, when the selection pressure is low, as with tournament with small k, the lower convergence speed is compensated by the possibility of exploring the parameter space more widely.
- **Recombination**: This operator produces new individuals by combining the information contained in the previously selected parents. A widely used technique for recombination is **k-point crossover**, where k points on both parents' chromosomes are picked randomly, and designate "crossover points". The bits in

between two points are swapped between the parent organisms. This results in two offspring for two parents, each carrying some genetic information from both of them. Differently, in a **uniform crossover**, each bit is selected randomly from one of the corresponding bit strings of the parents. Although usually one chooses from either parent with equal probability, other mixing ratios are sometimes used, resulting in offspring which inherit more genetic information from one parent than the other (**nonuniform crossover**).
For real-valued representation, the **geometrical crossover** breeds only one offspring from two parents by calculating the following representation:

$$\theta^{(new)} = (\theta_{p1})^{u}(\theta_{p2})^{1-u}$$

where u is a number between 0 and 1.

- **Mutation**: This operator may alter an individual randomly, thus changing one or more bits in the case of binary representation or applying some stochastic disturbance in the case of floating points. The **mutation step** is the probability of mutating a variable, and the **mutation rate** is the effective mutation applied to that variable. One widely used mutation method is called **inversion**, which defines a probability for bits in encoding to be flipped (0–1, or 1–0). Usually, this probability is such that, on average, the number of mutated bits is very small compared to the total number of bits that represent an individual. There are several more sophisticated methods, mutating more than one bit simultaneously and looking for deeper analogies with mutations in the genetic information of an organism.

Stopping Criterion: The iterative process is repeated until a candidate solution achieves sufficient quality defined by the user (for example, the objective function is equal or less than some value), or a predefined computational limit is reached (total number of iterations, for instance).

There are different advanced strategies in the literature to improve the performance of Genetic Algorithms. An **epidemic genetic operation** renews the population if its fitness does not improve after some number of generations, thus preserving only best-fitness individuals. Two parameters are needed: one to determine when the strategy will be activated, and another to determine the amount of surviving individuals.

7.4.2.4 Differential Evolution

The Differential Evolution algorithm involves maintaining a population of candidate solutions subjected to iterations of recombination, evaluation, and selection [30]. The recombination approach involves the creation of new candidate solution components based on the weighted difference between two randomly selected population members added to a third population member. This perturbs population members relative to the spread of the broader population. In conjunction with selection, the perturbation effect self-organizes the sampling of the problem space, bounding it to known areas of interest.

7.4.2.5 Other Methods

In this section, the main methods to guide the reader in the development of a parameter estimator for a problem modeled by reaction–diffusion networks were discussed. There is a vast literature of techniques for deepening the details of the area, from novel metaheuristics, Bayesian estimation (to take into account the uncertainties in the parameters found), and hybrid methods.

Novel metaheuristics are usually adaptations of strategies to traditional metaheuristics, such as Genetic Algorithms, including terms to better explore potential solutions to the optimization problem. They intend to increase the efficiency of the estimation while keeping robustness and reducing the computation time by orders of magnitude [31–33].

Bayesian estimation is an alternative to traditional optimization techniques, considering both the system and measurement noise during the estimation [5, 27]. Specifically, it calculates the posterior density of the parameters vector conditioned on the observed data. However, the calculation of this posterior involves high-dimensional integration, for which no analytical solution is generally available. Therefore, numerical approximations have to be made for this posterior probability density [21, 22, 29, 34–38]. Hybrid methods use a global search method to identify promising regions of the search space that are further explored by a local optimizer [16, 20]. This is motivated by the fact that global search methods, such as evolutionary algorithms, are very efficient in exploring parameter space. In contrast, local search, by using more detailed information about the neighborhood of a solution, can converge with better precision, given that it is close to the minimum.

7.4.3 Constrained Optimization

The algorithms discussed above are unrestricted and often create solutions that fall outside the feasible region. Therefore, a viable constraint-handling strategy is important to ensure that the criteria obeyed by the system are present, leading to a greater possibility of the solution generated having biological significance. As an example, one may want solutions for a balance between two biochemical species A and B such that the concentration of A is always higher than B.

Among the different approaches available to take into account the constraints of an optimization problem, the most traditional is through the use of **penalty functions** and **barrier methods** [39, 40]. In addition, there are procedures that explicitly use knowledge about restrictions and then use search operators to maintain the **feasibility of solutions** [41]. Besides these techniques, other methods involve working with specific parameter encodings, as well as possible combinations between different approaches, resulting in hybrid methods.

7.4.3.1 Penalty Methods

The most common way to incorporate constraints into search algorithms is by converting the constrained problem into an approximately equivalent unconstrained problem. Therefore, a term called *penalty* is added to the objective function prescribing a higher value if one or more constraints are violated. Hence, the original objective function is replaced by

$$\underline{J}(\theta) = J(\theta) + \sum_{i=1}^{n_g} \sigma_{g_i}(g_i) + \sum_{j=1}^{n_h} \sigma_{h_j}(h_j)$$

where c_{g_i} and c_{h_j} are *penalty coefficients* for each equality constraint g_i or inequality constraint h_j, and σ_g and σ_h are *penalty functions.* The most employed penalty function in the literature is the quadratic loss function: $\sigma(f) = \max(0, f)^2$ for inequality constraints, and $\sigma(f)^2$ for equality constraints. In general, a single parameter is defined for all constraints, in order to simplify the algorithm. However, there are several procedures to manipulate it through the evolution of the method in the search for an optimal solution.

The **static penalty method** does not depend on the iterations of the optimization method, remaining constant throughout the search process. The **dynamic penalty method**, on the other hand, is user-defined and tends to increase as the algorithm evolves, in order to favor exploration at the beginning, and exploitation in the final stages of the search. Although there is no consensus on the best dynamic method, a successful penalty applied in the literature is the **annealing penalty method**, following a logic analogous to simulated annealing, and defining a temperature parameter that decreases slowly during the iterations. Finally, it is possible to develop **adaptive penalty methods**, the parameter of which is updated after each iteration according to the fitness values found. Thus, if several prohibited solutions are being found during the search, the penalty parameter may increase, for example.

There is also a simple method called the **death penalty**, which rejects infeasible solutions from the algorithm, useful in population-based methods as genetic algorithms and differential evolution. In this case, there will never be any forbidden solutions in the population, being a useful technique if the feasible search space is convex or a reasonable part of the whole search space [26]. However, when the problem is highly constrained, the algorithm tends to spend a lot of time finding feasible solutions, therefore, hindering the search and impairing the variability of the population.

7.4.3.2 Barrier Methods

Rather than adding penalty terms that cancel out at the boundaries of the restrictions, the barrier method forces the algorithm to find values increasing to infinity as the point approaches the boundary of the feasible region of the optimization problem.

Note that this method may hinder the search for solutions that are close to the boundaries of the constraints, however. In addition, it is optimized for the existence of inequality restrictions only. The main barrier function is logarithmic, changing the original objective function, thus giving the following term:

$$\underline{J}(\theta) = J(\theta) - \mu \sum_{j=1}^{n_h} \log(-h_j)$$

where $\mu > 0$ is a *barrier parameter* for inequality constraints h_j.

7.4.3.3 Feasibility Preserving Constraint-Handling Strategies

Constraint-handling strategies repair infeasible solutions by bringing them back into the search space and explicitly preserve the feasibility of the solutions. Although several constraint-handling strategies have been proposed to bring solutions back into the feasible region when constraints manifest as variable bounds, some of these strategies can also be extended in the presence of general constraints.

The existing constraint-handling methods for problems with variable bounds can be broadly categorized into two groups [41]: techniques that perform feasibility check **variable-wise**, and techniques that perform feasibility check **vector-wise**. In the first group, for every solution, each variable is tested for its feasibility with respect to its boundaries and made feasible if the corresponding bound is violated; thus only the variables violating constraints are altered while other variables are kept unchanged. In the second group, if a solution is found to violate any of the defined constraints, it is brought back into the search space along a vector direction into the feasible space; consequently, variables that explicitly do not violate constraints may also get modified. In addition, there are approaches that, instead of modifying not factible solutions, **directly modify the fitness function**, ensuring that they are always worse than feasible solutions.

7.4.4 Software Availability

As much as an intermediate or advanced computer user may implement reaction networks and search algorithms using programming languages such as Python, C++, or MATLAB, there are several software available for such purposes.

COPASI is a software used for the creation, modification, simulation, and computational analysis of kinetic models in various fields. It is open-source, available for all major platforms, and provides a user-friendly graphical user interface, but is also controllable via the command line and scripting languages [42].

DEAP is an evolutionary computation framework developed in Python for rapid prototyping and testing of ideas. It seeks to make algorithms explicit and data structures transparent, working with parallelization mechanisms such as multiprocessing and SCOOP. DEAP includes many discussed methods, such as genetic algorithms, genetic programming, multi-objective optimization, and differential evolution [43].

7.5 Bistability in Developmental Biology

We ended the first section of this chapter by suggesting that modeling cell signaling in *Drosophila* is a good choice because of the strong conservative character of many signaling networks in *Drosophila*, mammalian, and human beings. It allows us to assume that a network in *Drosophila* as a good functional simplification of a mammalian network, i.e., a simplification that is still able to reproduce the key biological aspects of the flux and interpretation of external cellular information. However, it is very well known that *Drosophila* and mammalian embryonic development are quite different (see more details below), *why should we assume Drosophila as a good model to understand pattern formation in more complex organisms?*

In order to answer the above question, we need to return a little in time. Since the past century, *Drosophila* has been an important source of data to understand cell signaling. This is related to the discovery of NFkB homolog Dorsal, which is responsible for embryonic pattern formation and innate immune response in *Drosophila*. As discussed in the first section, this signaling network is conserved in many species. In fact, the general framework *External signal → Protein receptor in the cytoplasm membrane → Nuclear localization of TF* shows great universality among many molecular mechanisms of cell signaling (although it was not first described in Dorsal). Even though a great diversity of molecular data about NFkB in mammalian and even cancer patients are available, studying Dorsal *Drosophila* still has the potential to play a role in the understanding of NFkB signaling. Returning to the previous question, as far as pattern formation is concerned, we are in a situation similar to the middle of the past century. We are still looking for the general framework of pattern formation, the general framework at the molecular level. This justifies and, in fact, makes *Drosophila* one of the best biological systems to understand the basic principles that drive pattern formation from gene regulatory networks. Besides that, being simpler than the development in mammals makes the development in *Drosophila* easier to understand. Assuming that the basic principles of development at the molecular level are the same, understanding the development in *Drosophila* has the potential to provide key elements to understand the development in mammalians.

7.5.1 A System Biology Approach

Model building, calibration, and simulation are interdependent steps with very specific challenges. In order to organize the different stages of the *systems biology approach* here adopted, we proposed a general scheme (Fig. 7.7). The first stage in this approach corresponds to collecting *experimental information.* It encompasses, but does not limits to, the definition of the biological system of interest. In this stage, we define the scale and the level of approximation to be used in system modeling. This is critically defined by the questions to be answered and by the experimental results available. For example, if we are trying to identify a TF binding mechanism binding to a gene regulatory region, our model definition may take into account the number of the TF binding sites in the gene regulatory region, the monomeric, and dimer state of occupancy, among others. Depending on the level of accuracy, and experimental information available, the distance between the binding sites should also be considered. Besides these problems' specific information, other of more general aspects could, as much as possible, be considered. For example, assuming a more realistic number of binding sites in a gene regulatory region may not change the dynamical behavior of a theoretical model or even makes its mathematical solution more difficult, however, if it makes the model biologically more realistic it should be considered and have the potential to contribute to subsequent modeling stages (see 7.5.4). In fact, we must think about the balance between being more realistic and having a mathematical simple/soluble model. We must never forget that the main motivation is not to build a model that can be mathematically solved but a model that does not loses its bounds with reality. A heavy price that one could need to pay

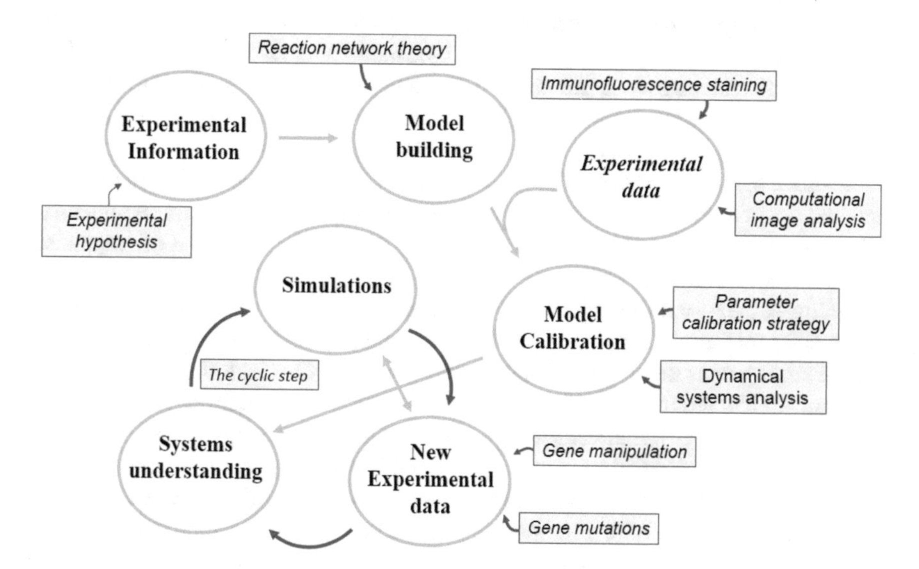

Fig. 7.7 A system biology approach

in order to make a realistic theoretical model has the potential to pay off when we are analyzing the experimental impact of its results or trying to convince the bench oriented researcher in the team about the results of your model; or and more importantly, to convince then to test your model orientated hypothesis. Short personal advice: *no matter how incredible or extraordinary a theoretical result could seem, it will be, at most, a hypothesis that needs experimental confirmation.* Also, the great majority of biological questions arise from the bench and this is there that any final answer will be verified.

Despite the extraordinary advances of experimental biology nowadays, the absence of sufficient experimental information is not rare. In this case, the set of experimental information to consider can be completed by an *experimental hypothesis*, i.e., hypothesis about the experimental behavior of the systems. One quite common experimental hypothesis is about regulatory interactions between specific genes, i.e., suppose that the protein codified by a gene *A* could activate/repress the expression of a gene *B*. In fact, the objective of an entire model could be check if this hypothesis could explain a gene expression level or even an entire gene expression pattern. Even though the kind of quantitative data available will influence the whole modeling strategy, in the *Experimental Information* stage of the approach, we are not dealing with the quantitative data itself. In fact, we can call this step as the *Qualitative Information* stage of the approach.

The *Model* step of the approach (Fig. 7.7) is one of the most diverse. We will focus here only on the model approach we are dealing with here: *Reaction Network Theory* and its application to biological systems that can be modeled by the *Law of Mass Action*. At this stage, a critical principle can be named as: *use all possible biological information available*. Remembering that, at this stage, biological information will have a dual impact on the model equations. At the same time, it can reduce the degree of freedom in the parameter space by establishing relations between model parameters; it can require including additional equations to describe a specific relation between the model state parameters. However, no matter what the case is, it will increase the model's biological meaningfulness.

In order to use all possible biological information available in the literature, a detailed and comprehensive literature review is critical; in fact, it is mandatory. We must think of it as an *error propagation problem.* Any mistake of imprecise information at this step will cause huge consequences in the next steps of your modeling strategy with the critical potential to compromise all committed efforts. Conversely, all time and effort spent on this step will pay off in the future. Also, if you are debugging your model, this is the first stage to be rechecked. Finally, a piece of personal advice: if during your literature search, you realize that the question you are looking for to answer was already solved, and the solution is *really consistent*, include this as additional information in your model and rephrase your model question.

In order to decide which *experimental data* to use, the third stage in the systems biology approach (Fig. 7.7), some key points must be kept in mind. First, the three first stages in the system biology approach we are discussing are strongly related. If we are willing to describe the biological systems by the *Law of Mass Action*, we must use only *experimental information* and *experimental data* that can contribute to building a mass action model. For example, if the biological system to be modeled is a *signaling network*, experimental data or information related to physiological conditions that cannot be directly considered in the model reaction cannot be considered. Even though this kind of information is critical to infer about the model applicability. We must respect the model scale we are dealing with. Even if a multiscale model is being built, to build a reaction network model, biological data at the molecular level are critically necessary. In other words, the most appropriate data to be used in a *reaction network* are molecular data. We are assuming here the strict sense definition of a reaction as the physical interaction between a pair of molecules. Therefore, it means that molecular data are the most appropriated to be used as experimental data to build a reaction network model. The good news is that, nowadays, the combination of experimental strategies with image analysis tools [44] has allowed an unprecedented generation of *statistically viable experimental* data. In other words, an amount of data enough for a statistical analysis that will reinforce the confidence in the model results.

The parameter calibration strategy is the heart of the model calibration stage in our approach. However, since we have dedicated the entire Sect. 7.4., *Parameter Estimation Strategies* to this issue, we reserve some space here only to discuss an inherent advantage of a specific strategy, the *genetic algorithm*: it provides a *family* of parameters that solves the problem. This is critical since it is almost inevitable to have a huge number of reactions in biologically realistic models, and parameter overfitting is frequently inevitable; we must be ready to deal with it. Apart from specific mathematical strategies to minimize this problem, combining a set of solutions with additional experimental data are good practical strategies. A statistical study of the different model behavior generated by a different set of parameters compared to experimental information to filter them can produce interesting results. For example, if in the activated state of signaling network the concentration of protein is expected to be high in the cell nucleus compared to the cytoplasm, all parameter sets that violate this experimental information must be ignored. Also, the capacity to generate a family of solutions is one of the major advantages of the genetic algorithm, besides its solution finding efficiency.

Despite how efficient the parameter calibration strategy could be, fitting a reaction network model to biological experimental is one of the most challenging stages in modeling. To minimize the inherent difficulties, we introduced a *pre-calibration step* in the system biology approach. This step can be done in two ways, a *dynamic system analysis* of the model reaction and a *parameter reducing strategy*. In the second one, we can use experimental information to establish constraints between the model parameters. It can reduce the dimension of the parameter space by reducing the number of parameters, or just reduce the parameter space volume [45] by establishing restrictions between the model parameters. For example, if we know the kinetic constant for monomer binding/unbinding, we can reduce the number of parameters to be calibrated; and, if we know that the DNA binding of a TF (k_{binding}) is more favorable than the unbinding ($k_{\text{unbinding}}$), we can set $k_{\text{binding}} > k_{\text{unbinding}}$, it reduces the parameter space volume.

Even though the parameter reduction strategy is very useful the most powerful strategy in the pre-calibration step is the *dynamic system analysis*. Let's suppose that we are looking for bistable solutions in a reaction kinetic model. A first, and sometimes useful, strategy is just using a parameter search strategy. However, if we do not find such a bistable solution, another critical question arises: *does the system not have any bistable solution or was my search strategy not good or lucky enough?* In order to solve this kind of problem, which can be very time spending, an especially useful analytical strategy was developed by Martin Feinberg, Bruce Clarke, F. Horn, R. Jackson, and others [12, 13]. Martin Feinberg, himself a pioneer in the *chemical reaction network theory*, developed a consistent set of strategies, based on the *Deficiency Zero Theorem, Deficiency One Theorem*, and a network structural property called concordance [46]. This strategy not only determines if the network has the ability to exhibit multiple stationary states but also allows us to find a set of kinetics parameters that produce such bistable solutions. This is a pretty powerful strategy that may be indispensable if we are dealing with a large reaction network model, a situation that in biology looks more to be the rule than the exception. For practical use, there is a convenient *Chemical Reaction Network Toolbox* freely available for download [47]. The applicability of this toolbox is exemplified in the section.

The *model calibration, systems understanding, simulations,* and *new experimental data* are another group of very connected stages in the systems biology approach (Fig. 7.7). The development of computational tools for parameter calibration and model simulation has been an area of great advances in recent years. For ordinary differential model, COPASI is most useful for a beginner and even for advanced users [42]. Success in the *model calibration* stages per se increases our understanding of the model's behavior. A good model with well-determined parameters comes from a well succeed *model calibration* stage, has the potential to improve our *understanding of the system*, and proposes answers to the biological questions. However, two critical points must be considered here. In the great majority of cases, the biological questions arise from the bench, and the verified solution will come from there. The first role of a theoretical model is to reduce the number of possible experiments needed to find or check a solution to the biological question, i.e., the model has the potential to reduce the great number of potential experimental routes to solve the problem. Theoretical models in advanced stages are also able to introduce new questions by itself, some of them will be solved only on the bench. However, one critical requirement must be fulfilled by any theoretical model before we can consider that it has proposed a reasonable route to solve a biological question or even to propose a new one: the model must be a good, reasonable, *synthetic reproduction* of the biological system. Reproducing an experimental curve, a set of experimental concentrations, or even a temporal evolution of a system is not, per se, sufficient for a model to be classified as a synthetic reproduction of the biological system. Even though not the unique, but a very efficient way to test a model quality is to submit it to the *cyclic step*. This step can be done by a combination of model simulation and a new set of data. Here, new data does not necessarily mean data that it wasn't known when the model began to be built. These data must be *new to the model*. It means that it wasn't used to build the model. Of course, the level of exigency in the cyclic step must be well dosed. It is *not fear* to require that a model simulation in this step be able to reproduce a quantitative data since the models were not calibrated to these new set of data. The model's ability to reproduce the qualitative aspect of the new data is sufficient to attest its quality and, maybe more important than anything, to build our confidence in the model's ability to be a *synthetic reproduction of the biological system*. In some very complex systems, to be this kind of representation is maybe even more important than solving a specific question as a first step.

7.5.2 Modeling Drosophila Embryonic Development

The embryonic development of *Drosophila melanogaster* has unique characteristics for theoretical modeling [48]. The egg fertilization triggers a cytoplasmic signal that leads to the duplication of genetic material, the initial stage of the process that culminates in mitosis, Fig. 7.8a. However, only the nucleus of the embryo duplicates itself; complete cell duplication does not occur as in most organisms. Thus, two nuclei share the same cytoplasm. This unusual nuclear duplication, called the mitotic cycle, is repeated 13 times in a period of about three hours after fertilization [49]. At the end of these cycles, the embryo, still unicellular, has around 2^{14} identical nuclei sharing the same cytoplasm. The time elapsed between each of these cycles increases progressively; the 14th and longest periods last approximately 60 min. At the end of

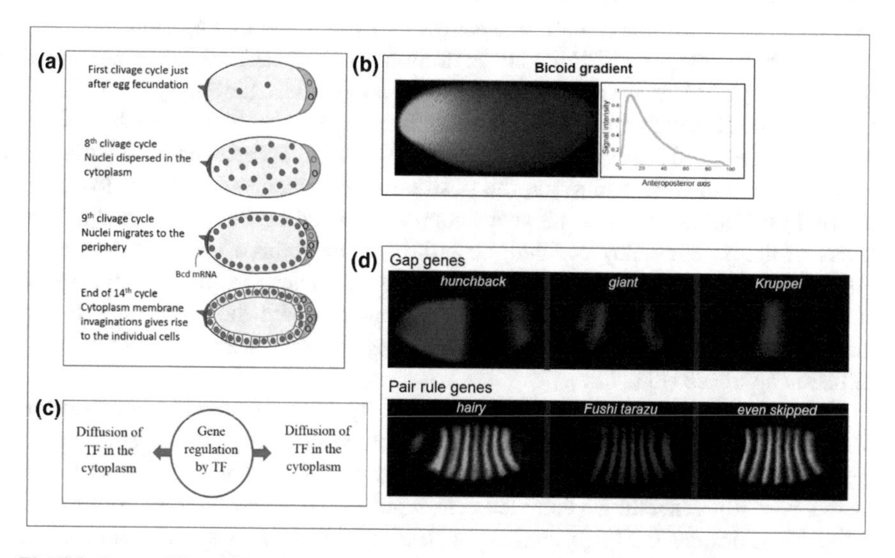

Fig. 7.8 Drosophila embryonic development. **a**. Mitotic cycles give rise to a syncytium. **b** Confocal images of Bcd antibody staining and fluorescence profiles versus anteroposterior position. **c** Reaction–diffusion schematic representation. **d** Some examples of the gap and pair-rule genes expression patterns [50]

this process, an invagination of the embryo's cytoplasmic membrane involves each nucleus giving rise to individual cells and thus embryonic tissue. Most experimental and modeling studies focus on this 60 min period, which corresponds to the 14th cleavage cycle. Throughout this period, no membrane separates the embryo nuclei. The presence of hundreds of nuclei sharing the same cytoplasm, called syncytium, allows us to use Fick's equation to describe protein diffusion. It also characterizes this stage of development of *Drosophila* ideal for studying the formation of patterns of gene regulation during embryonic development.

Another aspect of great theoretical importance in the embryonic development of *Drosophila* is related to the origin of the anteroposterior axis asymmetry of the embryo. In this organism, as in several others, asymmetry is generated by factors of maternal origin, that is, asymmetry is not acquired spontaneously as a result of the dynamics of the system. In the case of *Drosophila*, the messenger RNA of the Bicoid protein (Bcd) is maternally anchored in the anterior region of the embryo (Fig. 7.8a). Right after fertilization, which triggers the beginning of embryonic development, the synthesis of Bcd protein, from the RNA anchored in the anterior region of the embryo, gives rise to an anteroposterior gradient (Fig. 7.8b). This anchoring of the Bcd protein RNA occurs back during the formation of the egg, which defines the Bcd gradient as a maternal one. Accurate measurements of this gradient show that it is an exponential distribution [51, 52], indicating that, despite all the viscosity of the cytoplasm and the enormous diversity of interacting molecules, the simplest and most effective description of diffusion is sufficient, and in fact, the most suitable: Fick's diffusion equation. Another important theoretical aspect of this result is that, since an exponential corresponds to the stationary solution of the Fick´s equation, the system reaches a steady state, at least with respect to the Bcd gradient.

Since Bcd protein is a TF, its concentration gradient establishes the positional information [53] along the anteroposterior developing axis in the embryo (Fig. 7.8b). Different nuclei are submitted to different Bcd concentrations. Since in each nucleus, we have copies of all *Drosophila* genes, each one of these copies is subject to different Bcd concentrations depending on the position of the nucleus it belongs to are in the Bcd gradient. As a result, all genes copies, activated by Bcd, produce different levels of the proteins they codified, according to the embryo region. It generates different local protein concentrations around each nucleus in the anteroposterior axis. The combination of reaction inside the nucleus with the diffusion between the cytoplasmatic region around them establishes a perfect scenario for a reaction–diffusion approach (Fig. 7.8c).

The three first sets of genes that are sequentially activated after the transcription of maternal mRNAs are the *gap*, *pair-rule*, *segment-polarity*, and *homeotic* genes (Fig. 7.8). Many of these genes are regulated by each other and by other maternal factors [48, 50], generating a complex gene regulation network far to be completely understood, despite the huge amount of detailed information about them. As an example, the structure of the developmental *gap* gene *hunchback* (*hb*), one of the most studied Bcd target genes, has been detailed described [54–56]. All TF that binds to its regulatory region has been potentially described. Also, many papers have been published about the gene regulatory mechanism responsible for *hb* expression patterns (we refer by gene spatial pattern as the spatial distribution of mRNA or protein codified by a specific gene). However, many old questions remain to have a consensus answer in the literature.

By the end of the 14th cleavage cycle, all nuclei are involved by the invagination of the cytoplasm membrane. Since there are different local protein concentrations around each nucleus, as a result of the gene regulatory networks described above, the membrane invagination gives rise to cells with different cytoplasmatic content. This is called cell differentiation. The identity of different tissues that will be generated from these initial cells depends critically on this cell differentiation process. It is remarkable the just before membrane invagination, a set of seven strips of seven different pair-rule genes are accommodated along the anteroposterior axis. Each strip is around two to three nucleus width. As a result, even the neighbor nucleus may have different local protein concentrations around them, giving rise to cells with different cytoplasm content and, or, in other words, differentiated cells.

The sharp borders in the expression patterns of pair-rule genes, just before membrane invagination, play a critical role in the accommodation of the seven strips of each pair-rule genes (Fig. 7.8). At this point, an important question naturally arises: *what is the mechanism behind these sharp borders*? Trying to respond to this unsolved question by studying a pair-rule gene is exceedingly difficult because not all interactions responsible for the regulation of any of the gap genes are completely described. Also, since several proteins are involved in the regulation of these genes, an inevitable complexity makes this study especially challenging, maybe more than we can deal at the present time. In order to circumvent, or at least minimize, this difficulty, we can look at early genes since, at the early stages of development, only a comparatively small number of genes are being expressed. This can be achieved by studying a *gap* since they are regulated only by maternal genes and by each other. However, which gap gene has a sharp border? In fact, although many gap genes can be included as an answer to this question, we decided to study the *hb* because it has one well define sharp border at the middle of the embryo (Fig. 7.8).

7.5.3 Modeling the Expression of the Hunchback Gene

The developmental gene *hb* plays a critical role during *Drosophila* embryonic and central nervous development [48, 57]. This is one of the most studied genes in *Drosophila*. The importance of this gene goes beyond its role during *Drosophila* development. It has been one of the most studied examples of the so-called positional information mechanism proposed by Wolbert in 1969 [58], see also Zadorin et al. [59] for recent results about the French flag model. Although the first Hunchback (Hb) protein detected comes from RNA of maternal origin, mutant experiments have shown that this early expression is not essential for its mature expression pattern [60]. The zygotic *hb* expression is positively regulated by Bcd, Hb itself, and Krüppel protein at low concentration and repressed by Knirps and Krüppel protein (at high concentrations) [61, 62]. However, since in the first half of cycle 14, the Knirps and

Krüppel protein concentration is very low and *Kr* mutant experiment has shown that early Hb pattern is not disrupted in *Kr* mutant, we are considering here only the role of Bcd and Hb regulation for Hb pattern. To this end, we are studying only the Hb expression at the middle of cycle 14, i.e., the first half of cycle 14.

Bcd and Hb proteins play two well-characterized regulatory mechanisms in the *hb* regulation: cooperative binding and *hb* self-activation, respectively. The discussion about the molecular mechanisms of the Bcd cooperative is of great theoretical discussion. Despite the huge impact and success of the concept of positional information to understand the first regulatory process during morphogen, the molecular mechanism behind the establishment of position information is yet to be determined. Many possible mechanisms have been proposed. One of the first ones, in the case of *hb* expression pattern, was cooperative binding. The role of Bcd cooperative bonding for *hb* expression pattern has been subject to intense discussion [2, 51, 63–70]. Even though Bcd shows a low affinity to bind to a promoter region containing only a single site, this affinity is greatly increased if more than one binding site is presented. It has been experimentally shown that this increased affinity results from a cooperative mechanism between the Bcd protein molecules. Even though some molecular details about this mechanism are still uncovered, here we are only interested in the main effect of this mechanism: it increases the Bcd affinity to bind to multiple sites in the regulatory region of its target genes. In the present case, the *hb* gene. At this point, we can define the first question our *hb* model should be able to propose an answer: *what is the role of Bcd cooperative binding for hb expression pattern*? This is, in fact, related to a more general question: *which would be the molecular mechanism of positional information?* In the Sect. 5.3 *Modeling Drosophila embryonic development*, we describe how Bcd cooperative binding can be reproduced in the model by algebraic relation between the reaction kinetic parameters.

7.5.4 The Expression Pattern of a Developmental Gene: A Practical Example

As explained in the previous section, we are describing here *hb* expression pattern in the middle of cycle 14. As a result, we are taking the advantage that, until this stage, *only* Bcd e Hb protein controls *hb* regulation. Thus, we are describing the *hb* regulatory region as containing two Hb and six Bcd sites, among many other already identified [70]. In fact, the *hb* gene has a complex structure; however, a detailed analysis is out of the scope of the present study. We list below all *experimental information* used in the stage of the *systems biology approach* here adopted (Fig. 7.7):

i. Six sites for Bcd are sufficient for hb regulation by Bcd [71].
ii. *hb* activates its own production [54, 72].
iii. Bcd binds DNA cooperatively [73, 74].
iv. Genes with a higher number of TF bound in the regulatory region have a higher transcriptional rate [48].

Items (*i*) and (*ii*) are used to build the reaction network while (*iii*) and (*iv*) are used as part of the *parameter reducing strategy* in the *pre-calibration step* of the *model calibration* stage of our approach. The number of Bcd and Hb binding sites in the *hb* regulatory region to be assumed in our model is determined by items (*i*) and (*ii*), respectively. In the *hb* self-regulatory model (Fig. 7.9), we describe the *hb* gene activation by a two-step reaction. In the first step, the Bcd protein reversibly binds to the *hb* regulatory region with all Bcd sites empty (reaction (7.2)). In the second step, this gene with one Bcd protein-bound generates a Hb protein. This reaction is irreversible, and the gene is obviously preserved. Reactions (7.4)–(7.12) represents the sequential steps leading to the full occupancy of all Bcd sites in the *hb* regulatory region. The activation of *hb* by Hb protein (*hb* self-activation) follows the same strategy as for the activation by Bcd protein. This is reflected in the similarity between reaction (7.14) and (7.15) with (7.2) and (7.3). Reaction (7.1) represents the Bcd source from the mRNA anchored in the anterior region of the embryo. Reactions (7.18) and (7.19) represents the Hb and Bcd protein degradation. The HSR model has a set of 19 reactions having 27 kinetic parameters. Mass Action Law allows determining the set 12 ordinary differential equation equations [2] describing the dynamic of the system. The huge complexity of this system of ODE is a good example of how useful is the pre-calibration step described in the *model calibration* stage of the system biology approach, Fig. 7.7. As previously described, this stage can be done by two steps, a *dynamic system analysis* of the model reaction and a *parameter reducing strategy*. For parameter reduction, we can take advantage of *experimental information* (*iii*) and assume that Bcd cooperative binding implies that the binding of a second Bcd molecule in the *hb* regulatory region has a higher affinity than the binding of the first one (see 0 for an explanation about cooperative binding). A convenient way to assume that could be taking $k_{b(n-1),bn} = factor_{n.kb0,b1}$ and $k_{bn,b(n-1)} = k_{b1,b0}$ for $n = 2,\ldots 6$. To introduce *experimental information* (*iv*) we can assume $k_{bn,H} = (1 + Synt\_Factor).kb_{(n-1),H}$ for $n = 2,\ldots 6$. By applying both *experimental information* above, we reduced the number of kinetic parameters since

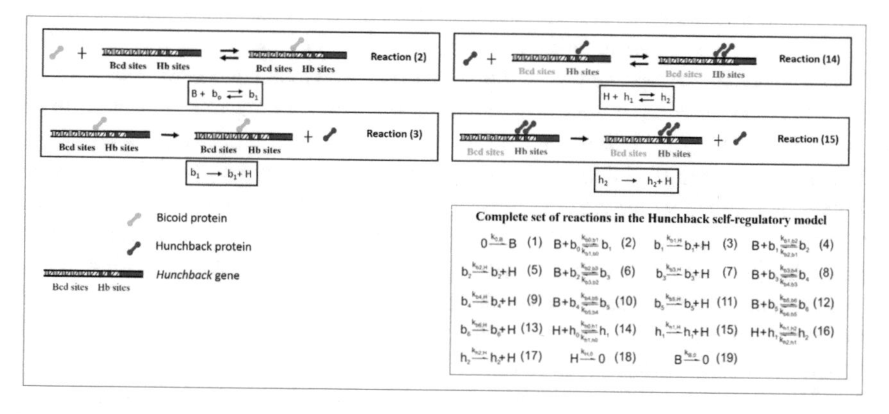

Fig. 7.9 hb self-regulatory (HSR) model

only parameters factor, *Synt_Factor*, $k_{b1,b0}$, and *factor* need to be determined. It is interesting to note that considering 6 Bcd sites in the *hb* regulatory region makes the model more biological meaningful but increases the number of kinetic parameters, however, using the *experimental information* (*iv*) allows to reduce the number of kinetic parameters. We can see that additional biological information plays a dual role here: increase the model complexity at the same time that allows the reduction in the number of kinetic parameters.

As part of the *pre-calibration step*, the *dynamic system analysis* has a huge impact on reducing *the parameter space volume*. As indicated in Sect. 7.5.1, by using the *Chemical Reaction Network Toolbox* [47], we found the following *Higher Deficiency Report*:

> Taken with mass action kinetics, the network DOES have the capacity for multiple steady states. That is, there are rate constants that give rise to two or more positive (stoichiometrically compatible) steady states—you'll see an example below—and also rate constants for which there is a steady state having an eigenvector (in the stoichiometric subspace) corresponding to an eigenvalue of zero. (To construct rate constants that give a degenerate steady state, use the Zero Eigenvalue Report.)

In order to construct your report, please follow the reference indicated above. The CRNT also shows a set of kinetic parameters that gives rise to the bistable solution and determines its stationary solutions. This set of parameters is the best start point for parameter calibration in order to find a good bistable that best fits the biological data. Of course, it will only produce effective results if a bistable model is required to reproduce the biological system behavior. In order words, if a bistable behavior is a good answer for the biological questions.

The parameter calibration strategy used in the *model calibration* stage of the system biology approach (Fig. 7.7), gives the data *versus* model plot as in Fig. 7.10a [2]. The model parameter calibration allows the determination of a bistable plot in the phase space Bcd versus Hb protein concentrations. The Bcd axis is inverted to facilitate comparison with Bcd concentration in the embryo that is high in the anterior region (Fig. 7.10a on the left). In order to explain how both Hb protein stable concentration generates a sharp border in the expression pattern of *hb* gene, Fig. 7.10b shows a schematic representation of seven embryo nuclei, very exaggerated in size. The same nuclei are represented in Fig. 7.10a. The Bcd total concentrations at nuclei (7.1)–(7.3) are located in the monostable region. The Hb concentration in all these nuclei, starting at zero initial concentration, evolves to the higher Hb concentration. In fact, any other initial Hb concentration on these nuclei has a higher Hb concentration as the stationary attractor. On the nucleus (7.4), the first is one located inside the bistable region, we see a very different behavior: for all zero or near zero Hb initial concentration, it evolves to the low Hb stationary state. The Hb concentration in this nucleus will be at low Hb stationary state. In this very particular region, where Bcd concentration enters the bistable region, the HSR model indicates a huge variation in the Hb stationary concentration as a result of a small variation in the Bcd concentration. This shift in the Hb concentration from the high to low stationary concentration generated the sharp border in the *hb* expression pattern. This result also indicates that the Bcd concentration threshold is located at the anterior border of the bistable

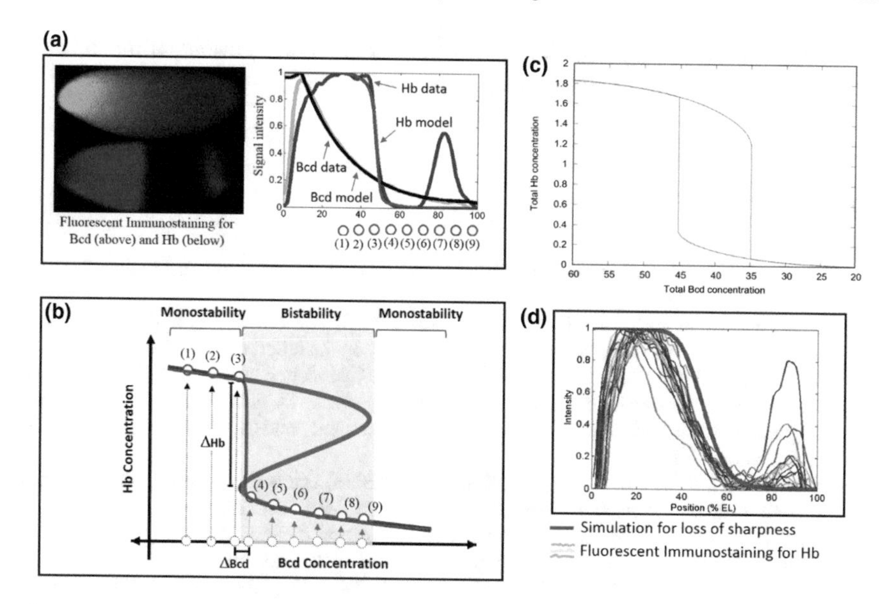

Fig. 7.10 HSR model compared to experimental data

region. In fact, Bcd has two threshold concentrations, which are both borders of the bistable region. The transition will be determined by the direction of variation in the Bcd concentration. If it comes from high to low concentration, we see a transition at a different position if it comes from low to high concentration; this generates a classical hysteresis curve, as seen in Fig. 7.10c.

Although the above results allow the HSR model to reproduce the *hb* sharp border, it just indicates that bistability is a mechanism that has the potential to generate *hb* sharp border. Is does not prove, per se, that this is *the mechanism.* Besides that, no answer has yet been proposed about the role of Bcd cooperative binding for *hb* border. In order to check these unanswered questions, we need to apply the *cyclic step* that involves the *simulation* and *new experimental data* stages (Fig. 7.7). considering that the HSR model reproduces both *Bcd cooperative binding* and *hb self-activation* mechanism, it would be appropriate to simulate mutation in only one of them. Fortunately, an experimentally generated *hb* mutant *Drosophila*, called hb^{14F}, expresses a truncated version of the Hb protein with lost transcriptional activity. In practical terms, it cannot activate itself. It corresponds to a model with mutated experimental information (*ii*); in terms of modeling, it means removing reaction (7.14)–(7.17), Fig. 7.9. Figure 7.10d shows the comparison between the model simulation and experimental data for *hb* self-regulation mutant. The qualitative concordance between these results allows us to go one step further from the previous conclusion to say that: *spatial bistability, produced by hb self-activation, is, with a reasonable degree of certainty, responsible for the Hb sharp border*.

Considering that the experimental information (*i*) and (*iii*) were not disrupted in the model simulation, the above result strongly indicates that Bcd cooperative binding cannot account for *hb* sharp border. In fact, this is supported by the experimental data since Bcd cooperative binding is not disrupted in the hb^{14F} mutant.

References

1. Rulands S, Klünder B, Frey E (2013) Stability of localized wave fronts in bistable systems. Phys Rev Lett 110:038102
2. Lopes FJP, Vieira FMC, Holloway DM, Bisch PM, Spirov AV (2008) Spatial bistability generates hunchback expression sharpness in the Drosophila embryo. PLoS Comput Biol 4
3. Graham TGW, Tabei SMA, Dinner AR, Rebay I (2010) Modeling bistable cell-fate choices in the Drosophila eye: Qualitative and quantitative perspectives. Development 137:2265–2278
4. Taniguchi K, Karin M (2018) NF-B, inflammation, immunity and cancer: coming of age. Nat Rev Immunol 18:309–324
5. Hoffmann A, Levchenko A, Scott ML, Baltimore D (2002) The IκB-NF-κB signaling module: temporal control and selective gene activation. Science 298:1241–1245
6. Araujo H, Fontenele MR, da Fonseca RN (2011) Position matters: variability in the spatial pattern of BMP modulators generates functional diversity. Genesis 49:698–718
7. Hong JW, Hendrix DA, Papatsenko D, Levine MS (2008) How the Dorsal gradient works: insights from postgenome technologies. Proc Natl Acad Sci USA 105:20072–20076
8. Brumby AM, Richardson HE (2005) Using Drosophila melanogaster to map human cancer pathways. Nat Rev Cancer 5:626–639
9. Rudrapatna VA, Cagan RL, Das TK (2012) Drosophila cancer models. Dev Dyn 241:107–118
10. Pires BRB, Mencalha AL, Ferreira GM, De Souza WF, Morgado-Díaz JA, Maia AM et al (2017) NF-kappaB is involved in the regulation of EMT genes in breast cancer cells. PLoS One 12
11. Horn F (1973) On a connexion between stability and graphs in chemical kinetics II. Stability and the complex graph. Proc R Soc London. A Math Phys Sci 334:313–30
12. Horn F, Jackson R (1972) General mass action kinetics. Arch Ration Mech Anal 47:81–116
13. Clarke BL (2007) Stability of complex reaction networks. In: Advances in chemical physics. Wiley, pp 1–215
14. Willamowski KD, Rossler OE (1978) Contributions to the theory of mass action kinetics I. Enumeration of second order mass action kinetics. Zeitschrift fur Naturforsch. Sect A J Phys Sci 33:827–33
15. Feinberg M (1987) Chemical reaction network structure and the stability of complex isothermal reactors-I. The deficiency zero and deficiency one theorems. Chem Eng Sci 42:2229–68
16. Ashyraliyev M, Fomekong-Nanfack Y, Kaandorp JA, Blom JG (2009) Systems biology: parameter estimation for biochemical models. FEBS J 276:886–902
17. Baker SM, Poskar CH, Schreiber F, Junker BH (2015) A unified framework for estimating parameters of kinetic biological models. BMC Bioinf 16
18. González J, Vujačić I, Wit E (2013) Inferring latent gene regulatory network kinetics. Stat Appl Genet Mol Biol 12:109–127
19. Hass H, Loos C, Raimúndez-Álvarez E, Timmer J, Hasenauer J, Kreutz C (2019) Benchmark problems for dynamic modeling of intracellular processes. Bioinformatics 35:3073–3082
20. Lillacci G, Khammash M (2010) Parameter estimation and model selection in computational biology. PLoS Comput Biol 6:e1000696
21. Rodriguez-Fernandez M, Rehberg M, Kremling A, Banga JR (2013) Simultaneous model discrimination and parameter estimation in dynamic models of cellular systems. BMC Syst Biol 7
22. Sun J, Garibaldi JM, Hodgman C (2012) Parameter estimation using metaheuristics in systems biology: a comprehensive review. IEEE/ACM Trans Comput Biol Bioinf 9:185–202

23. Vanlier J, Tiemann CA, Hilbers PAJ, van Riel NAW (2013) Parameter uncertainty in biochemical models described by ordinary differential equations. Math Biosci 246:305–314
24. Aster RC, Borchers B, Thurber CH (2013) Parameter estimation and inverse problems. Elsevier Inc
25. Dreossi T, Dang T (2014) Parameter synthesis for polynomial biological models. In: HSCC 2014—proceedings of the 17th international conference on hybrid systems: computation and control (Part of CPS week). Association for computing machinery, page 233–42
26. Michalewicz Z, Dasgupta D, Le Riche RG, Schoenauer M (1996) Evolutionary algorithms for constrained engineering problems. Comput Ind Eng 30:851–870
27. Reali F, Priami C, Marchetti L (2017) Optimization algorithms for computational systems biology. Front Appl Math Stat 3
28. Van Laarhoven PJM, Aarts EHL (1987) Simulated annealing. In: Simulated annealing: theory and applications. Springer Netherlands, Dordrecht, p 7–15
29. Whitley D (1994) A genetic algorithm tutorial. Stat Comput 4:65–85
30. Onwubolu GC, Davendra D (eds) (2009) Differential evolution: a handbook for global permutation-based combinatorial optimization. Springer, Berlin, Heidelberg
31. Abdullah A, Deris S, Anwar S, Arjunan SNVV (2013) An evolutionary firefly algorithm for the estimation of nonlinear biological model parameters. PLoS ONE 8:e56310
32. Abdullah A, Deris S, Mohamad MS, Anwar S (2013) An improved swarm optimization for parameter estimation and biological model selection. PLoS ONE 8:e61258
33. Stražar M, Mraz M, Zimic N, Moškon M (2014) An adaptive genetic algorithm for parameter estimation of biological oscillator models to achieve target quantitative system response. Nat Comput 13:119–127
34. Hussain F, Langmead CJ, Mi Q, Dutta-Moscato J, Vodovotz Y, Jha SK (2015) Automated parameter estimation for biological models using Bayesian statistical model checking. BMC Bioinf 16:S8
35. Mansouri MM, Nounou HN, Nounou MN, Datta AA (2014) Modeling of nonlinear biological phenomena modeled by S-systems. Math Biosci 249:75–91
36. Rodriguez-Fernandez M, Banga JR, Doyle FJ (2012) Novel global sensitivity analysis methodology accounting for the crucial role of the distribution of input parameters: application to systems biology models. Int J Robust Nonlinear Control 22:1082–1102
37. Rodriguez-Fernandez M, Egea JA, Banga JR (2006) Novel metaheuristic for parameter estimation in nonlinear dynamic biological systems. BMC Bioinf 7:483
38. Sequential monte carlo methods in practice. Springer, New York; 2001
39. Banga JR (2008) Optimization in computational systems biology. BMC Syst Biol 2:47
40. Boyd S, Vandenberghe L. (2004) Convex Optimization. Cambridge University Press
41. Padhye N, Mittal P, Deb K (2015) Feasibility preserving constraint-handling strategies for real parameter evolutionary optimization. Comput Optim Appl 62:851–890
42. Mendes P, Hoops S, Sahle S, Gauges R, Dada J, Kummer U (2009) Computational modeling of biochemical networks using COPASI. Methods Mol Biol 500:17–59
43. De Rainville FM, Fortin FA, Gardner MA, Parizeau M, Gagné C (2012) DEAP: a python framework for evolutionary algorithms. In: GECCO'12—proceedings of the 14th international conference on genetic and evolutionary computation companion. ACM Press, New York, USA, pp 85–92
44. De Sousa DJ, Arruda Cardoso M, Bisch PM, Pereira Lopes FJ, Nassif Travençolo BA (2013) A segmentation method for nuclei identification from sagittal images of Drosophila melanogaster embryos
45. Chaves M, Sengupta A, Sontag ED (2009) Geometry and topology of parameter space: investigating measures of robustness in regulatory networks. J Math Biol 59:315–358
46. Shinar G, Feinberg M (2013) Concordant chemical reaction networks and the species-reaction graph. Math Biosci 241:1–23
47. The chemical reaction network toolbox | chemical reaction network theory [Internet]. https://crnt.osu.edu/CRNTWin. Accessed 17 Apr 2020

48. Gilbert SF, Barresi MJF (2017) Developmental biology, 11th edn 2016. Am J Med Genet Part A 173:1430–1430
49. Campos-Ortega JA, Hartenstein V (1985) The embryonic development of Drosophila melanogaster. Springer, Berlin Heidelberg
50. FlyEx, the quantitative atlas on segmentation gene expression at cellular resolution [Internet]. https://www.ncbi.nlm.nih.gov/pmc/articles/PMC2686593/. Accessed 16 Apr 2020
51. Gregor T, Tank D, Wieschaus E, Cell WB (2007) Undefined. Probing the limits to positional information. Elsevier
52. Abu-Arish A, Porcher A, Czerwonka A, Dostatni N, Fradint C (2010) High mobility of bicoid captured by fluorescence correlation spectroscopy: implication for the rapid establishment of its gradient. Biophys J 99:L33–L35
53. Wolpert L (1989) Positional information revisited. Development 3–12
54. Margolis JS, Borowsky ML, Steingrímsson E, Shim CW, Lengyel JA, Posakony JW (1995) Posterior stripe expression of hunchback is driven from two promoters by a common enhancer element. Undefined
55. Schröder C, Tautz D, Seifert E, Jäckle H (1988) Differential regulation of the two transcripts from the Drosophila gap segmentation gene hunchback. EMBO J 7:2881–2887
56. Holloway DM, Lopes FJP, da Fontoura Costa L, Travencolo BAN, Golyandina N, Usevich K et al (2011) Gene expression noise in spatial patterning: hunchback promoter structure affects Noise amplitude and distribution in Drosophila segmentation. Plos Comput Biol 7:e1001069
57. Novotny T, Eiselt R, Urban J (2002) Hunchback is required for the specification of the early sublineage of neuroblast 7-3 in the Drosophila central nervous system. Development 129:1027–1036
58. Wolpert L (1969) Positional information and the spatial pattern of cellular differentiation. J Theor Biol 25:1–47
59. Zadorin AS, Rondelez Y, Gines G, Dilhas V, Urtel G, Zambrano A et al (2017) Synthesis and materialization of a reaction-diffusion French flag pattern. Nat Chem 9:990–996
60. Lehmann R, Nüsslein-Volhard C (1987) Hunchback, a gene required for segmentation of an anterior and posterior region of the Drosophila embryo. Dev Biol 119:402–417
61. Papatsenko D, Levine MS (2008) Dual regulation by the Hunchback gradient in the Drosophila embryo. Proc Natl Acad Sci U.S.A 105:2901–2906
62. Holloway DM, Spirov AV (2015) Mid-embryo patterning and precision in drosophila segmentation: Krüppel dual regulation of hunchback. PLoS One 10
63. Lopes FJP, Vanario-Alonso CE, Bisch PM (2005) A kinetic mechanism for Drosophila bicoid cooperative binding. J Theor Biol
64. Lopes FJP, Spirov AV, Bisch PM (2012) The role of bicoid cooperative binding in the patterning of sharp borders in Drosophila melanogaster. Dev Biol
65. Lebrecht D, Foehr M, Smith E, lopes FJP, Vanario-Alonso CE, Reinitz J et al (2005) Bicoid cooperative DNA binding is critical for embryonic patterning in Drosophila. Proc Natl Acad Sci U.S.A
66. Burz DSD (2001) Biology SH-J of molecular, 2001 U, Hanes SD. Isolation of mutations that disrupt cooperative DNA binding by the Drosophila Bicoid protein. J Mol Biol 305:219–230
67. Tran H, Desponds J, Perez Romero CA, Coppey M, Fradin C, Dostatni N et al (2018) Precision in a rush: trade-offs between reproducibility and steepness of the hunchback expression pattern. PLoS Comput Biol 14
68. Perry MW, Bothma JP, Luu RD, Levine M (2012) Precision of hunchback expression in the Drosophila embryo. Curr Biol 22:2247–2252
69. Crauk O, Dostatni N (2005) Bicoid determines sharp and precise target gene expression in the Drosophila embryo. Curr Biol 15:1888–1898
70. Ochoa-Espinosa A, Yucel G, Kaplan L, Pare A, Pura N, Oberstein A et al (2005) The role of binding site cluster strength in bicoid-dependent patterning in Drosophila. Proc Natl Acad Sci U.S.A. 102:4960–4965
71. Struhl G, Struhl K, Macdonald PM (1989) The gradient morphogen bicoid is a concentration-dependent transcriptional activator. Cell 57:1259–1273

72. Treisman J, Desplan C (1989) The products of the Drosophila gap genes hunchback and Krüppel bind to the hunchback promoters. Nature 341:335–337
73. Nature WD (1989) Undefined. Determination of zygotic domains of gene expression in the Drosophila embryo by the affinity of binding sites for the bicoid morphogen. ci.nii.ac.jp
74. Ma X, Yuan D, Diepold K, Scarborough T, Ma J (1996) The Drosophila morphogenetic protein bicoid binds DNA cooperatively. Development 122:1195–1206

Part II
Disease and Pathogen Modeling

Chapter 8
Challenges for the Optimization of Drug Therapy in the Treatment of Cancer

Nicolas Carels, Alessandra Jordano Conforte, Carlyle Ribeiro Lima, and Fabricio Alves Barbosa da Silva

Abstract Personalized medicine aims at identifying specific targets for treatment considering the gene expression profile of each patient individually. We discuss the challenges for personalized oncology to take off and present an approach based on hub inhibition that we are developing. That is, the subtraction of RNA-seq data of tumoral and non-tumoral surrounding tissues in biopsies allows the identification of up-regulated genes in tumors of patients. Targeting connection *hubs* in the subnetworks formed by the interactions between the proteins of up-regulated genes is a suitable strategy for the inhibition of tumor growth and metastasis in vitro. The most relevant protein targets may be further analyzed for drug repurposing by computational biology. The subnetworks formed by the interactions between the proteins of up-regulated genes allow the inference by Shannon entropy of the number of targets to be inhibited according to the tumor aggressiveness. There are common targets between tumoral tissues but many others are personalized at a molecular level. We also consider additional measures and more sophisticated modeling. This approach is necessary to improve the rational choice of therapeutic targets and the description of network dynamics. The modeling of attractors through Hopfield Network and ordinary differential equations are given here as examples.

Keywords Personalized oncology · Shannon entropy · Hopfield network · Boolean modeling · Differential equations · Drug repurposing

---

N. Carels (✉) · A. J. Conforte · C. R. Lima
Plataforma de Modelagem de Sistemas Biológicos, Center of Technology Development in Health, Oswaldo Cruz Foundation, Rio de Janeiro, Brazil
e-mail: nicolas.carels@gmail.com; nicolas.carels@cdts.fiocruz.br

A. J. Conforte
e-mail: conforteaj@gmail.com

C. R. Lima
e-mail: carlylelima@hotmail.com

A. J. Conforte · F. A. B. da Silva
Laboratório de Modelagem Computacional de Sistemas Biológicos, Scientific Computing Program (PROCC), Oswaldo Cruz Foundation, Rio de Janeiro, Brazil
e-mail: fabs.fiocruz@gmail.com; fabricio.silva@fiocruz.br

F. A. B. da Silva et al. (eds.), *Networks in Systems Biology*, Computational Biology 32,
https://doi.org/10.1007/978-3-030-51862-2_8

8.1 The Personalized Medicine of Cancer

The worldwide estimate of people diagnosed with cancer was 18.1 million in 2017, leading to 9.6 million deaths (https://ourworldindata.org/cancer) and is predicted by the *World Health Organization* (WHO) to be 27 million new cases worldwide by 2030. In Latin America, the cancer incidence is lower than in developed countries, but the mortality is significantly higher [1], which in part is due to diagnosis at later stages of the disease.

Cancer is a disease whose main feature is the uncontrolled division of malignant cells [2]; it affects genome replication due to dysfunctions in cell cycle checkpoints leading to the accumulation of mutations, indels, gene duplications, somatic crossing-over, chromosome number aberration, methylation errors among other molecular disorders. There are over 100 types of cancers, located in different organs and subtissues [3]. More efficient development of novel disease classifications based on molecular subtyping is likely to extend even more the diversity of cancer types [4].

As just emphasize, cell cycle dysfunctions are responsible for a huge genotype and phenotype plasticity of cancer cells, which continuously produce new cell clones as the tumor is growing. As a consequence, the tumors developed by two different patients for the same tissue are different according to their molecular profiles, in particular when considering the proteins encoded by their up-regulated genes [5]. The genetic heterogeneity at intratumor and interpatient levels, as well as the clonal evolution of tumors over time, remains among the major obstacles for precision medicine to materialize [6]. However, standard treatments of chemotherapy or radiation are one-size-fits-all, with the consequence that they are effective in only a subset of the patient population [3].

The emergence of drugs for specific targets and immunotherapy as made it possible to envisage the move from one-size-fits-all chemotherapies essentially based on cytotoxic drugs to *precision* or *personalized* treatments. In this text, we intend by *personalized medicine* a patient-centered care, which takes into account the necessity of therapies better corresponding to patient needs, i.e., the consideration of patient characteristics to optimize pharmaceutical treatment (specific medicines and associated devices) and/or overall care (individualized treatment that matches a patient's specific characteristics). By contrast, the commonly accepted meaning of *precision medicine* is the stratification of patients in subpopulations according to their specific features, i.e., the classification of individuals according to their susceptibility or prognosis for the diseases they may develop, or in their response to a specific treatment [4]. The difference between these definitions is very small since both involve decisions on patient-specific molecular signatures to perform diagnosis, prognosis, treatment, and prevention of cancer [7]. For several reasons that we present below, although genomic testing is available for patients at academic oncology centers, personalized medicine has not yet achieved widespread use in cancer care [8].

8.1.1 Benefits of Personalized Oncology

The perspective of cancer treatments by personalized oncology resulting in huge patient benefits has promoted companies and regulatory agencies to invest in this approach [3]. As outlined above, genetic information can enable one to distinguish between patients who are likely to respond strongly to pharmacologic treatment and those who will receive no benefit [4].

More specifically, the benefits of personalized oncology can be classified in: (i) Delivering an optimized therapy to patients more quickly with the consequence of promoting a better outcome and a reduced risk of negative collateral effects. (ii) Delivering benefits to healthcare systems and society by decreasing palliative costs and useless hospitalization. (iii) Increasing the development efficiency of new drugs and medicines. For instance, in 676 trials that occurred between 1998 and January 2012, biomarker targeted therapies had a 62% cumulative success rate, which is almost six times greater than the 11% cumulative success rate for any drug entering a phase 1 clinical trial. These results suggest that the use of biomarkers and personalized therapies may enable uncovering therapeutic mechanisms and design strategies that can decrease the amount of risk during drug development [9]. Unfortunately, the access of most low-income populations, such as the South American ones, to next-generation anti-cancer targeted drugs and personalized care is limited, mainly due to economic conditions when compared with higher income nations [10].

8.1.2 Personalized Cancer Therapies

There are four main types of standard cancer treatments: surgery, radiation therapy, chemotherapy, and immunotherapy. Cancer chemotherapy has always used generic cytotoxic drugs that aimed at inhibiting rapid cellular proliferation. These chemotherapies, although effective at controlling malignant proliferation by inhibiting cellular division, often produce high-risk side effects, i.e., development of resistance and immune suppression. A study conducted by Ramalho et al. [11] showed that 13% of drug therapy problems were the result of unnecessary drug prescriptions and inappropriately high dosages. Actually, due to the small specificity of cytotoxic drugs, the contribution of chemotherapy to overall cancer patient survival in the USA has been estimated to only 4.3% [12]. More recently, it has been estimated that any particular class of cancer drugs is ineffective in as much as 75% of patients [13].

By contrast to chemotherapies, personalized anti-tumor therapies are selected and combined based on their efficacy for particular cancer types [14], i.e., stratifying cancer treatments based on the tumor-specific features [7, 15].

In addition to targeting protein with specific drugs, immunotherapy is now used in personalized oncology to harness the own immune system of patients for fighting cancer. Immunotherapy treatments include monoclonal antibodies, checkpoint inhibitors, cytokines, vaccines, stem cell transplants, and *Chimeric Antigen Receptor* (CAR) T-cell therapies [16].

The genetic mixture should also be taken into account when profiling genes based on their mutations since a wide spectrum of polymorphisms have been identified in several oncogenes according to ethnic groups. In Latin American populations, for instance, treatment response differences are expected to be observed from one population to another and this should be evaluated before establishing a precision therapy [1].

As stated by Ersek et al. [17], "the health care environment and the practice of oncology are quickly changing. The traditional approach of prescribing chemotherapy for most patients with advanced or metastatic tumors is decreasing as the number of biomarker-driven treatments and immunotherapy drugs are increasing. For many cancers, multiple treatment options now exist, from traditional chemotherapy to biomarker-driven targeted therapies and immunotherapy. Oncologists today face the challenge of keeping up with novel therapeutics as they are approved. At the same time, oncologists must consider the pros and cons of various treatment regimens across many domains (e.g., efficacy, safety, costs) and weigh the options carefully with their patients, to select the most appropriate regimen".

Combination therapies are often needed for effective clinical outcomes in the management of complex diseases, but presently they are generally based on empirical clinical experience. To adapt clinical trial to this challenge, *adaptive design* has been shown to increase the efficiency of drug assessment by facilitating the dose selection, reducing the number of patients exposed to ineffective or potentially toxic doses, aiding the precise calculation of sample size, and reducing the duration and costs of clinical development. Adaptive therapy is a personalized approach in which clinicians continuously adapt their treatment strategy to the patient's anatomy and physiology. Adaptive therapy adjusted to the parameters of individual patients, can be advantageous when based on analytical tools, as provided by mathematical models [18], however, they often lead to more complex statistical assessment of clinical results.

New schemes, made possible by adaptive design, in early drug development allow (i) the integration of preclinical data, (ii) the incorporation of information beyond the traditional dose-limiting toxicity period, (iii) the integration of findings from other trials, and (iv) the alliance with emerging safety data, thereby increasing the likelihood of accurately determining any benefit of a new treatment and complying more quickly with regulatory requirements for efficacy and safety [19].

Many questions are raised by the personalized oncology approach such as listed by Lavi et al. [18]: (i) When several drugs are available to treat a cancer patient, how many drugs should be used to prevent treatment failure? (ii) What are the properties needed for this decision? (iii) What is the optimal drug administration in this case? (iv) Are the drug effects synergistic or sub-additive? (v) Should the drugs be administered simultaneously or sequentially?

These questions have been extensively studied in mathematical literature and partly addressed by clinical protocols. For instance, a series of clinical trials [20] confirmed the larger efficiency of the sequential schedule proposed by Norton and Simon [21] over the alternating schedule proposed by Goldie and Coldman [22].

8.1.3 OMIC Tests

The imprecise approach of one-size-fits-all treatments to patients has been undergoing a paradigm shift for several years, with the identification of molecular pathways predicting both tumor biology as well as response to therapy. It has been proposed to define a "new taxonomy" of disease based on molecular and environmental determinants rather than signs and symptoms [23]. The paradigm revolution lies in the change from a clinician selecting a generic therapy on a heuristic basis to one based on molecular facts.

At the moment, genetic information is used to evaluate patients at risk of developing particular diseases, or who have mutations that can be targeted by specific medicines. Specific biomarkers developed for each cancer type are used by physicians to stratify, diagnose, or take the best therapeutic options for each patient depending on the features of a specific tumor. When the molecular diagnosis is performed, targeted therapy is designed for acting on specific molecular targets supposed to be relevant to the tumor type under consideration [4]. The question of the parameters that may define how relevant a drug target may be is not obvious and many criteria were pursued in that quest [24].

Precision oncology has been associated with *Companion Diagnostics* (CDx) based on identifying a single biomarker. In this case, the idea is to apply a specific therapy expected to be efficacious for a given patient's condition [25]. CDx allows the selection of a treatment that is more likely to be effective for each individual based on the specific characteristics of its tumor. CDx is also developed for better predicting how patients will respond to a given treatment. According to the *U.S. Food and Drug Administration* (FDA), each CDx is associated with a particular drug therapeutic that, in turn, is associated with a specific genetic abnormality for which it is more effective [26]. Gaining insight into the molecular makeup of each tumor eliminates the misuse of ineffective and potentially harmful drugs. CDx assists healthcare providers in determining if a product's benefit outweighs its risks for patients. The FDA has therefore begun compiling a table of genomic biomarkers that they consider valid in guiding the clinical use of approved drugs [27].

At the moment, assays based on *Polymerase Chain Reaction* (PCR) currently have still the largest market share among the most common molecular diagnostic assays for detecting genetic abnormalities. Other established methods include DNA microarrays, and *Fluorescent* In Situ *Hybridization* (FISH) [28].

There are two types of omics tests: (i) prognostic test, which predicts a clinical outcome and (ii) a therapy guiding test (theranostic), which identifies subgroups of patients that are unique in their response to a particular therapy [29].

A variety of multigene assays are in clinical use or under investigation, which further defines the molecular characteristics of the cancers' dominant biologic pathways. These gene panels can be of diagnostic as well as prognostic power such as are the cases of Oncotype DX (www.genomichealth.com/) and Mamaprint [30], for instance. Even if there has been a growing use of biomarkers in clinical trials, and ~55% of all oncology clinical trials in 2018 involved the use of biomarkers, as compared with ~15% in 2000 [31], the use of single-marker and panel tests has become standard of care for only a very limited number of cancers [8]. This fact is mirrored by the market growth of global precision medicine that is forecasted to reach $ 141 billion by 2026 [32].

According to professional interviews, pathologists reported 75% inclusion of oncology biomarkers compared to 26% for oncologists. Even so, a majority of oncologists (68%) compared to pathologists (44%) reported high confidence in biomarkers [33].

As assay sensitivities improve, one development that holds much promise is the use of liquid biopsies to test for genetic abnormalities. Grail is currently running a series of clinical trials aiming at early stage disease detection of a range of common cancers through liquid biopsies [28]. However, the popularity of liquid biopsy among clinical professionals is still low with only 34% of oncologists and 14% of pathologists who strongly agree that sufficient evidence exists for this technology to be included in routine clinical practice [33].

Despite numerous technical challenges with liquid biopsies, there are at least four justification for its development (i) early disease detection, (ii) monitoring of disease progression, (iii) early detection of markers for drug resistance, and (iv) non-invasive technique of biopsy [34]. Unfortunately, the biggest challenge in implementing liquid biopsy is the rarity of tumor representative biomarkers [35]. In that regard, it seems that methylation abnormalities would offer a better promise than sequence mutations [36].

Interestingly, 30% of oncologists also strongly agree that full genomic sequencing should be conducted on all patient tumor biopsies upon initial diagnosis. The reason is that full genomic sequencing is considered as a memory of early tumor stage in case of future disease relapse where it is difficult to ask patients rebiopsying given their health condition and the invasivity of such process. However, 52% of pathologists strongly disagree with systematic full tumor sequencing [33].

Laboratory Developed Tests (LDT) that are a type of in vitro diagnostic test designed, manufactured and used within a single laboratory are neither supported by oncologists (8%) nor by pathologists (12%), which means that biomarkers and CDx strongly depend on regulation by official organizations for their acceptance by health decision-makers. From a general standpoint, the decision-maker confidence in oncology biomarker is variable with 68% of oncologists and 44% of pathologists reporting high confidence versus only 25% for payers [33].

According to decision-makers, key issues with a biomarker test is whether (i) its predictive power is acceptable (66% strongly agree), (ii) it is recommended in guidelines (64%), (iii) its companion diagnostic is mandated in the therapeutic's FDA labeling (52%), (iv) its clinical validity is acceptable (72% of payers strongly agree), (v) it is of clinical utility (58%), (vi) it is indicative of patient management (60%), (vii) its performance is good enough (58%), (viii) its price is affordable for the patient, (ix) it is reimbursed or not, and (x) its time delay is not larger than 10–14 days [33].

8.1.4 Drugs

The drug lifecycle can be divided into five distinct stages: discovery, preclinical research, clinical research, FDA review, and FDA post-market safety monitoring. This lifecycle is commonly agreed to be larger than 10 years, however, it is speculated that data-driven drug development should accelerate it and promote access to more effective and affordable treatments [37]. The number of FDA-approved precision drugs and indications is growing, and more are in clinical trials [38, 39]. According to the data stored in the web site of ClinicalTrials.gov, the study of the biomarker used in clinical trials of cancer performed between 2000 and 2018 let to a list of ~2,500 drugs based on more than 260,000 trials [31]. Approximately 73% of oncology drugs in development are based on personalized medicine approaches [13]. At the same time, the therapeutic armamentarium is rapidly increasing and the number of new drugs (including immune-oncology agents) entering drug development continues to rise. According to Vadas et al. [31], oncology drug approvals of new molecular entities by the FDA increased by 5% from 2014 to 2018. Furthermore, the proportion of personalized medicines among all new molecular entities gaining FDA approval has been doubling in the same time interval. Biomarkers are critical in the development of personalized medicines and have added more complexity to clinical trial design, execution, and data analysis.

The ongoing approach of drug development is the *High-Throughput Screening* (HTS) of vast reference libraries of molecules of known chemical properties aiming at observing their effects on specific proteins [40]. These methods allowed the development of 78 of the 113 first-in-class drugs (drugs that use novel methods of action, or specific biochemical interactions, rather than modified methods of existing therapies) approved by the FDA from 1999 to 2013.

However, successful oncology drug development has been sub-optimal as a result of the very low percentage of new agents that ultimately achieve regulatory approval. This high failure is partially due to the inability of standard models to accurately predict clinical response or tailor chemotherapy to specific high benefit patient groups. The search for more effective targeted therapies and immune checkpoint inhibitors are ongoing. However, resistance to these treatments is still a significant clinical problem [34].

One of the most important challenges for translational medicine (considering the whole chain from laboratory research to the patient bed) is to ensure the optimal use of pharmaceuticals [41]. The goal of this multidisciplinary activity is to provide comprehensive medication management services. The number of variables involved is huge encompassing (i) the identification and solution of drug therapy problems and (ii) the maximization of drug therapy efficiency while reducing adverse drug events and increasing cost savings [42]. These factors added to strong collaboration with regulatory agencies enabled the evolution in the design of early stage clinical trials and progressive incorporation of precision oncology methodologies.

Combining therapeutic strategies (e.g., targeted, chemo-, radiation-, and immunotherapy, as well as surgery) has become standard of care for the treatment of the majority of cancers. In particular, breast cancer patients who receive systemic delivery of targeted and cytotoxic drugs (given in parallel or sequentially) for various subgroups of breast cancers [43]. For instance, it has been shown that the simultaneous delivery of paclitaxel and trastuzumab drugs is not optimal. Regimens, where trastuzumab (a targeted monoclonal antibody) was applied first followed by paclitaxel (a cytotoxic drug that causes cell death by stabilizing microtubules during mitosis), had the greatest overall reduction in tumor cells compared to the opposite sequence [44]. This evidence suggested that there exists a spectrum of synergistic effects dependent on the sequence of drug administration. This is the kind of situation where in silico mathematical modeling may alleviate the challenge of investigating a myriad of combinations of experimental regimens in vitro [45].

Mathematical modeling has the strength of providing an analytic way of integrating and synthesizing individual components into a comprehensive picture, which allows the understanding of a system as a whole. This procedure is commonly referred to as *systems biology*. The analytical understanding of a specific mechanism can be used as a way to interpret the experimental data without reference to a heuristic receipt. Mathematical modeling may also provide insight into the dynamics of a system when experimental methodologies are not available, which itself may enable us to determine the most suitable measurement timing for a given problem. In addition, when validated, a mathematical model can serve for new experimental inferences [18].

Multiscale data integration requires specific model adaptation to accommodate heterogeneous information and/or indirect information because the required ones are often unavailable [46]. Hybrid models are well adapted for this function since these models are made of different subparts heterogeneously defined mathematically to adequately describe the events occurring at the many different spatiotemporal scales. For instance, this kind of modeling is required when: (i) A drug fails to reach its tumor targets due to vascular network bypasses [47]; (ii) The oxygen delivery is non-optimal as it occurs when part of the tumor becomes hypoxic and constitutive cells become quiescent [48]; (iii) The drug uptake is altered by membrane permeability and efflux pump activity with the effect of reducing intracellular drug accumulation, thereby preventing drug–target interactions; and (iv) A malignant cell normally sensitive to a drug physically escape to its action due to pharmacologic sanctuary. Cell-based Chaste [49, 50] (http://www.cs.ox.ac.uk/chaste/) is a software useful in these cases since it allows the heterogeneous modeling with objects varying from regulation networks, to mechanical constraints.

The resistance mechanisms outlined in the preceding paragraph can affect the treatment effectiveness with a single drug or multiple drugs. In the case of Multiple Drug Resistance (MDR), the cell becomes resistant to a variety of structurally and mechanistically unrelated drugs in addition to the drug initially administered [51]. Metastatic cancers challenged with targeted agents promote the growth of resistant cell populations and consequently targeted therapies are often only transiently effective. The active transport by drug efflux is the primary mechanism in MDR cell lines, while diffusion and exocytosis are not fast enough to account for the rapid efflux observed experimentally. This suggests that MDR reversal could be obtained with agents blocking molecular pumping activity [18].

The challenge of coping with drug resistance has promoted the development of several mathematical (control theory) and engineering (*artificial intelligence*, AI) approaches. The control theory supposes that the optimal drug dose can be calculated by reverse engineering. Typically, AI mimics *cognitive* functions by successive processes of *learning* according to a given algorithm and *solving problems* following the scheme that was learned. The purpose of AI in drug treatment is to reduce the data to a size the physicians can interpret and give them access to information they could not previously consider. The most common applications of AI in drug treatment have to do with matching patients to their optimal drug or combination of drugs [52]. Google developed a machine learning system that can identify from medical scans when breast cancer has metastasized with an 89% accuracy [53]. However, even if drug development has been boosted by mathematical modeling and AI, it is still a "lengthy, expensive, difficult and inefficient process with a low rate of successful therapeutic discovery" [54], which typically costs $ 2.56 billion [55].

In the case of dysfunction of complex biological networks, therapeutic interventions on multiple targets are required. Because the effect of drugs depends on the dose, several doses need to be studied at the same time and the number of possible combinations to be analyzed to find the optimal therapy rises quickly. For example, many cancer chemotherapy regimens are composed of six or more drugs from a pool of more than 100 clinically used anticancer drugs [56]. Using a network model of cell death, Calzolari et al. [57] found that the optimal combinations of interventions ensuring an 80–90% success rate were 6–9. In comparison, the success rate obtained with an equivalent random search was only 15–30%. These authors used a tree representation with drug combinations as nodes linking to all possible additions of one drug to the next level. Individual drugs form the base of the tree and combinations of maximum size are at the top. When the algorithm explores the drug combination tree going from smaller to larger combinations, more weight is given to lower order drug interactions. The best combination found can be of any size and typically searching within all possible combinations of different doses of ten drugs is applied. Drug interactions are known to cause adverse events, but multitherapy is common and essential for many patients and many hospitalized patients receive at least six drugs [58]. The trend of personalized therapy is evolving toward drug combinations since the different personalized drugs cannot be developed for each patient and the one-drug treatments are not effective.

Based on these results and considerations, precision therapy could be rationally designed by iteratively modifying apoptosis in silico networks [56]. By submitting malignant cells to drugs over a threshold concentration, it would enter apoptosis [59]. On the contrary, normal cells should not be significantly affected because their metabolism would not depend on the targeted genes. Of course, the success of this rationale depends on how much the idealized treatment may have non-specific off-target activity eventually toxic to normal cells as well.

Prediction of drug-target interactions is a smaller part of the monotherapy optimization problem, but it is no less important. Predicting drug-target interactions is vital to the repurposing of already existing drugs and the search for novel drugs along with the understanding of the signaling and metabolic pathways involved. Drug repurposing may lead FDA to approve therapies in a different tumor type (off-label) [60]. As stated above, the combinatorial explosion resulting from the large and growing number of drugs necessitates the development of in silico methods for selecting optimized combination therapies with minimal toxicity. For instance, Campillos et al. [61] observed similar drug-target interactions for drugs with similar therapeutical side effects, but which are not structurally similar.

Although *Drug–Drug Interaction* (DDI) is screened as part of the FDA approval process, many of them go unnoticed until after clinical trials due to the large number of possible combinations [62]. Li et al. [63] have employed several similarity scores to infer synergistic and antagonistic drug combinations. Even if several AI-based models were developed for the a priori detection of DDIs, none have reached the clinical implementation stage.

8.1.5 Preclinical Trial

A major challenge in precision medicine is to establish the relationship between laboratory models suitable for clinical translation. This is routinely obtained by screening in vitro culture, ex vivo tissues, or animals for thousands of molecular targets aiming at capturing the response of a complex system over time and constructing predictive models of human physiology for use in clinical trial development [64, 65].

Among other approaches, preclinical validation can be obtained by recapitulating individual tumors in vitro for determining the safest and most effective treatment before administration to a patient. 90% consistency of somatic mutations and DNA copy number profiles were observed between in vitro cultures of tumor organoids and patient original biopsies or surgical samples [66].

Xenograft tumor models are usually established by subcutaneous inoculation of a certain number of tumor cells into the flank of mice or rats. With xenograft, successful treatments for 12 individual patients were identified by screening a library of 63 drugs over 232 treatment regimens, which supports the use of personalized xenograft models for guided treatment platforms [67]. Hence, organoids hold large potential for the development of therapies in precision oncology. Considering the sets of variables of (i) tumor regression and therapeutic outcome and (ii) the same treatments applied to patients as well as to mice grafted with their tumors, a significant correlation was observed between both sets of variables in 87% of cases [68]. Izumchenko et al. [68] concluded that *patient-derived xenograft* models enable them to faithfully predict the outcome of chemical therapies before clinical presentation to patients, which enables them to test multiple options of ex vivo treatments (based on mutational profiling) before clinical implementation.

Unfortunately, the different microenvironment and physiology of mouse tissues may alter the genomic evolution of the implanted cells of patient-derived xenograft models, which explains the differences of therapy response compared to humans [69, 70]. Other challenges limiting the widespread use of patient-derived xenograft include the time taken to generate each model, the difficulty of result analysis induced by the necessity of using immune-compromised mice, the low success rate of xenografting, the significant resources required to create and maintain these models as well as the lack of standardized clinically relevant criteria [65].

8.1.6 Clinical Trial

One has to recognize that there are no prospective randomized controlled trials, at the moment, establishing oncology precision medicine-based treatment decisions as superior to routine care [60]. However, the concept of precision oncology has not yet been fully tested in a sufficient number of trials and many different ways of implementing it are still to be assessed [71]. There is considerable controversy in the oncology community regarding the value and future of precision oncology due to its often-exaggerated benefits [60]. Clinical trials validating the efficacy and utility of biomarkers and omics tests are not yet commonly successful. Also, the development and assessment of novel and robust statistical analysis methods are not following the same rigorous development as bioassays [3].

Oncologists require tools to assist them in selecting the most appropriate treatment for each patient. Oncologists and pathologists claim three factors as being keys in the selection of oncology biomarkers: (i) the predictive power of the test for identifying treatment responders vs. non-responders, (ii) the availability of *clinical guidelines* for patient applicability, and (iii) the existence of recommendations from governmental organizations such as FDA labeling [33]. Clinical guidelines are "statements that include recommendations intended to optimize patient care that is informed by a systematic review of evidence and an assessment of the benefits and harms of alternative care" [72]. *Clinical pathways* are detailed evidence-based treatment protocols within guideline sets that consist of the most efficacious and cost-effective regimens that minimize toxicity. Clinical pathways are typically disease and stage-specific and list drug names, doses, and administration schedules [73]. The purpose of clinical pathways is to support the best practice of patient care while considering efficacy, safety, and costs. However, as stated by Ersek et al. [17], "the development and implementation of clinical pathways are highly variable and reliant on the intent of the clinical pathways developer. Those who oppose the use of clinical pathways argue that they dictate patient care and lead to cookbook medicine. Those who are in favor of clinical pathways view them as resources to guide best practices and to reduce variation in care while taking efficacy, safety, and costs into consideration".

The proportion of trials using pharmacogenetics or *Pharmacogenomic Biomarkers* (PGX) to target specific patients continually increases [4]. At the moment, approximately 180 cancer clinical trials are open or in follow-up, with more than 1,100 patients enrolled [60]. Precision medicine has resulted in changing clinical trial methodology, which now focuses on individual, rather than average responses to therapy [74]. The traditional drug development track, where drugs were evaluated for safety in phase 1, early signs of efficacy in phase 2, and finally evaluated against standard therapy in a randomized phase 3 clinical trial is gradually evolving to a rapid phase 1 dose-escalation trial followed by large expansion cohorts. In that context, the emergence of new trials such as adaptive studies with basket and umbrella designs aim at optimizing the biomarker–drug co-development process. In parallel, with the growing complexity of clinical trials, new frameworks for stronger and faster collaboration between all stakeholders in drug development, including clinicians, pharma companies, and regulatory agencies are being established [19, 75].

As stated by Garralda et al. [6], "an adaptive design is defined as one that includes planned opportunities to modify one or more specified elements of the study design and hypothesis based on data analysis". For example, for assessing the predictive power of gene expression signatures, a Bayesian process of continuous learning is applied to recursively update the model with information of patients enrolled earlier [76, 77]. As a result, adaptive models aim at increasing the weights of good predictors and decreasing the weights of unstable predictors, while improving the overall performance of the classifier and selecting the best therapy according to current patients' characteristics.

Summarizing the main clinical trial types used to assess precision oncology, one has: (i) An *umbrella trial* is a master protocol for which the patient's eligibility is defined by the presence of a tumor type that is sub-stratified according to specific molecular alterations matched to different anticancer therapies [78]. (ii) A *basket trial* includes patients with different tumor types whose common molecular alteration is treated with the same matched therapy [79, 80]. This approach is expected to provide access to experimental therapies for patients across a wide range of tumor types including those rare tumors not studied in other clinical trials [81].

In Latin America, the situation is that 65% of all clinical trials are sponsored by the pharmaceutical industry, but public and private hospitals are often not adapted to perform clinical research, and the time length for regulatory approval of clinical trial applications is one of the longest in the world, which limits the interest of pharmaceutical companies to conduct clinical trials in Latin America [82].

8.1.7 Survival

As well as biomarkers need validation for the disease they represent, omics tests also require validation. Clinical viability and utility of biomarkers, as well as omics tests, must be established, which means that the use of a biomarker should be motivated by better *survival outcomes* [83]. By survival outcomes, one generally means *overall survival* (the length of time from either the date of disease diagnosis or the treatment start after which patients remain alive) and *progression-free survival* (the length of time during and after the disease treatment that a patient lives with the disease but it does not get worse). Von Hoff et al. [84], Radovich et al. [85], and Haslem et al. [86, 87] effectively reported gains in progression-free survival through the use of molecular-based therapy. In England, 5-year age-standardized net survival for breast cancer in women has increased from 71% in 1990–1999 to 87% in 2010–2011, with the greatest increase following the introduction of targeted therapies [4]. However, survival outcomes are not the only clinically meaningful endpoint to patients [17]. Other meaningful endpoints are (i) targeted and personalized intervention that identifies patients with the best likelihood to respond positively [88], (ii) reduced adverse events, suffering, and anxiety for patients and their families [8, 89, 90]. Actually, since scheduled precision medicine treatments are specific to individual patients, the theory is that it allows patients to receive treatments that have a higher likelihood of success and avoid therapies that will not work, carry high risk, or bring deleterious collateral effects [91–93].

Therapy optimization for survival maximization has been addressed mathematically as well. For instance, the population model of Monro and Gaffney [94] based on ordinary differential equations predicted that reduced chemotherapy protocols could lead to longer survival times due to competition between resistant and sensitive tumor cells. Castorina et al. [95] stated that the optimal chemotherapeutic dose should be determined by the balance of the two populations and their specific growth rates. These studies, conducted from the standpoint of control theory, provide a way to determine the minimal dose of a drug that warrants the asymptotic decay of the tumor cell population [18, 96].

Several models predict that continuous infusion (in the particular case of cell cycle phase-specific drugs) is more effective than short pulses of drug administration (see Ref [18]). The problem of continuous infusion is that if a drug is applied too quickly, the cells that are in a resistant part of their cycle may escape lethal exposure. If, on the other hand, the drug is applied too slowly by continuous infusion, drug resistance may develop. Gardner [97] modeled this tradeoff and used his model to provide insight on how much the chance of a cure is connected with the dose and the type of infusion.

8.1.8 Regulation

Managing adverse effects of precision treatments is also challenging because molecularly targeted drugs and immunotherapies have their toxicities that require new interventions. When making pathway or formulary decisions for new targeted therapies, oncologists and pathologists agree on the importance of (i) phase-three study results (82%), (ii) third-party guidelines, (iii) clinical pathways or compendia (72%), (iv) peer-reviewed journal articles (54%), (v) third-party value assessments (40%), and (vi) payer coverage decisions (~36%) [33].

Biomarker clinical trial data are unstructured and lack standardization for biomarker description and use [31]. Thus, the regulatory landscape needed to adapt and has been changing quickly. The enactment of the Precision Medicine Initiative in 2015 has helped in that process by promoting the FDA to develop a new platform for the assessment of the diagnostics and therapies produced by precision medicine [98]. The 21st Century Act (Cures Act) issued in 2016 accelerated medical product development by incorporating the patients' perspective and also modernizing clinical trial designs [99].

Payers traditionally consider many factors when making coverage decisions. Clinical validity remains the most important consideration at 72% with the two most important decision factors being: the capacity of test results to change patient management and the clinical utility of the test (60% and 58%, respectively). The information most widely utilized by decision-makers includes the (i) *National Comprehensive Cancer Network* (NCCN) biomarker compendium, (ii) NCCN evidence blocks, and (iii) guidance from regulatory bodies such as the FDA [33].

For oncologists, the top four factors for utilizing new targeted therapies are the following: (i) improved survival compared to the current standard of care, (ii) availability of a predictive biomarker or companion diagnostic, (iii) FDA approval of indication, and (iv) improved progression-free survival. Nearly half of all payers surveyed (40%) strongly agree that they have already or are likely to adopt the CMS decision to automatically cover FDA-approved *Next-Generation Sequencing* (NGS) companion diagnostics [33].

8.1.9 Costs

A 2015 study showed that the rate of inflationary pricing of cancer therapeutics was maintained at 10% per year between 1995 and 2013 [100]. In the USA, cancer care costs are projected to rise to $ 157.7 billion by 2020 [17]. The palliative costs of adverse collateral drug reactions were more than $ 30 billion in 2013 [101]. Thus the implementation of precision medicine can be of key importance for healthcare sustainability in the long term, especially in Latin American countries where healthcare resources are scarce. In these countries, a reduction of therapy use in patients for whom the treatment is not effective and the decrease in hospital stays can make a huge difference [102].

Many supporters of precision medicine believe they can reduce healthcare spending through the identification of therapy responders and non-responders. When asking whether healthcare spending is decreased through the use of precision strategies, skeptics point out that by stratifying patients and targeting therapies, the pharmaceutical companies are shrinking their available market share, which could result in a price increase to offset reduced volume [103]. However, we believe that this argumentation is wrong because precision medicine will just redistribute drugs for a more rational use, which could boost rather than shrink the market.

According to current healthcare practice, precision oncology approaches are used only after the failure of other therapies. However, conventional treatments, such as radiation and chemotherapy, usually leave patients exhausted, weakened, and unprepared for later treatments, which bias the assessment of the benefit that can be drawn from precision therapies.

Payers are reluctant to support new and untested treatments without overwhelming evidence that they are effective. Because precision oncology is a new and emerging field, a sufficient amount of data and pieces of evidence has not yet been accumulated to support widespread reimbursement for such treatments in the eyes of the payers. Because of these constraints, precision oncology requires collaborative efforts between regulatory agencies and industry to take off [3].

The oncology drug market is unique in the fact that the traditional notion of price competition between multiple drugs does not often apply to cancer therapeutics. In the treatment of most cancers, there exist few circumstances in which two biosimilar agents are directly competing in the same therapeutic space. Even when a particular biological pathway (e.g., EGFR, VEGF, mTOR) is being modulated by drug therapy, there are often a variety of different types of therapeutic agents that may be used and each has a specific mechanism of action and distinct rôle. The emerging field of immunotherapy is one of the few instances in which such pricing competition may occur since multiple agents are currently approved by FDA for similar situations [17].

Typically, a newly introduced drug or biopharmaceuticals is priced at a 10–20% larger rate than the most recent similar drug on the market [104]. Oncology drugs are priced at high initial levels, yet they rarely have associated price decreases with market expansion [17]. Because of these high drug prices, more than 80% of oncologists in the USA and Canada are influenced by patient out-of-pocket costs.

Concerning OMIC tests, oncologists, pathologists, and payers most readily agreed that the price would need to be comparable to two to four single biomarkers or CDx tests. When asked for an exact dollar value, oncologists are keen to agree with $ 1,233, pathologists with $ 1,356 and payers with $ 2,070. As concluded by the Novartis report of 2019, “oncologists, pathologists, and payers must work with pharmaceutical and diagnostic companies to develop products, services, and coverage policies that improve patient outcomes”. Payers are ahead of providers in evaluating oncology biomarkers for determining their overall cost-effectiveness. About 40% of payers are currently conducting this type of assessment, and another 30% are looking to implement this evaluation by 2020.

Innovations can take a decade or longer to become standard practice. As analysis published in the Journal of the Royal Society of Medicine identified a 17-year lag in translating medical research into practice [105]. This lag-time results in a vicious circle (Fig. 8.1) where payers are reluctant to cover the costs of a test until they have clear pieces of evidence for its benefits. Since oncologists are not ordering this test, the data to build a case for genomic testing, as part of an evidence-based standard of care, is not being accumulated. Because many new breaking technologies in the field of precision oncology are not standard of care, payers feel justified in not covering them [8].

Fig. 8.1 The vicious circle of innovation success on the market (modified from [8])

8.2 What the Molecular Phenotype Can Tell Us?

The development of personalized medicine is directly related to high-throughput technologies that became available in recent years. High-throughput techniques, such as *RNA sequencing* (RNA-seq), micro-array, and nanoString (https://www.nanostring.com/), are important tools for the characterization of tumors and their adjacent non-malignant tissues. These techniques allow a better understanding of tumor biology. In particular, RNA-seq demonstrated that each tumor is unique considering the protein profile of their up-regulated genes [5].

While the reference to the genome is generally performed to characterize indirect relationships between tumor development and gene alterations, such as indels, mutations, hyper- or hypo-methylation, transcriptome, proteome, and metabolome description allow the characterization of phenotypic events. However, most CDx on the market are based on mutation profiling.

An approach recently proposed was the identification of most relevant protein targets for personalized therapeutic intervention in breast cell lines [5]. This strategy combined *protein–protein interactions* (PPI) and RNA-seq data to infer the topology of the signaling network of malignant cell lines. It had the advantage to allow the association of a drug to the entropy of a target and to rank drugs according to their respective entropy by reference to their targets [106].

Three concepts were considered in the approach followed by Carels et al. [5]: (i) A vertex with a high expression level is more influential than a vertex with a low expression level. (ii) A vertex with a high connectivity level (hub) is more influential than a vertex with a low connectivity level. (iii) A protein target must be expressed at a significantly higher level in tumor cells than in the cells used as a non-malignant reference to reduce harmful side effects to the patient after its inhibition. It is worth mentioning that each combination of targets that most closely satisfied these conditions was found to be specific for its respective malignant cell lines.

The signaling network of a biological system is scale-free [107], which means that few proteins have high connectivity values and many proteins have low connectivity values. As a consequence, the inhibition of proteins with high connectivity values has a greater potential for signaling network disruption than randomly selected proteins [107].

As underlined above, the impact of node removal can be evaluated by the use of the Shannon entropy, which has been proposed as a network complexity measure and applied by many authors to determine a relationship with tumor aggressiveness. Breitkreutz et al. [108] for instance, found a negative correlation between the entropy of networks made of genes documented in the Kyoto Encyclopedia of Genes and Genomes (KEGG, http://www.genome.jp/kegg/) database considering cancer types and their respective 5-year survival. The Shannon entropy (*H*) is given by formula 1

$$H = -\sum_{k=1}^{n} p(k)\log_2(p(k)) \tag{1}$$

where *p*(*k*) is the probability that a vertex with a connectivity value *k* occurs in the analyzed network.

The approach of considering proteins with high connectivity values as a potential target of inhibition was in vitro validated with MDA-MB-231, a triple-negative cell line of invasive breast cancer [109]. It was shown that the inactivation, by interference RNA (*small interfering RNA*, siRNA), of the five top-ranked hubs of connection identified for this cell lineage resulted in a significant reduction of cell proliferation, colony formation, cell growth, cell migration, and cell invasion. Inhibition of these targets in other cell lines, such as MCF-7 (non-invasive malignant breast cell line) and MCF-10A (non-tumoral cell line used as a control), showed little or no effect, respectively [109]. Also, the effect of joint target inhibition was greater than the one expected from the sum of individual target inhibitions, which is in line with the buffer effect of regulatory pathway redundancy in malignant cells.

The transcriptome data from tumors and their paired non-malignant tissue (taken from the tumor surrounding) considered as control samples were used to determine their up-regulated genes, construct their corresponding subnetworks, and calculate their respective Shannon entropy among nine cancer types. The results confirmed the existence of a negative correlation (Fig. 8.2a) between the entropy of tumors' PPI subnetwork and the survival rate of the corresponding patients using data from bench experiments (*The Cancer Genome Atlas*, TCGA, now hosted by the *Genomic Data Commons Data Portal*, GDC Data Portal: https://portal.gdc.cancer.gov/) [110]. These authors also proposed a method to infer the suitable number of targets for inhibition according to the 100% patient survival in the 5 years after treatment (Fig. 8.2b). This method concerns the number of connectivity hubs that should be inactivated in a tumor to lower its subnetwork entropy to a level that maximizes patient survival [110].

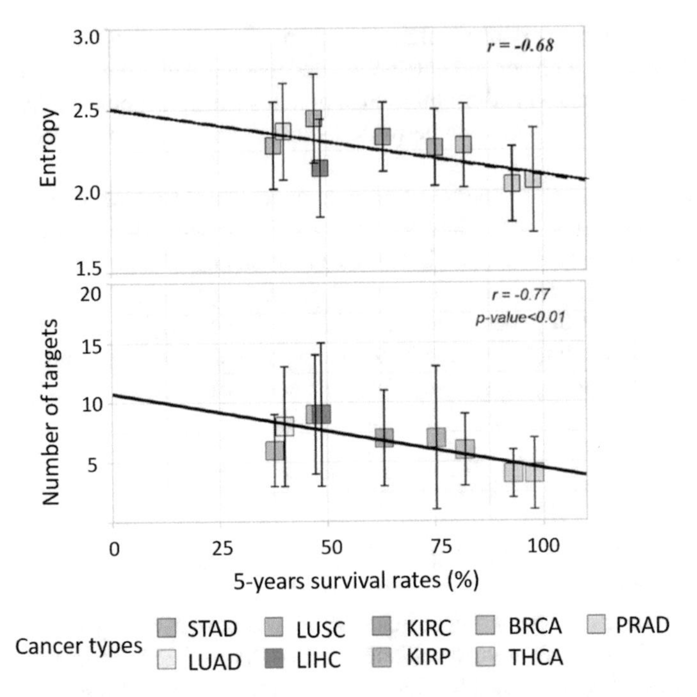

Fig. 8.2 Correlations between (**a**) entropies of up-regulated genes' subnetworks and their respective 5-year survival rates and (**b**) the number of targets to inhibit for a 100% survival in the 5-years after treatment for nine cancer type: Breast Cancer (BRCA), Kidney Renal Clear Cell Carcinoma (KIRC), Kidney Renal Papillary Cell Carcinoma (KIRP), Liver Hepatocellular Carcinoma (LIHC), Lung Adenocarcinoma (LUAD), Lung Squamous Cell Carcinoma (LUSC), Prostate Cancer (PRAD), Stomach Adenocarcinoma (STAD), and Thyroid Cancer (THCA). The vertical bars indicate standard deviations (modified from [110])

8.2.1 *Tumor Modeling*

As underlined above, from a general perspective, computational modeling facilitates the generation of hypotheses and tests. It enables the mitigation of time and expenses in the investigation of biological processes. Among biological networks, the modeling of cancer-related processes, such as the cancer hallmarks, is of great interest for unraveling disease dynamics and identify new biomarkers as well as for targeting proteins with therapeutic potential.

The use of up-regulated hubs in tumor tissues in a tentative of avoiding potential side effects is indicated to disrupt biological scale-free networks [5, 106]. Yet, they are other measures that could be explored as well [111–113].

The betweenness centrality of a node v (formula 2) is the ratio of the number of paths between two nodes s and t that passes through v divided by the number of all paths between s and t [114]. This measure is largely explored in metabolic networks aiming at finding essential metabolites [115].

$$g(v) = \sum_{s \neq v \neq t} \frac{\sigma_{st}(v)}{\sigma_{st}} \quad (2)$$

Knowledge of the clustering coefficient (formula 3) is also important when investigating alternative pathways that may preclude treatment success. The clustering coefficient measures how much the neighbors of one node are connected between them. The higher the coefficient, the more connected the nodes are [116].

$$C_i = \frac{2L_i}{k_i(k_i - 1)} \quad (3)$$

In formula 3, C_i is the clustering coefficient or the probability that two neighbors of a node link to each other, where L_i represents the number of links between the k_i neighbors of node i.

After determining the parameters for the selection of the most relevant nodes that should be included in a personalized therapeutic, the effect of node inhibition on the signaling network of malignant cells can be measured by the associated decrease in Shannon entropy [110]. The entropy has also been used by other authors as a measure of systems randomness [117], disorder [118], and signaling heterogeneity [119]. However, those approaches do not consider the dynamics of biological networks, which is an intrinsic trait of biological systems. As the inactivation (even transient) of a node initiates a signaling cascade, it is necessary to infer its consequences for the cell destiny through network dynamic modeling.

For exploring the signaling network dynamics, Huang et al. [120] proposed that cancer could be a pre-existing attractor in the cell epigenetic landscape. This landscape is a space including the state of all possible cell gene expression profiles. The hills of this landscape are considered unstable states, i.e., transient states, while wells indicate stable states related to cell phenotypes with low-energy levels, which are also called attractors. As biological systems are stochastic, attractors are surrounded by basins of attraction composed of similar gene expression profiles resulting in the same cell phenotype. For this reason, one does not need to perform many interventions to bring the energy state of a cell to its minimum. Interventions through gene manipulations should just need to bring the cell state to the border of a desired basin of attraction in such a way that the biological system would naturally evolve toward its corresponding minimum energy level [121].

Yet, one needs to consider the available interventions for such manipulations. For instance, a patient treatment usually involves the inhibition of therapeutic targets through different inhibitors, such as drugs and antibodies or *Fragment-Antigen Binding* (FAB), i.e., the present-day approaches, as well as RNA interference or CRISPR/Cas9, i.e., next-generation approaches [122, 123]. Consequently, it is necessary to search the tools among the pharmacopeia that will enable to bring the malignant cell state within the basin of attraction corresponding to its death with the fewest gene interventions as possible to cause the fewest toxic collateral effects to the patient as possible.

These concepts have been explored by many authors with different approaches, such as Boolean networks, ordinary differential equations, endogenous network theory, and Hopfield network.

The Boolean network approach was used by Crespo et al. [124], who proposed that the attractor stability is related to the presence of positive feedback loops. However, different phenotypes may present the same positive feedback loop with mutually exclusive expression profiles. In this context, those positive circuits may have determinant nodes with high interface out-degree, which refers to the number of genes that are directly regulated by these nodes; it is these determinant nodes that potentially regulate the cell phenotype and are keys for cellular reprogramming [125].

In short, Crespo et al. [124] reconstructed the regulatory network through literature search, enriched it with available miRNA data, combined the network with gene expression data and used a Boolean model to search for attractors and positive feedback loops with mutually exclusive gene expression profile associated to different phenotypes. Superconnected components (which occur when every vertex of a component can be reached from every other vertex of that component) were identified among the positive feedback loops and organized as a directed acyclic graph to analyze the hierarchy between them. This process allowed the identification of nodes with higher interface out-degree in the first level of the regulatory hierarchy.

This strategy was validated through a literature search considering three biological cases: (i) T helper phenotypes TH1 and TH2, (ii) myeloid and erythroid cell types, and (iii) fibroblast and hepatocyte cell types. For all cases, the nodes identified as determinants were previously described as essential for the phenotype considered. Moreover, an in vitro experimental validation was also performed in breast cancer considering the epithelial-mesenchymal transition [126], where the inhibition of SNAI1 with miRNA (miR203) was successful to obtain the transition from mesenchymal to epithelial phenotypes.

The differential equations approach was used to quantify the epigenetic landscape for cell development and cancer formation [127]. The network used was constructed based on a literature search for cancer and cell development markers. It included ZEB and OCT4 as biomarkers of cell development, p53 and MDM2 as biomarkers of cancer formation as well as two microRNAs that regulate the selected biomarkers. The modeling of the epigenetic landscape with stochastic differential equations were used to calculate the evolution of the probabilistic distribution for each variable and quantify the potential landscape. Li and Wang [127] found four stable states and their respective basins of attractions representing normal, cancer, cancer stem cells, and stem cell phenotypes. The transitions between stable states were calculated and characterized as irreversible, with different trajectories between two attractors.

There is a considerable variation in the landscape surface based on the set up of parameters (initial conditions). The decrease of the self-activation parameter changed the landscape from four stable states to a bistable cancer-related state, and then to a monostable cancer state. On the other hand, the increase of the repression strength parameter induces cancer state disappearance [127]. This experiment highlights the importance of parameter inference to improve accuracy in mathematical modeling.

As largely known, cancer may relapse in the 5–10 years after chemotherapy or radiotherapy. The disease recurrence may result from the maintenance of cancer stem cells after cancer cell removal. In this case, the cancer stem cells will eventually dynamically evolve toward a cancer stable state through transition paths. Consequently, efficient treatment should consider the network alteration to exclude both cancer stem cells and cancer stable states. In this context, the network proposed by Li and Wang [127] contains only six nodes and it is a huge simplification of the real situation, which does not decrease the merit of the methodology but underlines the size of the challenge to make that methodology applicable to real cases. This limitation was imposed by the mathematical method applied, which refers to the difficulty of defining parameters when describing a biological system with differential equations.

For therapeutic purposes, a complete network closer to the biological reality should be considered, which needs a modeling method that does not limit the network size due to the biological knowledge required concerning PPI or parameters. The endogenous network theory assumes that an endogenous network, including oncogenes and tumor suppressors as well as covering almost all aspects of molecular and cellular functions [128], exists. The nonlinear dynamical interactions among the endogenous agents result in multiple stable and unstable states, and their stochasticity allows the transition between stable states [128]. The application of this theory in cancer showed that the attractors found matched gene expression data for different cell types present in the heterogeneous tumor microenvironment [129, 130].

The Hopfield network approach is also not limited by biological knowledge requirements. This method allows the evaluation of hundreds of genes and the overall behavior of a gene regulatory network during progression from a normal to a disease state. It is defined as a recurrent artificial neural network, fully connected with undirected edges and ensures that sample states converge toward stored attractor patterns [131].

In the Hopfield method, the weight matrix (*Wa*) is defined as in formula 4, where P is the attractor's gene expression profile, P^T is its transpose, and I is the identity matrix necessary to impose symmetric behavior with diagonal equal to zero. Since W may be composed of more than one stored pattern, its value is equal to the sum of all weight matrices (formula 5).

$$W_a = \left(\mathrm{PP}^T\right) - I \tag{4}$$

$$W = W_{a1} + W_{a2} \tag{5}$$

The dynamic trajectory of each sample ($P(t + 1)$, defined in formula 6), was predicted by following the synchronous approach as shown by formula 6, where W is the final weight matrix, and $P(t)$ is the sample gene expression profile at time t. The sgn(x) function (formula 7) determines the binarized output pattern.

$$P_{(t+1)} = \mathrm{sgn}\left(P_{(t)} W\right) \tag{6}$$

$$\operatorname{sgn}(x) = \begin{cases} 1 : x \geq 0 \\ 0 : x < 0 \end{cases} \quad (7)$$

By analogy to a physical system, the discrete Hopfield network energy (E) is calculated using the Lyapunov function, which guarantees convergence to a low-energy attractor state (see formula 8, [132])

$$E[P(s)] = \frac{-\mathrm{PWP}^T}{2} \quad (8)$$

where $E[P(s)]$ is the energy of network state s for sample vector P and time t.

The Hopfield method has been widely used to better understand cancer-related attractors and basin of attraction in the epigenetic landscape (Fig. 8.3), such as cancer subtypes and stages [132–134]. Among the results that were obtained, it was found that a cancer basin of attraction is larger than the one associated with normal cells, indicating that it is likely to attract a more heterogeneous set of samples. In the therapeutic context, Szedlak et al. [135] used asymmetric Hopfield networks to test densely connected nodes as therapeutic targets, while Cantini and Caselle [136] adapted this approach to identify molecular similarities among transcriptomic data and stratify cancer patients to improve their therapy.

Although a lot has been achieved, some challenges still made difficult to perform the translation of mathematical models to personalized medicine approaches. For

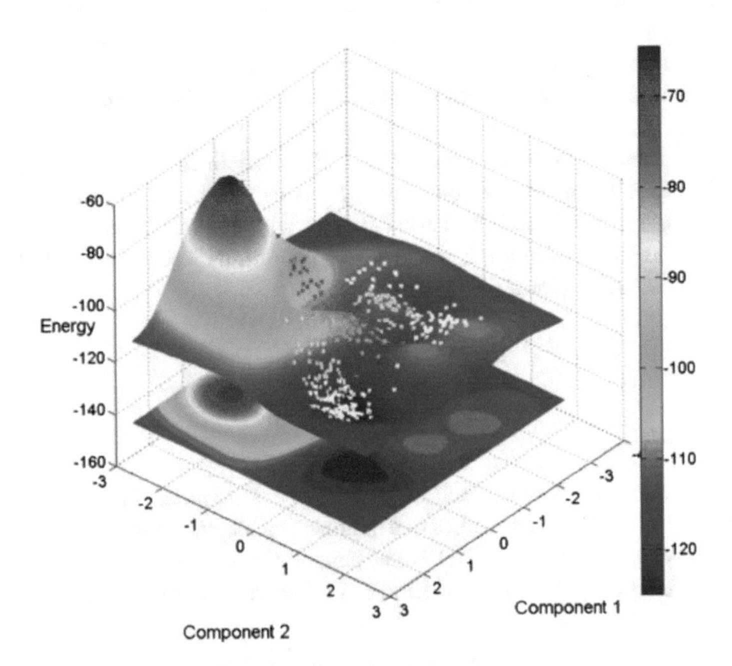

Fig. 8.3 Waddington's epigenetic landscape recovered using HopLand (*source* [137])

instance, many hypotheses and studies explore the state transition from cancer to normal cells as an option to identify the best therapeutic approach. However, this transition is impossible due to mutations accumulation and acquirement of deregulated processes during cancer development, which might result in topological differences in the regulatory network of different cell types. Yet, one can focus on cancer hallmarks modeling aiming at controlling cell proliferation and leading cancer cells toward death activation.

8.3 Drug Development

Since the dawn of humanity, natural products of plants have been used as traditional medicines, remedies, potions, and oils without any knowledge about their bioactive compounds [138].

From a historical standpoint, plant natural products were the most important source for drug development as demonstrated by their wide range of medicinal activity as well as structure and function diversity. However, aside from the single molecule-based drugs that dominated the scene in the past, the new trend is focused on the development of drugs based on natural mixtures, supposed to be more effective due to possible synergistic effects [139]. Particularly prominent examples of plant-derived natural compounds that have become indispensable for modern pharmacotherapy can be found in the field of anti-cancer agents, e.g., paclitaxel and its derivatives from yew (*Taxus*) species, vincristine, and vinblastine from Madagascar periwinkle (*Catharanthus roseus* (L.) G. Don), and camptothecin and its analogs initially discovered in the Chinese tree *Camptotheca acuminata* Decne [138, 140]. Despite their uncontested unique structural diversity [141], plant natural products still have several drawbacks associated with their low solubility, metabolic instability, and unknown exact mechanisms of action, which especially hamper the development of similar drugs [142]. Because of these drawbacks, the annual number of new chemical entities from plants has declined in the past decades [143].

The conduction of rigorous clinical trials needed for approval of natural products as drugs represents another major difficulty. While such clinical trials can only be performed with industrial support due to their high costs, the interest of pharmaceutical companies in natural products that are not synthetically modified is limited due to controversies regarding their patentability, e.g., curcumin [144], which explains why the interest in natural product-based drugs has been gradually declining. Even in the last decade, many big and medium-sized pharmaceutical companies, which were still active in the area in the 1990s, terminated their natural product programs, leaving them to the academic sector and start-up companies [145].

With modern advancements in computer hardware and the availability of free software packages, abstract concepts from general, organic, and physical chemistry started to be applied to realistic biomolecular systems, e.g., DNA, proteins. The use of *Computer-Aided Drug Discovery* (CADD) techniques and methods of molecular docking based on classical mechanics force fields enables the exploration of large

DNA–ligand systems [146]. This strategy has been shown to provide reliable results in predicting DNA–ligand binding modes and energies [147]. CADD has become an integral part of common protocols for initial screening and assessment of potential drug candidates [148–150]. The use of CADD actually started with the simulation of protein–ligand interactions because the force field of amino acids was obtained before that of nucleic acids [151]. There are two major types of drug design techniques: *Structure-Based Drug Design* (SBDD) and *Ligand-Based Drug Design* (LBDD). SBDD is used when the *three-dimensional* (3D) structure of a target protein is available, while LBDD is employed in cases where the 3D structure is unknown. CADD methods are dependent on biophysical models, mathematical algorithms, computer programs, and data from databases [152].

The information concerning a protein target is usually obtained experimentally by X-ray crystallography, *Nuclear Magnetic Resonance* (NMR), or *Cryo-Electron Microscopy* (cryo-EM). When neither of this information is available, modeling by comparison with homologous proteins may be used to predict the 3D structure of a protein target. Knowing the 3D structure of a protein target allows the docking of ligands from large in silico repository of chemical compound structures such as ZINC (https://zinc.docking.org/) by high-throughput virtual screening. The affinity of ligands for targets can be assessed by computing various measures of binding free energies. After subsequent steps of filtering and optimization, the lead (drug candidate) molecules are tested in vitro for their activity. When the 3D structure of a target is not available or is not predictable using computational methods, ligand-based approaches are often used as an alternative [153].

Both methods had great contributions to drug discovery; Zolmitriptan, produced by the LBDD approach, is used as a treatment of migraine [154], and Norfloxacin, which is a drug that is used in urinary tract infections [155]. In the case of SBDD, Saquinavir and Amprenavir were developed to target HIV-1, and Dorzolamide is used in the treatment of glaucoma.

For the discovery of new drugs, using LBDD, three main strategies must be considered: (i) molecular similarity approaches, (ii) *Quantitative Structure-Activity Relationship* (QSAR), and (iii) pharmacophore modeling. One selects novel compounds based on chemical and physical similarities to drugs known to inhibit the target.

Ligand similarity search methods are simple but effective; they are based on the observation that structurally similar molecules tend to have similar binding properties [156].

QSAR is based on the statistics of drug-target interactions based on various molecular descriptors. The hypothesis underlying the QSAR methodology is that structurally similar molecules tend to show similar biological activities. QSAR models mathematically describe the response activity of a ligand-target complex according to the structural features of the ligand. QSAR is obtained by calculating the correlation between experimentally determined biological activity and various properties of small ligands [157].

A pharmacophore is a molecular framework that defines the essential features responsible for the biological activity of a compound. When structural information about a drug target is limited or not known, pharmacophore models may be built using the structural characteristics of active ligands that bind to the target. When 3D information of the target structure is known, the information of the binding site can be used for generating pharmacophore models of higher precision. Pharmacophore models that use chemical features, such as acidic/basic residues and hydrogen bond acceptors and donors, are found to be the most effective. Pharmacophore modeling is also used for the virtual screening of ligands in large databases [158].

The information relative to the 3D structure of proteins and DNA started to be used for drug design almost three decades ago. The *Protein Databank* (PDB) is the largest repository of biomolecule structure information determined mostly by X-ray crystallography and NMR. In 1998, 2,058 structures were deposited in PDB. Since then, there has been a ~7.5% increase each year, resulting in a total of 161,002 structures by the end of February 2020. The use of this abundant structural information has been the cornerstone for structure-based drug design for the past years in academia as well as in the pharmaceutical industry [159]. When the structure of a protein target for a drug design project has not yet been solved, determining structures from sequences using computational methods is the only available mean [153].

After choosing the 3D model of interest, the next step is the identification of a potential drug and its orientation in the cavity of the active site. Depending on the method, there are different considerations concerning the flexibility of either the ligand or the protein during the docking process [160]. Docking and pharmacophore modeling (Fig. 8.4) are widely used in virtual screening of ligands to identify novel compounds able to inhibit protein targets.

According to the method of de novo design, a systematic search by trial and error is performed by the algorithm in the cavity of the target to best accommodate chemical groups on the ligand to improve its affinity and gradually optimize its conformation. The algorithm probes the energy landscape of the conformational space of the target

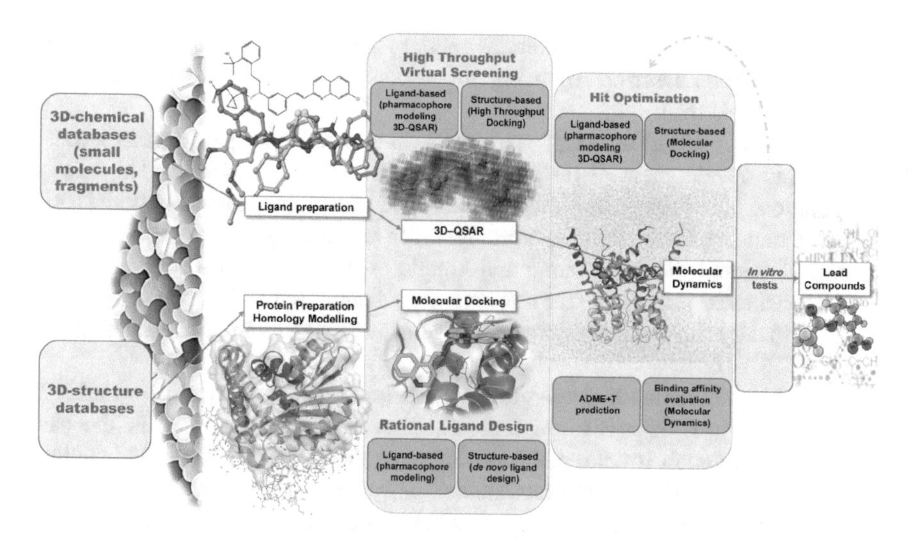

Fig. 8.4 Example of pipeline for assisted computational drug design (*source* [161])

and, after numerous search and assessment cycles, converges to the minimum energy solution corresponding to the most likely binding mode. Although the method is effective in exploring the conformational space, it can converge to a local minimum rather than the global minimum. The ligand uncover by this strategy can also be difficult to produce by chemical synthesis in the laboratory [162].

There are several examples of hit compounds that were successful in becoming leads through this process [159]. Typically, the ligand poses in the protein active site are assessed through a scoring function and the group of the compounds associated with the largest score values is chosen for the next step of the experimental screening.

To confirm a potential drug candidate, the orientation and indication of the best ligand in the 3D model of the target must be completed by data from *Molecular Dynamics* (MD). MD simulations nowadays allow the implementation of SBDD strategies that fully account for structural flexibility of the overall drug-target model system [163].

Classical MD regards atoms as solid spheres and the bonds connecting them as springs. This allows the atoms in the system to only oscillate within a specified distance. Classical MD is based on Newton's equations of motion (formula 9), where F_i is the component of the net force acting on atom i with a mass and mi. r_i denotes the position of the atom i at time t.

$$F_i = m_i \frac{\mathrm{d}^r_{r_i}}{\mathrm{d}t^2}. \tag{9}$$

The force F_i can then be computed as in formula 10

$$F_i = -\frac{\mathrm{d}U(r_1, r_2, \ldots r_n)}{\mathrm{d}r_i} \tag{10}$$

where $U(r_1, r_2, \ldots r_n)$ is the potential energy function of the specific conformation, which can be described by using the concept of a force field with predefined parameters [164, 165].

The impact of SBDD on drug discovery has intensified in the past decade because of the rapid development of faster hardware and better algorithms for high-level computations able to run complex problems in an acceptable interval of time [166]. Receptor and ligand flexibilities are crucial for correctly predicting drug binding as well as related thermodynamic and kinetic properties. As a result, MD is no longer considered prohibitive for effective drug design, which has enabled to push computationally driven drug discovery ahead of the frontiers in both academia and industry [167–169].

Since then, numerous research groups were using snapshots extracted from MD trajectories to provide a discrete representation of target plasticity. In the past decade, MD protocols for ensemble docking have shown great potential in handling targets governed by significant structural flexibility. These ensembles include protein kinases and G-protein-coupled receptors, whose function is characterized by remarkable conformational plasticity [170–172]. Today, a full dynamical description of the protein–ligand binding event can be obtained, with various degrees of accuracy. This is due to increasing computer power thanks to the advent of *Graphical Processor Unit* (GPU) architectures and MD software that can efficiently run on these innovative hardware infrastructures. In their pioneering studies, Buch et al. [173] and Shan et al. [174] investigated the binding of a ligand to a target protein by multiple replicas of microsecond-long MD simulations.

Even if the simulations today manage to reach milliseconds, some conformational changes of proteins do not occur in that time interval, which makes it necessary to have new approaches aiming at accelerating the process of MD. Among these developments, one may cite the umbrella sampling [175], replica exchange [176], metadynamics [177], and others [178].

Quantum mechanics (QM) is another MD method that simulates the electronic structure of the system under consideration, which allows an accurate investigation of its chemical reactivity. QM systems generally take into account only the closest protein environment that is directly involved in a chemical reaction but ignore the effects originating in the most distant protein environment. But even with these abstract representations of enzyme systems, very accurate predictions are possible for specific metabolic reactions [179].

Since QM methods deliver complementary information it comes as no surprise that the combination of both approaches, referred to as *Quantum Mechanical/Molecular Mechanical* (QM/MM) methods, has become a key technology to investigate enzyme reactions. The idea of the QM/MM hybrid approach is to tackle large systems by describing the region where a chemical reaction takes place by a QM method while accounting for the effects of the environment by MM methods [180, 181]. Knowing the chemical structure of ligand-target complexes is of immediate relevance to drug discovery as it allows the rational design of molecules with specific binding properties (in particular, substrates or inhibitors). This is an area of research in which QM/MM methods can be effective [182].

Acknowledgements This study was supported by a fellowship from Fundação de Amparo à Pesquisa do Estado do Rio de Janeiro (FAPERJ) to AC and a fellowship from Coordenação de Aperfeiçoamento de Pessoal de Nível Superior/Fiocruz (CAPES/Fiocruz) to CL.

References

1. Calderón-Aparicio A, Orue A (2019) Precision oncology in Latin America: current situation, challenges and perspectives. eCancer Med Sci 13:920
2. Hanahan D, Weinberg RA (2011) Hallmarks of cancer: the next generation. Cell 144:646–674
3. Krzyszczyk P, Acevedo A, Davidoff EJ, Timmins LM, Marrero-Berrios I, Patel M et al (2018) The growing role of precision and personalized medicine for cancer treatment. Technology (Singap World Sci) 6(3–4):79–100
4. Wilsdon T, Barron A, Edwards G, Lawlor R (2018) The benefits of personalised medicine to patients, society and healthcare systems. Charles River Assoc 2018:1–72

5. Carels N, Tilli T, Tuszynski JA (2015) A computational strategy to select optimized protein targets for drug development toward the control of cancer diseases. PLoS ONE 10:e0115054
6. Garralda E, Dienstmann R, Piris-Giménez A, Braña I, Rodon J, Tabernero J (2019) New clinical trial designs in the era of precision medicine. Mol Oncol 13(3):549–557
7. Ciriello G, Miller ML, Aksoy BA, Senbabaoglu Y, Schultz N, Sander C (2013) Emerging landscape of oncogenic signatures across human cancers. Nat Genet 45(10):1127–1133
8. Vuckovic N, Vuckovic BM, Liu Y, Paranjape K (2016) Accelerating clinical genomics to transform cancer care. Intel 1:8. https://www.intel.com/content/dam/www/public/us/en/documents/white-papers/accelerating-clinical-genomics-to-transform-cancer-care-paper.pdf. Accessed by Feb 2020
9. Falconi A, Lopes G, Parker JL (2014) Biomarkers and receptor targeted therapies reduce clinical trial risk in non-small-cell lung cancer. J Thorac Oncol 9:163–169
10. Morel CM, McGuire A, Mossialos E (2011) The level of income appears to have no consistent bearing on pharmaceutical prices across countries. Health Aff 30(8):1545–1552
11. Ramalho OD, Brummel AR, Miller DB (2010) Medication therapy management: 10 years experience in a large integrated health care system. J Manag Care Pharm 16(3):185–195
12. Morgan G, Ward R, Barton M (2004) The contribution of cytotoxic chemotherapy to 5-year survival in adult malignancies. Clin Oncol (R Coll Radiol) 16:549–560
13. PMC (Personalized Medicine Coalition) (2014) The case for personalized medicine, 4th edn. http://www.personalizedmedicinecoalition.org/Userfiles/PMC-Corporate/file/pmc_case_for_personalized_medicine.pdf. Accessed by Feb 2020
14. Coyle K, Boudreau J, Marcato P (2017) Genetic mutations and epigenetic modifications: driving cancer and informing precision medicine. BioMed Res Inter 2017:9620870
15. West J, Bianconi G, Severini S, Teschendorff AE (2012) Differential network entropy reveals cancer system hallmarks. Sci Rep 2:802
16. Maciejko L, Smalley M, Goldman A (2017) Cancer immunotherapy and personalized medicine: Emerging technologies and biomarker-based approaches. J Mol Biomark Diagn 8(5):350
17. Ersek JL, Nadler E, Freeman-Daily J, Mazharuddin S, Kim ES (2017) Clinical pathways and the patient perspective in the pursuit of value-based oncology care. Am Soc Clin Oncol Educ. Book. 37:597–606. https://doi.org/10.14694/EDBK_174794
18. Lavi O, Gottesman MM, Levy D (2012) The dynamics of drug resistance: a mathematical perspective. Drug Resist Updates 15(1–2):90–97
19. Harrington JA, Hernandez-Guerrero TC, Basu B (2017) Early phase clinical trial designs e state of play and adapting for the future. Clin Oncol (R Coll Radiol) 29:770–777
20. Citron ML, Berry DA, Cirrincione C, Hudis C, Winer EP, Gradishar WJ et al (2003) Randomized trial of dose-dense versus conventionally scheduled and sequential versus concurrent combination chemotherapy as postoperative adjuvant treatment of node-positive primary breast cancer: first report of Intergroup Trial C9741/Cancer and Leukemia Group B Trial 9741. J Clin Oncol 21:1431–1439
21. Norton L, Simon R (1977) Tumor size, sensitivity to therapy, and design of treatment schedules. Cancer Treat Rep 61:1307–1317
22. Goldie JH, Coldman AJ, Gudauskas GA (1982) Rationale for the use of alternating non-cross-resistant chemotherapy. Cancer Treat Rep 66:439–449
23. Collins FS, Varmus H (2015) A new initiative on precision medicine. N Engl J Med 372(9):793–795
24. Catharina L, de Menezes MA, Carels N (2018) System biology to access target relevance in the research and development of molecular inhibitors. In: da Silva FAB, Carels N, Paes Silva Junior F (Eds) Theoretical and applied aspects of system biology. Computational biology, 1st edn. Springer International Publishing, Cham, pp 221–242
25. Verma M (2012) Personalized medicine and cancer. J Pers Med 2:1–14
26. FDA (Food and Drug Administration) (2019) Policy for device software functions and mobile medical applications—Guidance for industry and food and drug administration staff. https://www.fda.gov/media/80958/download. Accessed by Feb 2020

27. FDA (Food and Drug Administration) (2020) Table of pharmacogenomic biomarkers in drug labeling. https://www.fda.gov/drugs/science-and-research-drugs/table-pharmacogenomic-biomarkers-drug-labeling. Accessed by Feb 2020
28. Gelifescience (2019) Delivering precision health: the role of molecular diagnostics. https://thepathologist.com/fileadmin/pdf/GE-app-note-0919-supplied.pdf. Accessed by Feb 2020
29. McShane LM, Polley MY (2013) Development of omics-based clinical tests for prognosis and therapy selection: the challenge of achieving statistical robustness and clinical utility. Clin Trials 10:653–665
30. van de Vijver MJ, He YD, van 'T Veer LJ, Dai H, Hart AA, Voskuil DW et al (2002) A gene expression signature as a predictor of survival in breast cancer. N Engl J Med 347(25):1999–2009
31. Vadas A, Bilodeau TJ, Oza C (2019) The evolution of biomarker use in clinical trials for cancer treatments: key findings and implications. L.E.K. Consulting 2019:1–25. https://www.lek.com/insights/sr/evolution-biomarker-use-clinical-trials-cancer-treatments. Accessed by Feb 2020
32. BIS Research (2017) Global precision medicine market to reach $141.70 billion by 2026, reports BIS Research. PR Newswire website. https://www.prnewswire.com/news-releases/global-precision-medicine-market-to-reach-14170-billion-by-2026-reports-bis-research-664364683.html. Accessed by Feb 2020
33. Novartis (2019). https://www.hcp.novartis.com/contentassets/4c6d6843d6cf4231804e0e7d7b865ec3/18-nvsonc-0005-poatr5_trend_report.pdf. Accessed by Feb 2020
34. Matchett KB, Lynam-Lennon N, Watson RW, Brown JAL (2017) Advances in precision medicine: tailoring individualized therapies. Cancers 9(11):146
35. Hong B, Zu Y (2013) Detecting circulating tumor cells: current challenges and new trends. Theranostics 3:377–94
36. Sheridan C (2019) Investors keep the faith in cancer liquid biopsies. Nat Biotechnol 37(9):972–974
37. New J (2019) The promise of data-driven drug development. Center for Data Innovation 1–33. http://www2.datainnovation.org/2019-data-driven-drug-development.pdf. Accessed by Feb 2020.
38. Cohen J (2018) Taking a wider view of precision oncology. https://www.forbes.com/sites/joshuacohen/2018/08/02/taking-a-wider-view-of-precision-oncology/#2dd6d94d2022 2018. Accessed by Feb 2020
39. Madhavan S, Subramaniam S, Brown TD, Chen JL (2018) Art and challenges of precision medicine: Interpreting and integrating genomic data into clinical practice. Am Soc Clin Oncol Ed Book, pp. 546–553
40. Croston GE (2017) The utility of target-based discovery. Expert Opin Drug Discov 12(5):427–429. https://tandfonline.com/doi/full/10.1080/17460441.2017.1308351. Accessed by February 2020
41. Conforte AJ, Magalhães M, Tilli TM, da Silva FAB, Carels N (2018) The challenge of translating system biology into targeted therapy of cancer. In: da Silva FAB, Carels N, Paes Silva Junior F (eds) Theoretical and applied aspects of system biology. Computational biology, 1st edn. Springer International Publishing, Cham, pp 175–194
42. Brummel A, Lustig A, Westrich K, Evans MA, Plank GS, Penso J et al (2014) Best practices: improving patient outcomes and costs in an ACO through comprehensive medication therapy management. J Manag Care Spec Pharm 20(12):1152–1158
43. Carels N, Spinassé LB, Tilli TM, Tuszynski JA (2016) Toward precision medicine of breast cancer. Theor Biol Med Model 13:7
44. Jarrett AM, Shah A, Bloom MJ, McKenna MT, Hormuth DA II, Yankeelov TE et al (2019) Experimentally-driven mathematical modeling to improve combination targeted and cytotoxic therapy for HER2 + breast cancer. Sci Rep 9:12830
45. Yankeelov TE, Quaranta V, Evans KJ, Rericha EC (2015) Toward a science of tumor forecasting for clinical oncology. Cancer Res 75(6):918–923

46. Jarrett AM, Lima EABF, Hormuth DA, McKenna MT, Feng X, Ekrut DA et al (2018) Mathematical models of tumor cell proliferation: a review of the literature. Expert Rev Anticancer Ther 18(12):1271–1286
47. Stéphanou A, McDougall SR, Anderson ARA, Chaplain MAJ (2005) Mathematical modelling of ow in 2d and 3d vascular networks: applications to antiangiogenic and chemotherapeutic drug stategies. Math Comput Model 41:1137–1156
48. Stéphanou A, Lesart AC, Deverchère J, Juhem A, Popov A, Estève F (2017) How tumour-induced vascular changes alter angiogenesis: insights from a computational model. J Theor Biol 419:211–226
49. Mirams GR, Arthurs CJ, Bernabeu MO, Bordas R, Cooper J, Corrias A et al (2013) Chaste: An open source c ++ library for computational physiology and biology. PLoS Comput Biol 9(3):e1002970
50. Pitt-Francis J, Pathmanathan P, Bernabeu MO, Bordas R, Cooper J, Fletcher AG, et al (2009) Chaste: a test-driven approach to software development for biological modelling. Computer Physics Communications. 180(12):2452–2471. 40 YEARS OF CPC: A celebratory issue focused on quality software for high performance, grid and novel computing architectures
51. Gillet JP, Gottesman MM (2010) Mechanisms of multidrug resistance in cancer. Methods Mol Biol 596:47–76
52. Yuaney G, Shah P (2018) Reinforcement learning with action-derived rewards for chemotherapy and clinical trial dosing regimen selection. In: Proceedings of the 3rd machine learning for health care conference, vol 85, pp 161–226. http://proceedings.mlr.press/v85/yauney18a.html. Accessed by Feb 2020
53. Liu Y, Gadepalli K, Norouzi M, Dahl GE, Kohlberger T, Boyko A et al (2017) Detecting cancer metastases on gigapixel pathology images. https://arxiv.org/abs/1703.02442
54. Patel J (2013) Science of the science, drug discovery and artificial neural networks. Curr Drug Discov Technol 10(1):2–7
55. Atkinson RD (2018) Drug price controls will be more pain than gain. The Hill. https://thehill.com/opinion/healthcare/416068-drug-price-controls-will-be-more-pain-than-gain. Accessed by Feb 2020; Cost to develop and win marketing approval for a new drug is $2.6 billion. Tufts Center for the Study of Drug Development. 2014; https://static1.squarespace.com/static/5a9eb0c8e2ccd1158288d8dc/t/5ac66adc758d46b001a996d6/1522952924498/pr-coststudy.pdf. Accessed by Feb 2020
56. Calzolari D, Paternostro G, Harrington PL, Piermarocchi C, Duxbury PM (2007) Selective control of the apoptosis signaling network in heterogeneous cell populations. PLoS ONE 2:e547
57. Calzolari D, Bruschi S, Coquin L, Schofield J, Feala JD, Reed JC et al (2008) Search algorithms as a framework for the optimization of drug combinations. PLoS Comput Biol 4(12):e1000249
58. Hardman JG, Limbird LE, Gilman AG (2001) Goodman & Gilman's the pharmacological basis of therapeutics. McGraw-Hill, New York
59. Pons-Salort M, van der Sanden B, Juhem A, Popov A, Stéphanou A (2012) A computational framework to assess the efficacy of cytotoxic molecules and vascular disrupting agents against solid tumours. Math Model Nat Phenom 7(1):49–77
60. Thompson MA, Godden JJ, Wham D, Ruggeri A, Mullane MP, Wilson A et al (2019) Coordinating an oncology precision medicine clinic within an integrated health system: lessons learned in year one. J Patient Cent Res Rev 6(1):36–45
61. Campillos M, Kuhn M, Gavin A, Jensen LJ, Bork P (2008) Drug target identification using side-effect similarity. Science 321(5886):263–266
62. Cheng F, Zhao Z (2014) Machine learning-based prediction of drug–drug interactions by integrating drug phenotypic, therapeutic, chemical, and genomic properties. J Am Med Inform Assoc 21(e2):e278–e286
63. Li X, Xu Y, Cui H, Huang T, Wang D, Lian B et al (2017) Prediction of synergistic anti-cancer drug combinations based on drug target network and drug induced gene expression profiles. Artif Intell Med 83:35–43

64. DeRose YS, Wang G, Lin Y-C, Bernard PS, Buys SS, Ebbert MT et al (2011) Tumor grafts derived from women with breast cancer authentically reflect tumor pathology, growth, metastasis and disease outcomes. Nat Med 17:1514–1520
65. Tentler JJ, Tan AC, Weekes CD, Jimeno A, Leong S, Pitts TM et al (2012) Patient-derived tumour xenografts as models for oncology drug development. Nat Rev Clin Oncol 9:338–350
66. Weeber F, van de Wetering M, Hoogstraat M, Dijkstra KK, Krijgsman O, Kuilman T et al (2015) Preserved genetic diversity in organoids cultured from biopsies of human colorectal cancer metastases. Proc Natl Acad Sci USA 112:13308–13311
67. Hidalgo M, Bruckheimer E, Rajeshkumar NV, Garrido-Laguna I, De Oliveira E, Rubio-Viqueira B et al (2011) A pilot clinical study of treatment guided by personalized tumorgrafts in patients with advanced cancer. Mol Cancer Ther 10:1311–1316
68. Izumchenko E, Paz K, Ciznadija D, Sloma I, Katz A, Vasquez-Dunddel D et al (2017) Patient-derived xenografts effectively capture responses to oncology therapy in a heterogeneous cohort of patients with solid tumors. Ann Oncol 28:2595–2605
69. Ledford H (2017) Cancer-genome study challenges mouse "avatars". Nat News. https://doi.org/10.1038/nature.2017.22782
70. Ben-David U, Ha G, Tseng Y-Y, Greenwald NF, Oh C, Shih J et al (2017) Patient-derived xenografts undergo mouse-specific tumor evolution. Nat Genet 49(11):1567–1575
71. Shin SH, Bode AM, Dong Z (2017) Precision medicine: the foundation of future cancer therapeutics. NPJ Precis Oncol 1(1):12
72. Graham R, Mancher M, Wolman DM, Greenfield S, Steinberg E (eds) (2011) Clinical practice guidelines we can trust. National Academies Press, Washington, DC
73. Zon RT, Frame JN, Neuss MN, Page RD, Wollins DS, Stranne S et al (2016) American Society of Clinical Oncology policy statement on clinical pathways in oncology. J Oncol Pract 12:261–266
74. Schork NJ (2015) Personalized medicine: time for one-person trials. Nature 520(7549):609–611
75. Weber JS, Levit LA, Adamson PC, Bruinooge SS, Burris HA 3rd, Carducci MA et al (2017) Reaffirming and clarifying the American Society of Clinical Oncology's policy statement on the critical role of phase I trials in cancer research and treatment. J Clin Oncol 35:139–140
76. Dienstmann R, Rodon J, Tabernero J (2015) Optimal design of trials to demonstrate the utility of genomically-guided therapy: putting precision cancer medicine to the test. Mol Oncol 9:940–950
77. Xiao G, Ma S, Minna J, Xie Y (2014) Adaptive prediction model in prospective molecular signature based clinical studies. Clin Cancer Res 20:531–539
78. Woodcock J, LaVange LM (2017) Master protocols to study multiple therapies, multiple diseases, or both. N Engl J Med 377:62–70
79. Carey LA, Winer EP (2016) I-SPY 2: toward more rapid progress in breast cancer treatment. N Engl J Med 375:83–84
80. Redig AJ, Jänne PA (2015) Basket trials and the evolution of clinical trial design in an era of genomic medicine. J Clin Oncol 33:975–977
81. Billingham L, Malottki K, Steven N (2016) Research methods to change clinical practice for patients with rare cancers. Lancet Oncol 17:e70–e80
82. Rolfo C, Caglevic C, Bretel D, Hong D, Raez LE, Cardona AF et al (2016) Cancer clinical research in Latin America: current situation and opportunities expert opinion from the first ESMO workshop on clinical trials, Lima, 2015. ESMO Open 1(4):e000055
83. McShane LM, Cavenagh MM, Lively TG, Eberhard DA, Bigbee WL, Williams PM et al (2013) Criteria for the use of omics-based predictors in clinical trials: Explanation and elaboration. BMC Med 11:220
84. Von Hoff DD, Stephenson JJ Jr, Rosen P et al (2010) Pilot study using molecular profiling of patients' tumors to find potential targets and select treatments for their refractory cancers. J Clin Oncol 28:4877–4883
85. Radovich M, Kiel PJ, Nance SM, Niland EE, Parsley ME, Ferguson ME et al (2016) Clinical benefit of a precision medicine based approach for guiding treatment of refractory cancers. Oncotarget 7:56491–56500

86. Haslem DS, Van Norman SB, Fulde G, Knighton AJ, Belnap T, Butler AM et al (2017) A retrospective analysis of precision medicine outcomes in patients with advanced cancer reveals improved progression-free survival without increased health care costs. J Oncol Pract 13:e108–e119
87. Haslem DS, Chakravarty I, Fulde G, Gilbert H, TudorBP, Lin K et al (2018) Precision oncology in advanced cancer patients improves overall survival with lower weekly healthcare costs. Oncotarget 9:12316–12322
88. Schwaederle M, Zhao M, Lee JJ, Lazar V, Leyland-Jones B, Schilsky RL et al (2016) Association of biomarker-based treatment strategies with response rates and progression-free survival in refractory malignant neoplasms: a meta-analysis. JAMA Oncol 2(11):1452–1459
89. Bellmunt J, De Wit R, Vaughn DJ, Fradet Y, Lee JL, Fong L et al (2017) Pembrolizumab as second-line therapy for advanced urothelial carcinoma. N Engl J Med 376(11):1015–1026
90. Levit LA, Kim ES, McAneny BL, Nadauld LD, Levit K, Schenkel C et al (2019) Implementing precision medicine in community-based oncology programs: three models. J Oncol Pract 15(6):325–329
91. Moscow JA, Fojo T, Schilsky RL (2018) The evidence framework for precision cancer medicine. Nat Rev Clin Oncol 15:183–192
92. Jameson JL, Longo DL (2015) Precision medicine—personalized, problematic, and promising. N Engl J Med 372:2229–2234
93. Schwartzberg L, Kim ES, Liu D, Schrag D (2017) Precision oncology: Who, how, what, when, and when not? Am Soc Clin Oncol Ed Book 37:160–169
94. Monro HC, Gaffney EA (2009) Modelling chemotherapy resistance in palliation and failed cure. J Theor Biol 257:292–302
95. Castorina P, Carco D, Guiot C, Deisboeck TS (2009) Tumor growth instability and its implications for chemotherapy. Cancer Res 69:8507–8515
96. Kapoor S, Rallabandi VP, Sakode C, Padhi R, Roy PK (2013) A patient-specific therapeutic approach for tumour cell population extinction and drug toxicity reduction using control systems-based dose-profile design. Theor Biol Med Model 10:68
97. Gardner SN (2000) Scheduling chemotherapy: catch 22 between cell kill and resistance evolution. J Theor Med 2:215–232
98. FDA (Food and Drug Administration) (2016) FDA advances precision medicine initiative by issuing draft guidances on next generation sequencing-based tests. https://www.fda.gov/newsevents/newsroom/pressannouncements/ucm509814.htm. Accessed by Feb 2020
99. FDA (Food and Drug Administration) (2020b) 21st century cures act. https://www.fda.gov/regulatory-information/selected-amendments-fdc-act/21st-century-cures-act. Accessed by Feb 2020
100. Howard DH, Bach PB, Berndt ER, Conti RM (2015) Pricing in the market for anticancer drugs. J Econ Perspect 29:139–162
101. Sultana J, Cutroneo P, Trifiro G (2013) Clinical and economic burden of adverse drug reactions. J Pharmacol Pharmacother 4:S73–S77
102. Katz G, Romano O, Foa C, Vataire AL, Chantelard JV, Hervé R et al (2015) Economic impact of gene expression profiling of early stage breast cancer patients in France. PLoS ONE 10(6):e0128880
103. Jakka S, Rossbach M (2013) An economic perspective on personalized medicine. HUGO J 7:1
104. ECML (Experts in Chronic Myeloid Leukemia) (2013) The price of drugs for chronic myeloid leukemia (CML) is a reflection of the unsustainable prices of cancer drugs: from the perspective of a large group of CML experts. Blood 121:4439–4442
105. Morris ZS, Wooding S, Grant J (2011) The answer is 17 years, what is the question? Understanding time lags in translational research. J R Soc Med 104(12):510–520
106. Carels N, Tilli TM, Tuszynki JA (2015) Optimization of combination chemotherapy based on the calculation of network entropy for protein-protein interactions in breast cancer cell lines. EPJ Nonlinear Biomed Phys 3:6

107. Albert R, Jeong H, Barabási A-L (2000) Error and attack tolerance of complex networks. Nature 406(6794):378–382
108. Breitkreutz D, Hlatky L, Rietman E, Tuszynski JA (2012) Molecular signaling network complexity is correlated with cancer patient survivability. Proc Natl Acad Sci USA 109(23):9209–9212
109. Tilli TM, Carels N, Tuszynski JA, Pasdar M (2016) Validation of a network-based strategy for the optimization of combinatorial target selection in breast cancer therapy: siRNA knock-down of network targets in MDA-MB-231 cells as an in vitro model for inhibition of tumor development. Oncotarget 7(39):63189–61203
110. Conforte AJ, Tuszynski JA, da Silva FAB, Carels N (2019) Signaling complexity measured by shannon entropy and its application in personalized medicine. Front Genet 10:1–14
111. Peng Q, Schork N (2014) Utility of network integrity methods in therapeutic target identification. Front Genet 5:12
112. Schramm G, Kannabiran N, König R (2010) Regulation patterns in signaling networks of cancer. BMC Syst Biol 4:162
113. Winterbach W, Mieghem PV, Reinders M, Wang H, de Ridder D (2013) Topology of molecular interaction networks. BMC Syst Biol 7:90
114. Freeman LCA (1977) Set of measures of centrality based on betweenness. Sociometry 40:35–41
115. Frainay C, Jourdan F (2017) Computational methods to identify metabolic sub-networks based on metabolomic profiles. Brief Bioinform 18:43–56
116. Watts DJ, Strogatz SH (1998) Collective dynamics of 'small-world' networks. Nature 393:440–442
117. Teschendorff AE, Banerji CRS, Severini S, Kuehn R, Sollich P (2015) Increased signaling entropy in cancer requires the scale-free property of proteininteraction networks. Sci Rep 5:1–9
118. West HJ (2016) Can we define and reach precise goals for precision medicine in cancer care? J Clin Oncol 34:3595–3596
119. Banerji CRS, Severini S, Caldas C, Teschendorff AE (2015) Intra-tumour signalling entropy determines clinical outcome in breast and lung cancer. PLoS Comput Biol 11:e1004115
120. Huang S, Ernberg I, Kauffman S (2009) Cancer attractors: a systems view of tumors from a gene network dynamics and developmental perspective. Semin Cell Dev Biol 20:869–876
121. Cornelius SP, Kath WL, Motter AE (2013) Realistic control of network dynamics. Nat Commun 4:1942
122. Bora RS, Gupta D, Mukkur TKS, Saini KS (2012) RNA interference therapeutics for cancer: challenges and opportunities (review). Mol Med Rep 6:9–15
123. Ehrke-Schulz E, Schiwon M, Hagedorn C, Ehrhardt A (2017) Establishment of the CRISPR/Cas9 system for targeted gene disruption and gene tagging. Methods Mol Biol 1654:165–176
124. Crespo I, del Sol A (2013) A general strategy for cellular reprogramming: The importance of transcription factor cross-repression. Stem Cells 31:2127–2135
125. Sgariglia D, Conforte AJ, de Carvalho VLA, Carels N, da Silva FAB (2018) Cellular reprogramming. In: da Silva FAB, Carels N, Paes Silva Junior F (eds) Theoretical and applied aspects of system biology. Computational biology, 1st edn. Springer International Publishing, pp 41–55
126. Moes M, Le Béchec A, Crespo I, Laurini C, Halavatyi A, Vetter G et al (2012) A novel network integrating a miRNA-203/SNAI1 feedback loop which regulates epithelial to mesenchymal transition. PLoS ONE 7(4):e35440
127. Li C, Wang J (2015) Quantifying the landscape for development and cancer from a core cancer stem cell circuit. Cancer Res 75:2607–2618
128. Ao P, Galas D, Hood L, Zhu X (2008) Cancer as robust intrinsic state of endogenous molecular-cellular network shaped by evolution. Med Hypotheses 70:678–684
129. Yuan R, Zhang S, Yu J, Huang Y, Lu D, Cheng R et al (2017) Beyond cancer genes: colorectal cancer as robust intrinsic states formed by molecular interactions. Open Biol 7(11)

130. Yuan R, Zhu X, Wang G, Li S, Ao P (2017) Cancer as robust intrinsic state shaped by evolution: a key issues review. Rep Prog Phys 80:042701
131. Hopfield JJ (1982) Neural networks and physical systems with emergent collective computational abilities. Proc Natl Acad Sci USA 79:2554–2558
132. Maetschke SR, Ragan MA (2014) Characterizing cancer subtypes as attractors of Hopfield networks. Bioinformatics 30:1273–1279
133. Taherian Fard A, Ragan MA (2017) Modeling the attractor landscape of disease progression: a network-based approach. Front Genet 8:48
134. Conforte AJ, Alves LD, Coelho FC, Carels N, da Silva FAB (2020) Modeling basins of attraction for breast cancer using Hopfield networks. Front Genet 11:314. https://doi.org/10.3389/fgene.2020.00314
135. Szedlak A, Paternostro G, Piermarocchi C (2014) Control of asymmetric hopfield networks and application to cancer attractors. PLoS ONE 9:e105842
136. Cantini L, Caselle M (2019) Hope4Genes: a Hopfield-like class prediction algorithm for transcriptomic data. Sci Rep 9:1–9
137. Guo J, Zheng J (2017) HopLand: single-cell pseudotime recovery using continuous Hopfield network-based modeling of Waddington's epigenetic landscape. Bioinformatics 33(14):i102–i109
138. Kinghorn AD, Pan L, Fletcher JN, Chai H (2011) The relevance of higher plants in lead compound discovery programs. J Nat Prod 74(6):1539–1555
139. Bernardini S, Tiezzi A, Laghezza Masci V, Ovidi E (2018) Natural products for human health: An historical overview of the drug discovery approaches. Nat Prod Res 32(16):1926–1950
140. Cragg GM, Newman DJ (2013) Natural products: a continuing source of novel drug leads. Biochem Biophys Acta 1830(6):3670–3695
141. Breinbauer Rolf, Vetter Ingrid R, Waldmann Herbert (2002) From protein domains to drug candidates: natural products as guiding principles in the design and synthesis of compound libraries. Angew Chem Int Ed Engl 41(16):2878–2890
142. Bauer Armin, Brönstrup Mark (2014) Industrial natural product chemistry for drug discovery and development. Nat Prod Rep 31(1):35–60
143. Newman DJ (2008) Natural products as leads to potential drugs: An old process or the new Hope for drug discovery? J Med Chem 51(9):2589–2599
144. Corson Timothy W, Crews CM (2007) Molecular understanding and modern application of traditional medicines: triumphs and trials. Cell 130(5):769–774
145. David B, Wolfender J-L, Dias DA (2015) The pharmaceutical industry and natural products: historical status and new trends. Phytochem Rev 14(2):299–315
146. Kholod Y, Hoag E, Muratore K, Kosenkov D (2018) Computer-aided drug discovery: molecular docking of diminazene ligands to DNA minor groove. J Chem Educ 95(5):882–887
147. Prato G, Silvent S, Saka S, Lamberto M, Kosenkov D (2015) Thermodynamics of binding of di- and tetrasubstituted naphthalene diimide ligands to DNA G-quadruplex. J Phys Chem B 119(8):3335–3347
148. Bajorath J (2015) Computer-aided drug discovery. Research 4:630
149. Chaudhary KK, Mishra N (2016) A review on molecular docking: novel tool for drug discovery. JSM Chem 3(4):1029
150. de Ruyck J, Guillaume B, Ralf B, Lensink MF (2016) Molecular docking as a popular tool in drug design, an in silico travel. Adv Appl Bioinform Chem 9:1–11
151. Pârvu L (2003) QSAR: a piece of drug design. J Cell Mol Med 7(3):333–335
152. Rao VS, Srinivas K (2011) Modern drug discovery process: An *in silico* approach. J Bioinform Seq Anal 2(5):89–94
153. Leelananda SP, Lindert S (2016) Computational methods in drug discovery. Beilstein J Org Chem 12(1):2694–2718
154. Buckingham J, Glen RC, Hill AP, Hyde RM, Martin GR, Robertson AD et al (1995) Computer-aided design and synthesis of 5-substituted tryptamines and their pharmacology at the 5-HT1D receptor: discovery of compounds with potential anti-migraine properties. J Med Chem 38(18):3566–3580

155. Koga H, Itoh A, Murayama S, Suzue S, Irikura T (1980) Structure-activity relationships of antibacterial 6,7- and 7,8-disubstituted 1-alkyl-1,4-dihydro-4-oxoquinoline-3-carboxylic acids. J Med Chem 23(12):1358–1363
156. Klopmand G (1992) Concepts and applications of molecular similarity, by Mark A. Johnson and Gerald M. Maggiora, Eds., John Wiley & Sons, New York, 1990, pp. 393. J Comp Chem 13(4):539–540
157. Verma RP, Hansch C (2009) Camptothecins: a SAR/QSAR study. Chem Rev 109(1):213–235
158. Lin S-K (2000) Pharmacophore perception, development and use in drug design. Edited by Osman F. Güner. Molecules 5(7):987–989
159. Katsila T, Spyroulias GA, Patrinos GP, Matsoukas M-T (2016) Computational approaches in target identification and drug discovery. Comput Struct Biotech J 14:177–184
160. Yuriev E, Agostino M, Ramsland PA (2011) Challenges and advances in computational docking: 2009 in Review. J Mol Recognit 24(2):149–164
161. Chemi G, Brogi S (2017) Breakthroughs in computational approaches for drug discovery. J Drug Res Dev 3(1):2470
162. Sousa SF, Fernandes PA, Ramos MJ (2006) Protein-ligand docking: current status and future challenges. Proteins 65(1):15–26
163. Durrant JD, McCammon JA (2011) Molecular dynamics simulations and drug discovery. BMC Biol 9:71
164. Mackerell AD Jr (2004) Empirical force fields for biological macromolecules: overview and issues. J Comput Chem 25(13):1584–1604
165. Ganesan A, Coote ML, Barakat K (2017) Molecular dynamics-driven drug discovery: leaping forward with confidence. Drug discovery Today 22(2):249–269
166. De Vivo M (2011) Bridging quantum mechanics and structure-based drug design. Optimization 7:8
167. Abagyan R, Totrov M (2001) High-throughput docking for lead generation. Curr Opin Chem Biol 5(4):375–382
168. Borhani DW, Shaw DE (2012) The future of molecular dynamics simulations in drug discovery. J Comput Aided Mol Des 26(1):15–26
169. Fischer M, Coleman RG, Fraser JS, Shoichet BK (2014) Incorporation of protein flexibility and conformational energy penalties in docking screens to improve ligand discovery. Nat Chem 6(7):575–583
170. Ivetac A, McCammon JA (2011) Molecular recognition in the case of flexible targets. Curr Pharm Des 17(17):1663–1671
171. Tarcsay Á, Paragi G, Vass M, Jójárt B, Bogár F, Keserű GM (2013) The impact of molecular dynamics sampling on the performance of virtual screening against GPCRs. J Chem Inf Model 53(11):2990–2999
172. Tian S, Sun H, Pan P, Li D, Zhen X, Li Y et al (2014) Assessing an Ensemble docking-based virtual screening strategy for kinase targets by considering protein flexibility. J Chem Inf Model 54(10):2664–2679
173. Buch I, Giorgino T, De Fabritiis G (2011) Complete reconstruction of an enzyme-inhibitor binding process by molecular dynamics simulations. Proc Nat Acad Sci USA 108(25):10184–10189
174. Shan Y, Kim ET, Eastwood MP, Dror RO, Seeliger MA, Shaw DE (2011) How does a drug molecule find its target binding site? J Am Chem Soc 133(24): 9181–9183
175. Torrie GM, Valleau JP (1977) Nonphysical sampling distributions in Monte Carlo free-energy estimation: umbrella sampling. J Comput Phys 23(2):187–199
176. Sugita Y, Okamoto Y (1999) Replica-exchange molecular dynamics method for protein folding. Chem Phys Lett 314(1–2):141–151
177. Laio A, Parrinello M (2002) Escaping free-energy minima. Proc Natl Acad Sci USA 99(20):12562–12566
178. De Vivo M, Masetti M, Bottegoni G, Cavalli A (2016) Role of molecular dynamics and related methods in drug discovery. J Med Chem 59(9):4035–4061

179. Korzekwa KR, Jones JP, Gillette JR (1990) Theoretical studies on cytochrome P-450 mediated hydroxylation: A predictive model for hydrogen atom abstractions. J Am Chem Soc 112(19):7042–7046
180. Warshel A, Levitt M (1976) Theoretical studies of enzymic reactions: dielectric, electrostatic and steric stabilization of the carbonium ion in the reaction of lysozyme. J Mol Biol 103:227–249
181. Shaik S, Cohen S, Wang Y, Chen H, Kumar D, Thiel W (2010) P450 enzymes: Their structure, reactivity, and selectivity modeled by QM/MM calculations. Chem Rev 110(2):949–1017
182. Kirchmair J, Göller AH, Lang D, Kunze J, Testa B, Wilson ID et al (2015) Predicting drug metabolism: Experiment and/or computation? Nature Rev Drug discov. 14(6):387–404

Chapter 9
Opportunities and Challenges Provided by Boolean Modelling of Cancer Signalling Pathways

Petronela Buiga and Jean-Marc Schwartz

Abstract Cancer is a multifactorial disease, which involves intricate signalling pathways leading to different cellular processes such as survival, proliferation, differentiation or apoptosis. This diversity significantly complicates cancer investigation and treatment. There have been studies on investigating signalling in cancer, but it is difficult to rely solely on molecular biology approaches in order to predict the properties of cancer pathways. Therefore, mathematical modelling has been increasingly used to understand tumour progression, direct the design of new experiments by introducing likely candidates for clinical investigation and present novel approaches to cancer therapy. Different types of modelling techniques, including static and dynamic methods, have been used to model processes that occur in cancer. Boolean models have attracted a large interest in cancer research due to their efficiency and scalability. We here review the use of Boolean modelling for cancer, highlighting the facts that this technique is efficient at generating predictions and that the gain brought by their scalability and flexibility often offsets the lack of quantitative values upon which they are based.

Keywords Boolean model · Cancer · Systems biology · Signalling pathway

9.1 Background

Cancer represents a major global public health issue and the second cause of death in humans after cardiovascular disease [1]. Being a multifactorial disease, understanding the cellular response to specific signals and control of survival, differentiation, proliferation or apoptosis in different cell conditions is essential to efficient therapy [2]. There have been many studies to explore the mechanisms and pathways related to cancer. These studies generally require long and expensive laboratory

---

P. Buiga
School of Dental Medicine, Harvard University, Boston, MA, USA

P. Buiga · J.-M. Schwartz (✉)
School of Biological Sciences, University of Manchester, Manchester, UK
e-mail: jean-marc.schwartz@manchester.ac.uk

F. A. B. da Silva et al. (eds.), *Networks in Systems Biology*, Computational Biology 32,
https://doi.org/10.1007/978-3-030-51862-2_9

experiments, which require extensive research facilities [3]. Cancer has the potential to develop in different tissues affecting nearly all organs, including breast, liver, prostate, colon and blood (leukaemia). Findings arising from one type of cancer cannot be readily applied to another [4]. This diversity significantly complicates the investigation of cancer, as well as the application of cancer treatments. Nevertheless, there are many common cellular pathways involved in human cancer. These pathways govern the cellular response to cancer and to its treatment, for example, proliferation pathways (MAPK) and survival pathways (PI3K/AKT) [5].

Malignant growth of cancer cells is considered to be a result of six main alterations in cell physiology (Table 9.1): evading programmed death, evading anti-growth signals, sustained growth signalling, limitless replicative potential, sustained angiogenesis, tissue invasion and metastasis [6].

Because of the complexity of biological systems, it is difficult to rely solely on experimental approaches in order to predict the behaviour of such systems. Therefore, mathematical modelling has been increasingly used to understand tumour progression, direct the design of new experiments by introducing likely candidates for further clinical investigation and present novel approaches to cancer therapy. Such complex biological processes across a variety of spatial and temporal scales can be predicted by models, which integrate experimental data into a consistent representation thus enabling simulation and prediction. Computational models have many advantages such as being cost effective, reproducible, applicable on different cancer types and time efficient [7].

Various types of modelling techniques, including static and dynamic methods (Fig. 9.1) [7], have been used to model processes that occur in cancer. Static network analysis relies on the topology of interaction networks in order to assess some of their properties. The global topological properties facilitate understanding of the system behaviour as a whole, for example, in the case of random errors such as gene mutation, and targeted treatment [8]. Local topological properties point towards important individual elements in the network like nodes, edges motifs and modules [9, 10].

On the other hand, dynamic modelling is designed to illustrate time-dependent properties and predict dynamic changes in the components of the system. Most of these approaches use ordinary differential equations (ODE) [11]. However, a major drawback of ODE-based models is that kinetic equations and parameters are often

Table 9.1 Hallmarks of cancer (adapted from [6])

Hallmarks of cancer
1. Evading programmed cell death
2. Evading anti-growth signals
3. Sustained growth signalling
4. Limitless replicative potential
5. Sustained angiogenesis
6. Tissue invasion and metastasis

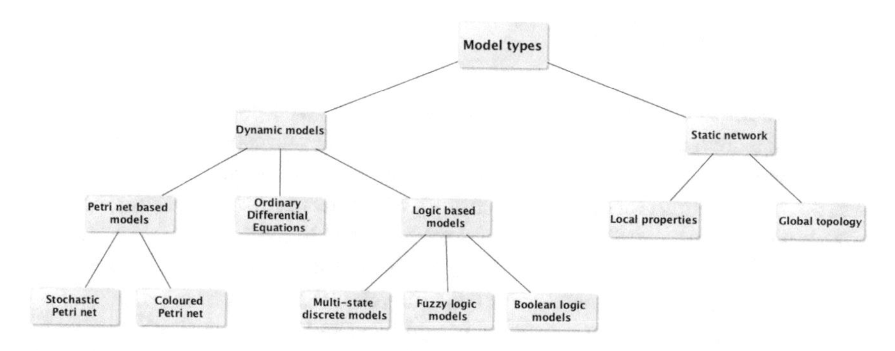

Fig. 9.1 Relations between different methods used for cancer modelling (adapted from [7])

difficult and expensive to derive experimentally and are highly dependent on specific cellular and experimental conditions. Therefore alternative methods of dynamic modelling have attracted much interest, including Petri-based models (Signalling Petri net SPN, Coloured Petri net CPN) and logic-based models [12]. The logic-based models are themselves represented by fuzzy, multistate and Boolean models; by far, the most commonly used type is Boolean modelling.

In a Boolean model, nodes (genes/proteins) are represented by two states ON or OFF, and edges (interactions) can be of two types, either activation or inhibition. A logical function determines the response of each node in response to incoming interactions, which can include combinations of "AND" and "OR" relations [13]. Boolean models can be applied to systems of hundreds of components and large-scale biological systems; one of the largest Boolean models of the p53 interactome is composed of over 200 components and 700 interactions [14]. These authors provide a qualitative dynamic description of the system's behaviour, which can be used for simulation and prediction in the same way as other types of models, including predicting effects of deletions, loss of function mutations, changes in input signals and treatments [15]. Boolean models have been used in several research areas, such as immune response [16], leukaemia [17], respiratory systems [18] and HIV infection [19].

The scope of this review is to present the use of Boolean models in cancer research. Boolean models are particularly efficient at providing insights into the dynamics of complex signalling pathways and gene regulatory networks, which are some of the most important factors of cancer. We highlight the fact that Boolean models are efficient at generating predictions and that the gain brought by their scalability and flexibility often offsets the lack of quantitative values in the data upon which they are based.

9.2 Methodology

The development of a Boolean model starts by identifying the network of genes or proteins that are required to understand the system under investigation, and how they interact with each other. Two complementary approaches can be followed (Fig. 9.2): first, direct modelling (forwards engineering), where the network is constructed based on knowledge from scientific literature, databases and biological hypotheses; second, network inference (reverse engineering), where the main source of knowledge is experimental data, and the network is constructed so that it can fit the data.

A Boolean network consists of a set of variables and functions. The variables are represented only by two values (0 and 1), which correspond to the logic values FALSE and TRUE. In a biological system, the variable entity is binary referring to OFF (inactive) and ON (active). The Boolean function is defined as a function on the set of variables, for instance, a function with k variables is a mapping B: $\{0, 1\}^k \rightarrow \{0, 1\}$ from the set of all k variables over $\{0, 1\}$ to a binary output. The Boolean output is determined by this function, on the basis of certain logical operations obtained from k binary inputs [20]. The functions AND, OR and NOT are the only logical operations used [15]. Figure 9.3 shows a simple Boolean interaction network and its representative Boolean functions and truth tables.

The Boolean functions are applied in an iterative way and the succession of states gives a dynamic time course for the system. There are different ways to update the Boolean function in the course of a simulation: synchronous and asynchronous. In synchronous update, the function is applied to all nodes simultaneously, while in asynchronous update it is applied to one node at a time. The update sequence may be random or determined according to a fixed order. For that reason, synchronous update is a deterministic method, while asynchronous methods can be determined or stochastic [21]. Synchronous simulations necessarily lead to a limit state, which is either a steady state or a cycle, while asynchronous simulations may not reach any limit state.

Several software programs have been developed to enable manipulation and simulation of Boolean models (Table 9.2). Most of them provide functionalities such as

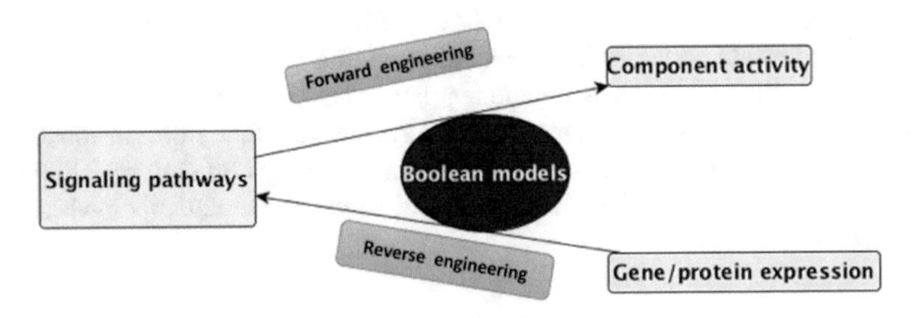

Fig. 9.2 Use of Boolean models in forwards and reverse engineering of signalling networks (adapted from [15])

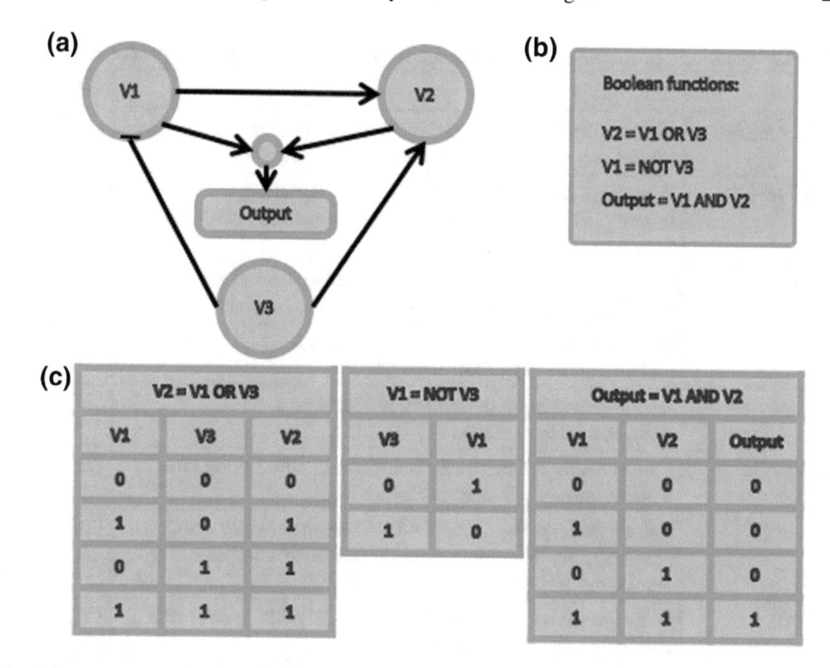

V2 = V1 OR V3		
V1	V3	V2
0	0	0
1	0	1
0	1	1
1	1	1

V1 = NOT V3	
V3	V1
0	1
1	0

Output = V1 AND V2		
V1	V2	Output
0	0	0
1	0	0
0	1	0
1	1	1

Fig. 9.3 Boolean interaction network. **a** Directed graph with its associated Boolean model. The edges with sharp arrows denote activating effect, while blunt arrows represent inhibiting effect. **b** Boolean functions in the model. **c** Truth tables of Boolean functions

dynamic simulation using synchronous or asynchronous update and detection of limit states. Many of them are open source and can be run on different platforms.

9.3 Application of Boolean Modelling to Oncogenic Pathways

Boolean models are useful to gain insight of the dynamics and end state of signalling pathways. This is particularly useful for cancer to understand the conditions that lead to survival and proliferation. Some breast cancers are characterised by overexpression of HER2 protein or amplification of *HER2* gene [31]. We illustrate how using a Boolean model we can predict situations in which HER2 overexpression leads to Survival and how this process can be reversed.

In the models shown in Fig. 9.4, we represented the components of HER2 signalling.

HER2 is known to activate the major MAP kinases (ERK, p38 and JNK). Dual specificity phosphatases (DUSPs) are part of the larger tyrosine phosphatase superfamily, representing key players in inactivating some MAP kinases by dephosphorylation [32]. This is represented by inhibition in the models and at the same time

Table 9.2 Examples of software used for Boolean modelling with their main properties

Software	Description	References
CellNetAnalyzer	MATLAB tool to perform structural and qualitative analysis of mass-flow- and signal-flow-based cellular networks	[5, 22]
CellNOptR	R package for building logic models of signalling networks by training networks derived from prior knowledge; cytoscape interface available	[23]
BooleanNet	Python package for simulation of Boolean networks, integrating synchronous and asynchronous methods	[24]
BoolNet	R package integrating methods for synchronous, asynchronous and probabilistic Boolean networks	[25]
GINsim	Java application for the construction and analysis of multi-valued logical models	[26]
MaBoSS	C++ software for stochastic Boolean modelling using kinetic Monte Carlo or Gillespie algorithm	[27]
PyBoolNet	Python package for manipulating Boolean networks, model checking, standard graph algorithms, visualisation and attractor detection	[28]
SQUAD	Software for the dynamic simulation of signalling networks and listing of steady states	[29]
BooleSim	In-browser tool for simulation and manipulation of Boolean networks	[30]

DUSPs are activated by these kinases. Some DUSPs are reported to be overexpressed in HER2 positive breast cancer, therefore, they are thought to be involved in survival and proliferation of these cancer cells [33].

The basal level of ERK, JNK and p38 kinases leads to proliferation in normal cells, while increased activity of both JNK and p38 leads to apoptosis and represses proliferation [34–36]. This is represented in the model by an AND function. One of the common treatments used against HER2 positive breast cancer is Herceptin (Trastuzumab), which targets HER2 protein by attaching to the 4th extracellular domain, represented in the model by inhibition.

In the condition of untreated HER2 positive breast cancer, we start the simulation by having HER2 ON, Herceptin OFF, JNK and p38 OFF, ERK ON and DUSP1 ON. The model correctly predicts that Survival remains ON in this condition, representing survival and proliferation of cancer cells (Fig. 9.4a).

In Fig. 9.4b, the same system is simulated with Herceptin ON. We can see that the model reaches a steady state where Survival is still predicted to remain ON, which indicates that the overexpression of DUSP1 can maintain survival and proliferation.

Another example is shown in Fig. 9.4c where DUSP4 is represented instead of DUSP1. In this case, Survival is still ON, but other components of the network are seen to oscillate between ON/OFF states. In this case, the system reached a limit cycle instead of a steady state.

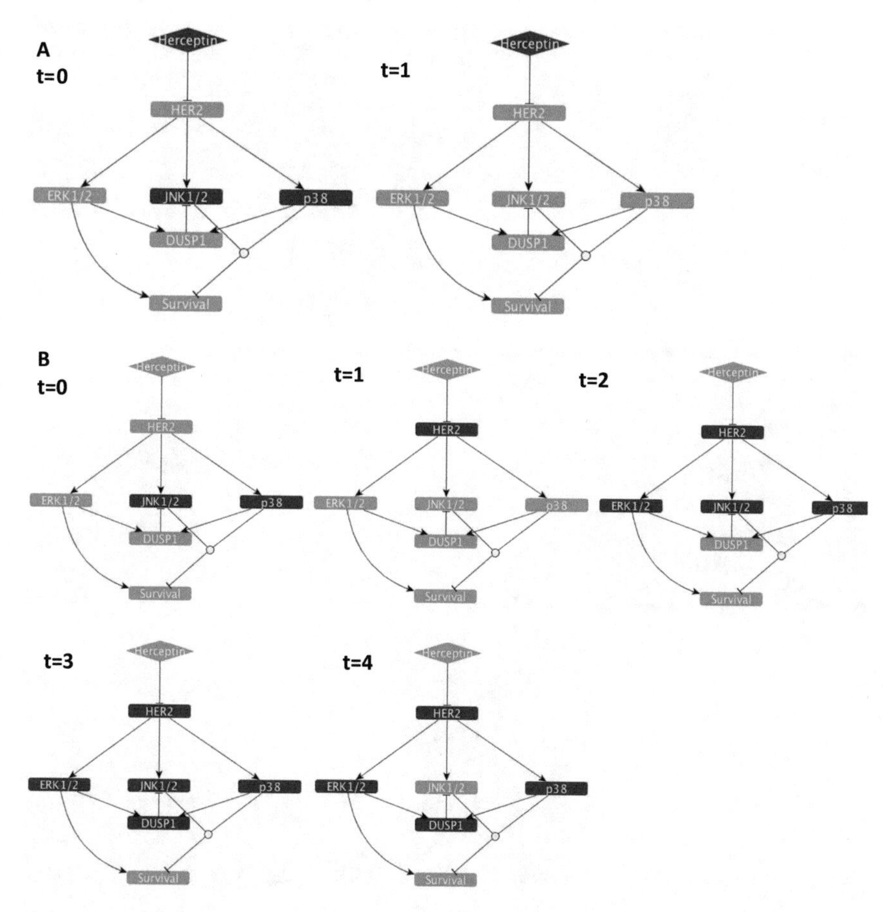

Fig. 9.4 Boolean simulations of HER2 signalling in breast cancer. **a** System simulations in the absence of Herceptin. **b** System simulations reaching a steady state in the presence of Herceptin and DUSP1 overexpression. **c** System simulations reaching a limit cycle in the presence of Herceptin and DUSP4 overexpression. **d** System simulations reaching a steady state in the presence of Herceptin and DUSP16 overexpression, where Survival is turned OFF

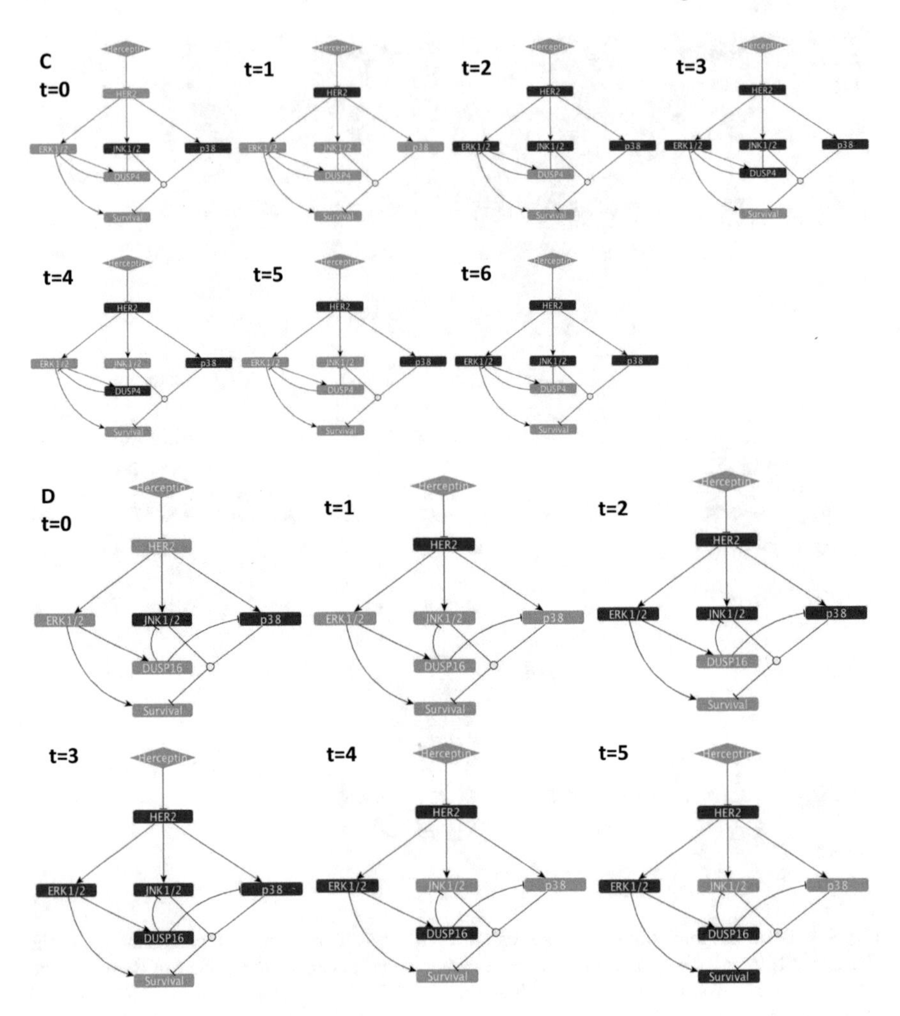

Fig. 9.4 (continued)

By testing all the known DUSPs we found that DUSP16 plays a potential role in reversing Survival. In this case, the model reaches a different steady state where Survival is turned OFF, which means that proliferation can be potentially reversed when DUSP16 is active (Fig. 9.4d).

This shows that Boolean models are an efficient method to test the system-level effects of a large number of conditions, which can then be tested experimentally and potentially lead to therapeutically relevant strategies.

9.4 Boolean Dynamics in Cancer Signalling

Boolean models have been developed and applied to several cancer types. Since the processes involved in cancer are often cell type specific, models need to take their unique characteristics into account. Therefore, cancer models are generally developed for a precise type of disease. In the literature, there are several examples of well-curated Boolean models for leukaemia, prostate cancer, colon and breast cancer, which are in good agreement with experimental data. In the following sections, we describe in detail these models and how they were used to gain new knowledge.

9.4.1 Leukaemia

Leukaemia has been one of the most widely studied types of cancer using Boolean models. Zhang et al. [37] developed one of the first Boolean models in cancer. They incorporated the signalling pathways involved in survival of T-cell large granular lymphocytes and obtained several predictions identifying key mediators of cytotoxic T lymphocyte survival in leukaemia, which were validated experimentally. Enciso and co-workers [38] used Boolean modelling to study the pro-inflammatory tumour microenvironment in acute lymphoblastic leukaemia to predict the breakdown of hematopoietic–mesenchymal communication networks. They developed a Boolean model to investigate the biological consequences of micro-environmental perturbation on crucial communication networks between stem and progenitor cells. The simulations performed by the authors on the effects of constitutive expression of NF-kB, support the hypothesis that the pro-inflammatory microenvironment perturbs CXCR4/CXCL12 communication. The model studied was built using GINsim, while its dynamic behaviour was analysed with BoolNet by applying both synchronous and asynchronous update schemes. The authors also applied knock-out and knock-in simulation techniques. This model is believed to be the first system wide study explaining the pathways involved in progression of haematological hyperproliferative diseases.

In another study, published by Alvarez-Silva and co-workers, Boolean modelling was used to investigate the regulatory mechanisms of chronic lymphocytic leukaemia and to analyse its regulatory genes [39]. The authors constructed a simplified Protein–Protein Interaction (PPI) network of the Immunoglobulin Heavy Chain Variable (IGVH) gene and its mutational status by analysing its structure and the related critical genes. In this study, the researchers developed a Boolean model based on the protein interaction network, using BoolNet. Their simulations allowed for the identification of several disease regulatory genes, for example, PTEN, EGR1, TNF, TGFBR3 and IFGR2. Through Boolean modelling the team had insights towards the understanding of different prognosis status of leukaemia.

Saadatpour and co-workers analysed the survival of T-cells in T-cell Large Granular Lymphocytic (T-LGL) leukaemia, for understanding the unknown mechanism leading to this disease [21]. The authors proposed components related to the programmed cell death that might lead to development of therapeutic targets. For this, a Boolean model was constructed using BooleanNet and considered all possible initial states to analyse the long-term behaviour of the disease. Asynchronous updating scheme and network reduction methods were applied in order to predict the status of individual network components. Also, node perturbation analysis was used to find the magnitude of each component's contribution to programmed cell death. The results suggested 19 key components in the network, intervention in any of which would lead to correct the abnormal signalling of the network. Hence, the study proposed these components as potential therapeutic targets.

9.4.2 Colon Cancer

Integrating considerable amount of database and literature data, Tian and co-workers constructed a large-scale logical model of the p53 interactome, which identified pathways that contribute to tumour growth and altered pathways in knock-outs leading to chemotherapy resistance [14]. The prediction capacity of in silico knock-outs and steady state model analysis was assessed in large-scale validation using gene expression and literature data. The authors found an upregulation of several pathways (Chk1, ATM and ATR) in p53 negative cells. Moreover, growth factors and their receptors (e.g. FGF2, IGF1R) were determined to be controlling factors of U2OS osteosarcoma and HCT116 colon cancer cell growth.

In a study published by Cho et al., the mechanism of colorectal tumour development was studied by modelling the human signalling pathways in a large-scale network [40]. The authors included relevant interactions leading to proliferation, apoptosis and metastasis, based on previous literature and databases. A simulation was performed to restore the normal phenotype from the cancerous state. As a result, they were able to identify a group of specific components in the network, such as beta-catenin, MEK and protein phosphatase 2A (PP2A), that are able to affect the steady state of the system and to shift it towards the desired normal proliferative state.

Chowdhury and co-workers explored new potential drug targets for three different types of cancer, namely, colon, glioma and pancreatic, through Boolean modelling of the human Hedgehog pathway [41]. The authors presented a master model based on literature and databases for the pathway. The team followed systematic perturbation analysis for potential drug target identification, considering either single protein targets or their combinations. The presented model was simulated using both the synchronous and asynchronous update schemes in *CellNetAnalyzer*. Their findings suggested an innovative approach towards the identification of new biomarkers and therapeutic marker identification in different types of cancers.

9.4.3 Prostate Cancer

Arshad & Datta integrated key signalling pathways involved in the development and progression of prostate cancer into a Boolean model to test the efficiency of different drug combinations in decreasing cancer growth [42]. The regulatory network consisted of the integral components of the main pathways, Androgen Receptor (AR), Mitogen-Activated Protein Kinase (MAPK) and PI3K/AKT/mTOR pathways, which are involved in the development and progression of prostate cancer. These pathways were modelled as a combinatorial network, and the resulting Boolean model was then used to propose therapies aimed to oppose the effect of each mutation and to predict potential combinatorial therapies.

9.4.4 Breast Cancer

Sahin and co-workers employed Boolean logic to define potential strategies to treat breast cancer patients that show acquired resistance to trastuzumab (Herceptin), using the GINsim software [43]. Their model included two major ErbB-dependent signalling pathways and regulation of the G1/S transition in the cell cycle. The resulting interaction network was used to identify new therapeutic targets. The authors found that targeting ERBB receptors and key signalling intermediates in different combinations did not improve treatment of acquired trastuzumab resistance in cells. Instead, c-MYC was selected as a new potential target.

Gómez Tejeda Zañudo et al. [44] presented a discrete model of apoptosis and proliferation in estrogen-receptor-positive breast cancer with the objective of determining resistance mechanisms to targeted drugs and effective drug combinations. The model was used to reproduce the response to PI3K inhibitors and describe their resistance mechanisms. Several proteins were identified that can potentially be targeted to mitigate resistance and increase cancer cell apoptosis.

Another study by Khan et al. [45] represented the regulatory network controlled by the E2F family of transcription factors, which is involved in breast and bladder cancers. They simulated scenarios in which input signals lead to a highly invasive phenotype, then determined the best combinations of perturbations that can lead to reduction of this phenotype. The predictions were validated in cell line models.

We built a series of Boolean models to study the initial response of HER2-positive breast cancer cells to Herceptin [46]. In these models, we studied the contribution of dual specificity phosphatases (DUSPs) in cell survival signalling pathways. We were able to explain the observed dynamics in the expression of several DUSPs involved in regulation of MAPK signalling and to predict new regulatory mechanisms for other DUSPs.

9.4.5 Other Cancer Types

Hetmanski et al. built a Boolean model using existing Epidermal Growth Factor signalling pathway and literature data, to understand the signalling network that controls Rab-Coupling Protein (RCP) driven invasive migration [47]. This model agreed with previous experimental data and further identified a negative feedback loop, which was unpredicted before. This loop includes MAPK-dependent control of the Rac1 activating Sos1-Eps8-Abi1 complex, which determines the RCP/integrin $\alpha 5\beta 1$ dependent Rac to RhoA switch. In this way, a targetable aspect of RhoGTPase activity control was reported, which could be of therapeutic benefit as an anti-invasive approach in mutant p53 expressing cancers. Expression of gain-of-function mutant p53 determines the cooperation of RCP with $\alpha 5\beta 1$ integrin. Cell types used for this study were ovarian and non-small cell lung cancer cell lines.

In another study [48], the authors used a cell-based model of bortezomib action in relation to the survival pathway proteins (NF-kB), antiapoptotic protein BclxL, and an apoptotic marker to simulate cell proliferation upon exposure to bortezomib in U266 human myeloma cells. In this study, a Boolean model was constructed using *CellNetAnalyzer*, which was then reduced using a model reduction algorithm to identify specific proteins that play a role in the signal transduction upon treatment with bortezomib. Their results showed agreement between the predicted simulation and their experimental data. The model provides a platform to study combinations of bortezomib with antimyeloma agents. It can be extended to predict responses in other cancer types with similar pathways.

In order to understand pathways that contribute to cell fate decision in urinary bladder cancer, Grieco et al. built a logical model for the MAPK signalling network, based on literature, using the GINsim software [49]. They introduced new algorithms for the compression of state transition graphs and for model reduction, to facilitate working with a large number of states. Specific roles of components, cross-talks and regulatory feedbacks in cell fate decision were determined by in silico simulations. Furthermore, they discovered that both proliferative and anti-proliferative mechanisms can be connected with bladder cancer deregulations, such as Epidermal Growth Factor Receptor (EGFR) overexpression and Fibroblast Growth Factor Receptor 3 (FGFR3) activating mutations.

Samaga et al. built a large-scale Boolean model and one of the largest existing mathematical models of EGFR/ErbB signalling using CellNetAnalyzer, based on a stoichiometric pathway map published earlier [50]. The authors implemented novel techniques for assessing high-throughput data with logical models, using phosphoproteomic data from primary hepatocytes and the HepG2 cell line. They were able to discover a contradiction between experimental results and qualitative knowledge. The results of the study show that the Rac/Cdc42 induced p38 and JNK cascades are independent of PI3K. Moreover, they discovered that the activation of JNK in response to neuregulin is dependent on the PI3K signalling pathway.

9.5 Discussion

An important characteristic of Boolean models is that they conceptually match intuitive reasoning used in biology. Biological knowledge is often established and recorded in qualitative ways: for example, many biological studies have identified pro- or anti-apoptotic factors [51]; the Gene Ontology contains numerous terms for biological processes that include positive and negative regulation [52, 53]; MAPK pathways are described in terms of activation and inhibition of several proteins [54]. This qualitative knowledge cannot be readily translated into quantitative models, but it can serve as the basis for Boolean modelling. In this way, more biological information can be incorporated into models, which would otherwise be ignored. This quality makes Boolean modelling better suited to integrate numerous factors contributing to a biological process of interest in complex systems.

A key question arising for the use of Boolean model is whether the discretisation of complex molecular amounts and interactions into binary states is a reasonable approximation to model biological processes. In fact, there are numerous situations in biology where interactors behave in a switch-like manner. For example, the relation between an activator and a target results in a sigmoidal shape which can be well approximated by a Boolean switch [13]. The same relation is found in biochemical reaction kinetics when the reaction rate is plotted against the substrate concentration. Comparisons between Boolean models and more precise kinetic models of the same system have shown that Boolean models can produce the same qualitative dynamic outputs and provide the same level of insights towards hypothesis testing and experimental design [13, 55]. The main restriction in their usage would be for situations where interactions are close to equilibrium, i.e. in the middle of the sigmoidal curve. For example, Boolean models are rarely used to model metabolic reactions since situations where metabolic reactions are close to equilibrium are more frequently encountered, then a small change in enzyme level can lead to an important change in flux value. But they are still useful to predict effects of gene deletions and intervention strategies in metabolic networks [56].

When the Boolean approximation is considered to be adequate, different strategies can be followed to discretise molecular levels into binary states (0 and 1). One approach is to compare gene expression levels against the background distribution formed by all expression values. Typical values can be obtained from quantitative gene expression databases such as Gene Expression Commons [57], Expression Atlas [58] or StemMapper [59]. For example, Okawa et al. [60] used statistical testing to determine the significance of gene expression levels against the expression background in stem cells and selected a threshold above which a gene was considered active ("1") in the Boolean model. Buiga et al. [61] compared qPCR levels of phosphatases in cells treated with Herceptin versus untreated cells, after which overexpressed proteins were determined by statistical comparison between the treated and untreated samples, and those overexpressed phosphatases were then considered active ("1"). Christensen et al. [62] tested the significance of gene expression fold changes between wild-type and deletion mutants in yeast to discretise gene expression levels and evaluate a Boolean model of glucose signalling.

Boolean models are particularly well suited for cancer, because most cancer pathways consist of dysregulated signalling processes [6]. Signalling pathways can be more easily represented by ON/OFF logic than other more continuous biological processes; when compared to precise molecular interaction kinetics simulated by Ordinary Differential Equations, these relations often behave in a switch-like manner [13]. Therefore, the gain in scale and efficiency allowed by Boolean modelling comes at a relatively limited cost compared to other computational methods, whose implementation is generally more time consuming and challenging [63].

It has been shown in many studies that Boolean modelling combined with experimental approaches helps to develop potential therapeutic strategies. This approach enabled to simulate and select drug candidates to be tested for future treatments in breast cancer [43], to contribute to the development of novel approaches for the possible therapy of androgen-refractory prostate cancer [42] and to identify combinations of drug targets in the Hedgehog pathway in different cancer cell lines [41]. Also, Boolean modelling was used for enhancing the understanding of the pathways involved in progression of haematological hyperproliferative diseases [38].

The increased interest in Boolean modelling provides a new set of opportunities for advancing cancer therapies. There is a strong need for further development in mathematical modelling of cancer to enable the design of refined cancer therapy approaches, and crucially to move towards more individualised therapeutic procedures [64]. This means that not only should specific types of mutations involved in different tumours be taken into account, but also the specific genotype, transcriptome, proteome and clinical features of individual patients. We believe that Boolean modelling provides a highly promising approach towards the integration of such wealth of data into predictive computational models, in order to advance patient-individualised cancer therapy [65]. It represents a powerful tool to help us understand integrative systems behaviour and can help to decrease the experimental effort and cost needed to achieve new discoveries.

Competing Interests The authors declare no competing interests.

References

1. Ferlay J, Soerjomataram I, Ervik M, Dikshit R, Eser S, Mathers C, Rebelo M, Parkin DM, Forman D, Bray F (2012) Estimated cancer incidence, mortality and prevalence worldwide in 2012. IARC Cancer Base No. 11. International Agency for Research on Cancer, Lyon, France
2. Chen WW, Schoeberl B, Jasper PJ, Niepel M, Nielsen UB, Lauffenburger DA, Sorger PK (2009) Input–output behavior of ErbB signaling pathways as revealed by a mass action model trained against dynamic data. Mol Syst Biol 5:239
3. Breitkreutz D, Hlatky L, Rietman E, Tuszynski JA (2012) Molecular signaling network complexity is correlated with cancer patient survivability. Proc Natl Acad Sci USA 109(23):9209–9212
4. Schneider G, Schmidt-Supprian M, Rad R, Saur D (2017) Tissue-specific tumorigenesis: context matters. Nat Rev Cancer 17(4):239–253
5. Morris MK, Saez-Rodriguez J, Sorger PK, Lauffenburger DA (2010) Logic-based models for the analysis of cell signaling networks. Biochemistry 49(15):3216–3224
6. Hanahan D, Weinberg RA (2000) The hallmarks of cancer. Cell 100(1):57–70
7. Zhou T-T (2012) Network systems biology for targeted cancer therapies. Chin J Cancer 31(3):134–141
8. Albert R, Jeong H, Barabási AL (2000) Error and attack tolerance of complex networks. Nature 406(6794):378–382
9. Shoval O, Alon U (2010) SnapShot: network motifs. Cell 143(2):326-e1
10. Barabasi AL, Oltvai ZN (2004) Network biology: understanding the cell's functional organization. Nat Rev Genet 5(2):101–113
11. Tyson JJ, Chen K, Novak B (2001) Milestones network dynamics and cell physiology. Nat Rev Mol Cell Biol 2(12):908–916
12. Ruths D, Muller M, Tseng J-T, Nakhleh L, Ram PT (2008) The signaling petri net-based simulator: a non-parametric strategy for characterizing the dynamics of cell-specific signaling networks. PLoS Comput Biol 4(2):e1000005
13. Wynn ML, Consul N, Merajverca SD, Schnell S (2012) Logic-based models in systems biology: a predictive and parameter-free network analysis method. Integr Biol 4(11):1323–1337
14. Tian K, Rajendran R, Doddananjaiah M, Krstic-Demonacos M, Schwartz J-M (2013) Dynamics of DNA damage induced pathways to cancer. PLoS ONE 8(9):e72303
15. Wang R-S, Saadatpour A, Albert R (2012) Boolean modeling in systems biology: an overview of methodology and applications. Phys Biol 9(5):055001
16. Thakar J, Pilione M, Kirimanjeswara G, Harvill ET, Albert R (2007) Modeling systems-level regulation of host immune responses. PLoS Comput Biol 3(6):e109
17. Zhang R, Shah MV, Yang J, Nyland SB, Liu X, Yun JK, Albert R, Loughran TP (2008) Network model of survival signaling in large granular lymphocyte leukemia. Proc Natl Acad Sci 105(42):16308–16313
18. Walsh ER, Thakar J, Stokes K, Huang F, Albert R, August A (2011) Computational and experimental analysis reveals a requirement for eosinophil-derived IL-13 for the development of allergic airway responses in C57BL/6 mice. J Immunol 186(5):2936–2949
19. Oyeyemi OJ, Davies O, Robertson DL, Schwartz J-M (2015) A logical model of HIV-1 interactions with the T-cell activation signalling pathway. Bioinformatics 31(7):1075–1083
20. Boole G (2009) The mathematical analysis of logic. Cambridge University Press (CUP)
21. Saadatpour A, Wang RS, Liao A, Liu X, Loughran TP, Albert I, Albert R (2011) Dynamical and structural analysis of a T cell survival network identifies novel candidate therapeutic targets for large granular lymphocyte leukemia. PLoS Comput Biol 7(11):e1002267
22. Klamt S, Saez-Rodriguez J, Gilles ED (2007) Structural and functional analysis of cellular networks with CellNetAnalyzer. BMC Syst Biol 1:2
23. Terfve C, Cokelaer T, Henriques D, MacNamara A, Goncalves E, Morris MK, van Iersel M, Lauffenburger DA, Saez-Rodriguez J (2012) CellNOptR: a flexible toolkit to train protein signaling networks to data using multiple logic formalisms. BMC Syst Biol 6:133

24. Albert I, Thakar J, Li S, Zhang R, Albert R (2008) Boolean network simulations for life scientists. Source Code Biol Med 3(16)
25. Müssel C, Hopfensitz M, Kestler HA (2010) BoolNet—an R package for generation, reconstruction and analysis of Boolean networks. Bioinformatics 26(10):1378–1380
26. Naldi A, Berenguier D, Fauré A, Lopez F, Thieffry D, Chaouiya C (2009) Logical modelling of regulatory networks with GINsim 2.3. Biosystems 97(2):134–139
27. Stoll G, Viara E, Barillot E, Calzone L (2012) Continuous time boolean modeling for biological signaling: application of Gillespie algorithm. BMC Syst Biol 6(1):116
28. Klarner H, Streck A, Siebert H (2017) PyBoolNet: a python package for the generation, analysis and visualization of boolean networks. Bioinformatics 33(5):770–772
29. Di Cara A, Garg A, De Micheli G, Xenarios I, Mendoza L (2007) Dynamic simulation of regulatory networks using SQUAD. BMC Bioinform 8:462
30. Bock M, Scharp T, Talnikar C, Klipp E (2014) BooleSim: an interactive Boolean network simulator. Bioinformatics 30(1):131–132
31. Slamon D, Godolphin W, Jones L, Holt J, Wong S, Keith D, Levin W, Stuart S, Udove J, Ullrich A et al (1989) Studies of the HER-2/neu proto-oncogene in human breast and ovarian cancer. Science 244(4905):707–712
32. Alonso A, Sasin J, Bottini N, Friedberg I, Friedberg I, Osterman A, Godzik A, Hunter T, Dixon J, Mustelin T (2004) Protein tyrosine phosphatases in the human genome. Cell 117(6):699–711
33. Lucci MA, Orlandi R, Triulzi T, Tagliabue E, Balsari A, Villa-Moruzzi E (2010) Expression profile of tyrosine phosphatases in HER2 breast cancer cells and tumors. Anal Cell Pathol 32(5–6):361–372
34. Woodgett J, Avruch J, Kyriakis J (1996) The stress activated protein kinase pathway. Cancer Surv 27:127–138
35. Loda M, Capodieci P, Mishra R, Yao H, Corless C, Grigioni W, Wang Y, Magi-Galluzzi C, Stork P (1996) Expression of mitogen-activated protein kinase phosphatase-1 in the early phases of human epithelial carcinogenesis. Am J Pathol 149(5):1553
36. Haagenson KK, Wu GS (2010) The role of MAP kinases and MAP kinase phosphatase-1 in resistance to breast cancer treatment. Cancer Metastasis Rev 29(1):143–149
37. Zhang R, Shah MV, Yang J, Nyland SB, Liu X, Yun JK (2008) Network model of survival signaling in large granular lymphocyte leukemia. Proc Natl Acad Sci USA 105:16308–16313
38. Enciso J, Mayani H, Mendoza L, Pelayo R (2016) Modeling the pro-inflammatory tumor microenvironment in acute lymphoblastic leukemia predicts a breakdown of hematopoietic-mesenchymal communication networks. Front Physiol 7:349
39. Alvarez-Silva MC, Yepes S, Torres MM, Barrios AF (2015) Proteins interaction network and modeling of IGVH mutational status in chronic lymphocytic leukemia. Theor Biol Med Model 12:12
40. Cho SH, Park SM, Lee HS, Lee HY, Cho KH (2016) Attractor landscape analysis of colorectal tumorigenesis and its reversion. BMC Syst Biol 10(1):96
41. Chowdhury S, Pradhan RN, Sarkar RR (2013) Structural and logical analysis of a comprehensive hedgehog signaling pathway to identify alternative drug targets for glioma, colon and pancreatic cancer. PLoS ONE 8(7):e69132
42. Arshad OA, Datta A (2017) Towards targeted combinatorial therapy design for the treatment of castration-resistant prostate cancer. BMC Bioinform 18(Suppl 4):134
43. Sahin Ö, Fröhlich H, Löbke C, Korf U, Burmester S, Majety M, Mattern J, Schupp I, Chaouiya C, Thieffry D, Poustka A, Wiemann S, Beissbarth T, Arlt D (2009) Modeling ERBB receptor-regulated G1/S transition to find novel targets for de novo trastuzumab resistance. BMC Syst Biol 3:1
44. Zañudo JGT, Scaltriti M, Albert R (2017) A network modeling approach to elucidate drug resistance mechanisms and predict combinatorial drug treatments in breast cancer. Cancer Converg 1:5
45. Khan FM, Marquardt S, Gupta SK, Knoll S, Schmitz U, Spitschak A (2017) Unraveling a tumor type-specific regulatory core underlying E2F1-mediated epithelial-mesenchymal transition to predict receptor protein signatures. Nat Commun 8:198

46. Buiga P, Elson A, Tabernero L, Schwartz JM (2018) Regulation of dual specificity phosphatases in breast cancer during initial treatment with Herceptin: a Boolean model analysis. BMC Syst Biol 12(Suppl 1):11
47. Hetmanski JH, Zindy E, Schwartz JM, Caswell PT (2016) A MAPK-driven feedback loop suppresses Rac activity to promote RhoA-driven cancer cell invasion. PLoS Comput Biol 12(5):e1004909
48. Chudasama VL, Ovacik MA, Abernethy DR, Mager DE (2015) Logic-based and cellular pharmacodynamic modeling of bortezomib responses in U266 human myeloma cells. J Pharmacol Exp Ther 354(3):448–458
49. Grieco L, Calzone L, Bernard-Pierrot I, Radvanyi F, Kahn-Perles B, Thieffry D (2013) Integrative modelling of the influence of MAPK network on cancer cell fate decision. PLoS Comput Biol 9(10):e1003286
50. Samaga R, Saez-Rodriguez J, Alexopoulos LG, Sorger PK, Klamt S (2009) The logic of EGFR/ErbB signaling: theoretical properties and analysis of high-throughput data. PLoS Comput Biol 5(8):e1000438
51. Wong RSY (2011) Apoptosis in cancer: from pathogenesis to treatment. J Exp Clin Cancer Res 30(1):87
52. Ashburner M, Ball CA, Blake JA, Botstein D, Butler H, Cherry JM, Davis AP, Dolinski K, Dwight SS, Eppig JT, Harris MA, Hill DP, Issel-Tarver L, Kasarskis A, Lewis S, Matese JC, Richardson JE, Ringwald M, Rubin GM, Sherlock G (2000) Gene ontology: tool for the unification of biology. Gene Ontol Consort Nat Genet 25(1):25–29
53. Consortium T.G.O. (2017) Expansion of the gene ontology knowledgebase and resources. Nucleic Acids Res 45(D1):D331–D338
54. Santen RJ, Songb RX, McPherson R, Kumarc R, Adamc L, Jeng MH, Yue W (2002) The role of mitogen-activated protein (MAP) kinase in breast cancer. J Steroid Biochem Mol Biol 80:239–256
55. Akman OE, Watterson S, Parton A, Binns N, Millar AJ, Ghazal P (2012) Digital clocks: simple Boolean models can quantitatively describe circadian systems. J R Soc Interface 9(74):2365–2382
56. Lu W, Tamura T, Song J, Akutsu T (2015) Computing smallest intervention strategies for multiple metabolic networks in a boolean model. J Comput Biol 22(2):85–110
57. Seita J, Sahoo D, Rossi DJ, Bhattacharya D, Serwold T, Inlay MA, Ehrlich LI, Fathman JW, Dill DL, Weissman IL (2012) Gene expression commons: an open platform for absolute gene expression profiling. PLoS ONE 7(7):e40321
58. Papatheodorou I, Fonseca NA, Keays M, Tang YA, Barrera E, Bazant W, Burke M, Fullgrabe A, Fuentes AM, George N, Huerta L, Koskinen S, Mohammed S, Geniza M, Preece J, Jaiswal P, Jarnuczak AF, Huber W, Stegle O, Vizcaino JA, Brazma A, Petryszak R (2018) Expression Atlas: gene and protein expression across multiple studies and organisms. Nucleic Acids Res 46(D1):D246–D251
59. Pinto JP, Machado RSR, Magno R, Oliveira DV, Machado S, Andrade RP, Braganca J, Duarte I, Futschik ME (2018) StemMapper: a curated gene expression database for stem cell lineage analysis. Nucleic Acids Res 46(D1):D788–D793
60. Okawa S, Angarica VE, Lemischka I, Moore K, Del Sol A (2015) A differential network analysis approach for lineage specifier prediction in stem cell subpopulations. NPJ Syst Biol Appl 1:15012
61. Buiga P, Elson A, Tabernero L, Schwartz JM (2019) Modelling the role of dual specificity phosphatases in herceptin resistant breast cancer cell lines. Comput Biol Chem 80:138–146
62. Christensen TS, Oliveira AP, Nielsen J (2009) Reconstruction and logical modeling of glucose repression signaling pathways in Saccharomyces cerevisiae. BMC Syst Biol 3:7
63. Garland J (2017) Unravelling the complexity of signalling networks in cancer: a review of the increasing role for computational modelling. Crit Rev Oncol Hematol 117:73–113
64. Weston AD, Hood L (2004) Systems biology, proteomics, and the future of health care: toward predictive, preventative, and personalized medicine. J Proteome Res 3(2):179–196

65. Zañudo JGT, Steinway SN, Albert R (2018) Discrete dynamic network modeling of oncogenic signaling: mechanistic insights for personalized treatment of cancer. Current Opinion Syst Biol 9:1–10

Chapter 10
Integrating Omics Data to Prioritize Target Genes in Pathogenic Bacteria

Marisa Fabiana Nicolás, Maiana de Oliveira Cerqueira e Costa, Pablo Ivan P. Ramos, Marcelo Trindade dos Santos, Ernesto Perez-Rueda, Marcelo A. Marti, Dario Fernandez Do Porto, and Adrian G. Turjanski

Abstract This review focuses on efforts toward target prioritization from several pathogenic bacteria applying integrated multi-omics approaches. In an integrative scheme, diverse layers of multi-omics data, genome-scale models (GSMs), and structural/functional data related to any pathogenic species can be used to prioritize genes and proteins with attractive target characteristics for the development of new antimicrobials agents. The reconstruction of genome-scale metabolic models

M. F. Nicolás (✉) · M. de Oliveira Cerqueira e Costa · M. T. dos Santos
National Laboratory for Scientific Computation, LNCC/MCTIC, Petrópolis, RJ, Brazil
e-mail: marisa@lncc.br

M. de Oliveira Cerqueira e Costa
e-mail: maiolivei@gmail.com

M. T. dos Santos
e-mail: msantos@lncc.br

P. I. P. Ramos
Center for Data and Knowledge Integration for Health (CIDACS), Instituto Gonçalo Moniz, Fiocruz-Bahia, Salvador, BA, Brazil
e-mail: pablo.ramos@fiocruz.br

E. Perez-Rueda
Instituto de Investigaciones En Matemáticas Aplicadas Y En Sistemas (IIMAS) Universidad, Nacional Autónoma de México, Unidad Académica Yucatán, Mérida City, Mexico
e-mail: ernesto.perez@iimas.unam.mx

M. A. Marti · D. F. Do Porto · A. G. Turjanski
Departamento de Química Biológica, Facultad de Ciencias Exactas Y Naturales, Universidad de Buenos Aires (FCEyN-UBA), Ciudad de Buenos Aires C1428EHA, Buenos Aires, Argentina
e-mail: marti.marcelo@gmail.com

D. F. Do Porto
e-mail: dariofd@gmail.com

A. G. Turjanski
e-mail: aturjans@gmail.com

M. A. Marti · A. G. Turjanski
Instituto de Química Biológica de La Facultad de Ciencias Exactas Y Naturales (IQUIBICEN) CONICET, Pabellòn 2 de Ciudad Universitaria, Ciudad de Buenos Aires C1428EHA, Buenos Aires, Argentina

F. A. B. da Silva et al. (eds.), *Networks in Systems Biology*, Computational Biology 32,
https://doi.org/10.1007/978-3-030-51862-2_10

(GSMMs) and transcriptional regulatory networks (TRNs) is described in detail. Also, we discuss the methods for the integration of GSMs and diverse web servers for drug targeting in pathogens. Structural approaches are also illustrated. We stress the clinical importance of the drug-resistant isolates related to severe nosocomial or community infections belonging to the species *Klebsiella pneumoniae* and *Staphylococcus aureus*, two of the six ESKAPE pathogens, as well as *Mycobacterium tuberculosis*, the causative agent of tuberculosis.

Keywords Drug target · Structural modeling · Metabolism reconstruction · Transcriptional regulatory network · Bacterial pathogens · *Staphylococcus aureus* · *Klebsiella pneumoniae* · *Mycobacterium tuberculosis*

List of Used Acronyms

AMR	Antimicrobial resistance
BEC	Brazilian Epidemic Clone
BIGG	Biochemical, Genetic and Genomic knowledge base
CAI	Community-acquired infections
CA-MRSA	Community-associated MRSA strains
cis-RE	*cis*-regulatory elements
CCR	Carbon Catabolite Repression
ChIP-seq	Chromatin immunoprecipitation followed by sequencing
CG	Clonal group
COBRA	Constraint-based reconstruction and analysis (COBRA), a wide class of methods to analyze possible flux distributions in metabolic network
CP	Choke-points
CRKp	carbapenem-resistant *Klebsiella pneumoniae*
CRISPR-Cas	Clustered Regularly Interspaced Short Palindromic Repeats
CWD	Crystallized With Drugs
D	Druggable
DBD	Domain-Based Druggable
DCW	Dry Cell Weight
DS	Druggability Score
ESKAPE	acronym encompassing the names of six bacterial pathogens commonly associated with antimicrobial resistance belonging to species *Enterococcus faecium, Staphylococcus aureus, Klebsiella pneumoniae, Acinetobacter baumannii, Pseudomonas aeruginosa, and Enterobacter spp*
FBA	Flux Balance Analysis

D. F. Do Porto
Instituto de Cálculo, Facultad de Ciencias Exactas Y Naturales, Universidad de Buenos Aires (FCEyN-UBA), Ciudad de Buenos Aires C1428EHA, Buenos Aires, Argentina

FVA	Flux Variability Analysis
GCNs	Gene Co-expression Networks
GENRE	Genome-Scale Reconstruction
GIMME	Gene Inactivity Moderated by Metabolism and Expression
GIM^3E	Gene Inactivation Moderated by Metabolism, Metabolomics and Expression
GPRs	Gene–Protein–Reactions
GSM	Genome-scale Model also referred to as GEM
GSMM	Genome-scale Metabolic Model also referred to as GEMM
GRN	Gene Regulatory Network also referred to as TRN
HA-MRSA	Healthcare-associated Methicillin-Resistant *Staphylococcus aureus*
HAIs	Healthcare-associated infections
HD	Highly Druggable
HIV	Human Immunodeficiency Virus
HMM	Hidden Markov Models
iFBA	integrated Flux Balance Analysis
KEGG	Kyoto Encyclopedia of Genes and Genomes
KL	K-Locus, referred to as *in silico* predicted capsular serotype
KPC	*Klebsiella pneumoniae* carbapenemase, also referred to as KPC-2
Kp13-MN	Kp13 metabolic network
LPS	Lipopolysaccharide
mAb	monoclonal antibody
MADE	Metabolic Adjustment by Differential Expression
MDR	Multidrug-resistant, an organism resistant to multiple drugs
MRSA	Methicillin-Resistant *Staphylococcus aureus*
MSSA	Methicillin-Susceptible *Staphylococcus aureus*
MTB	*Mycobacterium tuberculosis*
Mtb-MN	*Mycobacterium tuberculosis* metabolic network
mRNA	messenger RNA
PDB	Protein Data Bank
PB	Polymyxin B
PFAM	Protein Family Database
PROM	Probabilistic Regulation Of Metabolism
PWM	Position Weight Matrix
rFBA	regulated Flux Balance Analysis
RNAi	RNA interference
RNA-seq	RNA sequencing
RNOS	Reactive Nitrogen and Oxygen Species
RSA	Regulatory Steady-state Analysis
SBML	Systems Biology Markup Language
SBML-qual	Systems Biology Markup Language Qualitative
ST	Sequence Type
SR-FBA	Steady-State Regulated Flux Balance Analysis
TB	Tuberculosis
TF	Transcription Factor

TP	Target-Pathogen database
TFBS	Transcription Factor Binding Site
TRN	Transcriptional Regulatory Network
TU	Transcription Unit
WGCNA	Weighted Gene Co-expression Network Analysis
WHO	World Health Organization's

10.1 Introduction

The emergence of bacterial clinical isolates with resistance to a wide range of antibiotic drugs represents a global health concern. The first reports of antibiotic-resistant bacteria date from the principle of clinical use of these wonder drugs in the first half of the twentieth century. For example, strains of sulfonamide-resistant *Streptococcus pyogenes* were reported as early as the 1930s, while the emergence of penicillin-resistant *Staphylococcus aureus* bacteria occurred immediately after the introduction of this antibiotic into clinical practice in the 1940s [1]. Thus, although the availability and use of antibiotics represented one of the greatest scientific achievements of the twentieth century, enabling the control of bacterial diseases considered deadly in the pre-antibiotic era, the emergence of resistant bacteria, often called "superbugs," is now a source of concern in the clinical practice. The World Health Organization's (WHO) latest report on antimicrobial resistance (AMR) warns that drug-resistant diseases could cause 10 million deaths each year by 2050 and damage to the economy as catastrophic as the 2008–2009 global financial crisis. By 2030, antimicrobial resistance could force up to 24 million people into extreme poverty (WHO, 29 April 2019) [2]. Therefore, the discovery of new antibacterial drugs becomes essential for the present and future control of bacterial pathogens presenting AMR, which are becoming increasingly common in clinical practice.

The high cost associated with prospecting research and the development of new drugs is considered one of the limiting factors in this area, which has held back investment even by large pharmaceutical industries [3]. However, the popularization of high-throughput DNA sequencing techniques, coupled with the growing number of important pathogen genomes deposited in public databases, enables large-scale genomic data mining. In this sense, In silico techniques are currently being used from the field of Bioinformatics, which make it possible to significantly reduce the costs of the initial bioprospecting phases for the identification of new target candidates. Thus, the initial detection of new targets can be done entirely computationally by integrating data and applying filters to select pathogen molecules with desirable druggability features. An example of a filter that can be used would be to consider only target proteins that have an inhibition site, where the possible drug would be placed (pocket), with the structural characteristics to bind a drug-like compound (druggable or bindable) and that it is also different from the pockets of similar human proteins. The latter is relevant because drugs with inhibitory capacity for the pathogen proteins could react undesirably with the human proteins, which would result in an adverse cytotoxic effect triggered by this drug.

For bioprospecting new targets, bioinformatic algorithms take as a starting point the genomic sequence of an organism in order to identify molecular targets (genes or proteins). During those phases, it is necessary to develop bioinformatics tools that perform both protein sequence and structural analysis [3–5]. Additionally, the development of new combined therapies can be done by studying metabolic, regulatory, and signaling networks, where the combined effect, that is, the set of metabolic and regulatory pathways most impacted by the use of a drug combination, is evaluated using systems biology [6, 7]. The three hierarchical levels mentioned (1: sequence → DNA/protein, 2: structure → protein, 3: enzymatic/regulatory reactions → metabolic/regulatory network) have generally been used alone to determine candidate therapeutic targets (e.g., [8–12]). While the application of computational techniques alone does not envisage a definitive identification of drug targets, it does permit shortlisting more plausible targets, effectively reducing the search space to candidates with an increased probability of serving as targets for either a new or a repositioned drug.

In general, the rational development of new antimicrobials requires the prioritization of specific molecular targets and the determination of drug-like compounds that are capable of binding and inhibiting those targets. These new antimicrobials can be used in polytherapy, a strategy capable of increasing bacterial death, via subpopulation and/or mechanistic synergy [13]. For instance, the polymyxin combination therapy with many drug combinations is increasingly used clinically. Substantial improvements in bacterial killing and reduction of the emergence of polymyxin-resistant subpopulations was observed when a polytherapy approach was applied, even for multidrug-resistant *Klebsiella pneumoniae* isolates [14].

It is important to remark that other innovative approaches that are not discussed in this chapter are currently emerging, such as those focused on the development and use of monoclonal antibodies (mAbs) as antimicrobial biopharmaceuticals. In studies with Gram-negative bacteria (e.g., *Escherichia coli* and *K. pneumoniae*), specific targets for the development of antibacterial mAbs are currently focused on the outer surface structures, such as capsular polysaccharides, lipopolysaccharides (LPS), and fimbriae [15]. However, epitopes from external structures, despite being more accessible to mAbs, present considerable variability among the isolates of these bacteria and are limited to few serotypes belonging to the most widespread clonal types [16, 17]. Thus, the balance between accessibility and diversity of the target epitope is crucial. The recent publication by Storek et al. [18], describing a single outer membrane protein containing a target epitope for the production of a mAb with proved bactericidal activity against *E. coli*, highlights several interesting options for the future development of antimicrobial mAbs.

Another innovative therapy is the RNA interference (RNAi), where the molecular targets are prioritized at the RNA level focusing on gene suppression. Few applications using silencing genes by RNAi against pathogen target genes have been described so far (e.g., for some intracellular species as *Legionella* spp. [19]). Alternatively to silencing genes by RNAi, the CRISPR-Cas technology is evolving as an effective tool to break double-stranded DNA and represents an opportunity to combat resistance in clinical pathogens. This system associated with phage therapy can be used to design and deliver antimicrobials with "precision" to fight bacterial pathogens in a DNA sequence-dependent manner. Thus, this system proves to be very prosperous by directing the selective silencing of genes that are involved in the antibiotic resistance, biofilm formation, or virulence in bacterial pathogens [20].

In this chapter, we discussed the generation, modeling, organization, and integration of data from the standpoint of bioinformatics, computational and systems biology methods to prioritize molecular targets for future application in novel therapeutic approaches against multidrug-resistant pathogenic bacteria.

10.2 How to Prioritize Targets in Pathogenic Bacteria?

Classically, the discovery of molecules with antimicrobial activities have relied on either experimental laboratory studies, involving complex designs and costly reagents and equipments, or through sheer serendipity—as was the discovery of penicillin, the first molecule with antibiotic activity, identified by Sir Alexander Fleming in the 1920s [21]. While the latter can be considered exceptions, the usual path to drug discovery is filled with experimental caveats and involves extensive trial-and-error (Fig. 10.1). In the so-called antibiotics 'golden era,' natural scaffolds and variations thereof were disclosed through the study of bacterial and fungal (particularly actinomycetes) metabolism, which produce these molecules as a competition mechanism, and by chemical modification of these scaffolds [22]. Many antibiotic classes were disclosed as a result of this strategy, yielding the current notion that most 'low-hanging fruit' was already picked, an analogy that reveals that the further pursuit of antibiotic molecules from natural sources mostly rediscovered already known molecules [23].

High-throughput experimental methodologies, where chemical libraries of the order of 10^5–10^6 molecules are screened to identify potential drug-like molecules, have also allowed the identification of "new" candidate antibiotic molecules, but the

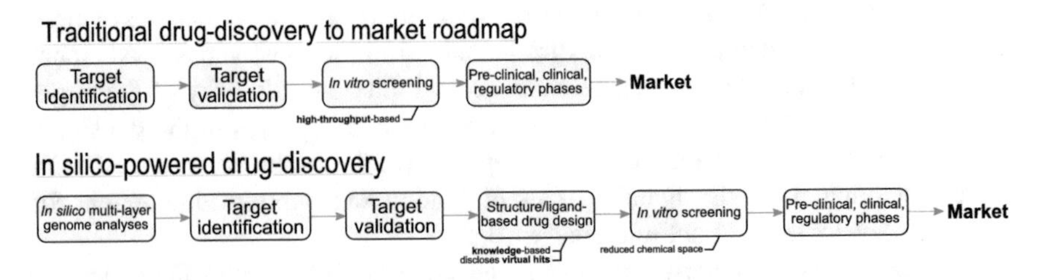

Fig. 10.1 Traditional and *in silico*-powered drug discovery. The traditional route to drug discovery focuses on extensive high-throughput screening against fragment libraries. *In silico*-powered drug discovery focuses on the extensive use of integrative approaches to disclose knowledge about targets, virtual screening, and ligand optimization to reduce the search space in which candidate drugs can be further developed

usefulness of this brute–force approach is currently questioned as the rate of novel lead discovery slumped [24]. Also, limitations in the representation of chemical libraries themselves hamper the usefulness of the HTS approach, especially considering that estimates of the total number of molecules interesting for drug discovery are of the order of 10^{60} molecules [25]. In this context, alternative methodologies, relying on the use of whole genome-level information available since the initial advent of Sanger sequencers, followed by next-generation sequencing instruments, were proposed early in the 1990s.

These omic technologies shifted the focus of novel drug discovery into a target-centric approach, in which molecular targets were explored based on genome-level information. This would allow for more informed decisions to be made regarding targets with higher probability, and to which further experimental and financial resources can be directed. For instance, several high-throughput screening campaigns were run by biopharmaceutical companies using candidate targets that were prioritized through genome-derived criteria; these were then expressed, purified, and inhibitors were sought for by searches against chemical libraries [26]. However, after extensive genome mining efforts, target-centric approaches were heavily criticized due to its low output in terms of generating novel leads or inhibitors [27]. Part of these criticisms related to the expectation that the genomics 'era' would open doors to a plethora of novel antimicrobials, a clearly naïve belief. Strict drug approval policies by regulatory agencies in recent years, unfavorable risk–benefit ratio to pharmaceuticals, and overall unattractiveness of the antibacterial market (compared to other diseases) were among the key factors contributing to the low output of research into novel antimicrobials during this period [28]. Despite the mentioned caveats, successful examples of target-based approaches are available in the literature, one of the first being the development of peptide deformylase inhibitors (based on N-alkyl urea hydroxamic acids) [29]. Other fruitful examples include the development of HIV inhibitors [30], the identification of inhibitors targeting proteins involved in wall teichoic acid biogenesis in *Staphylococcus aureus*, which strongly potentiates β-lactam antibiotics against methicillin-resistant *S. aureus* [31], the discovery of inhibitors of the MvfR-regulon (involved in quorum-sensing) in *Pseudomonas aeruginosa* [32], and the recent development of mAbs targeting the β-barrel assembly in *E. coli*, with notable bactericidal activity [18]. The rapid advance of antibiotic-resistant strains and the lack of novel compounds against bacterial pathogens has reinstalled the idea that target-based approaches, particularly when integrating multiple layers of information (e.g., 3D structure, essentiality, subcellular localization, off-target criteria, sequence conservation and functional properties), still offer a promising research avenue in tackling infectious diseases. In the following sections, we discuss how these layers of data can be generated, modeled, and used for bacterial protein target identification.

10.2.1 Metabolic Network Modeling

Metabolism refers to the set of reactions and (bio)chemical transformations performed within the cell that together contribute to its homeostasis, energy acquisition, and adaptation to the surrounding medium. The main parts that compose an organism's metabolism are enzymes, proteins that possess catalytic activity by converting a set of substrates into products (chemical compounds), and regulatory proteins, which orchestrate the activation and suppression of these components (see Sect. 2.2). The organism's genome exerts a fundamental role upon its metabolic repertoire since it holds the necessary information for enzyme synthesis. In this sense, it is possible to reconstruct metabolic model for a given organism using as input the information encoded in the genome.

In recent years, the reconstruction of metabolic models has been of fundamental importance for model-driven discovery, metabolic engineering, and studies of evolutionary processes [33, 34], with direct applications in the search of treatments for many diseases, such as infections from multidrug resistant bacteria, Alzheimer's disease, and cancer [35–38]. However, the myriad of models recently produced have an enormous variability in both quality and coverage, which can jeopardize their predictive capacities [39]. Therefore, to improve a model's predictability it is essential to collect knowledge from the physiology, biochemistry, and genetics of the organism of interest.

In this regard, metabolic model can be represented by its underlying enzymatic reactions, which can be considered one of its fundamental units, with the respective substrate and product chemical compounds. Reactions and metabolites organize to form coordinated networks, responsible for the functioning of the basic cellular apparatus, such as energy transduction phenomena. Similarly, to traditional phenotypic traits used in bacterial identification and classification schemes (e.g., CO_2 production, usage of carbon sources), an organism expresses a certain profile of enzymes that yield a "metabolic phenotype," which is usually distinct from that of other species or even within a species [40, 41]. The dynamics of a metabolic network can, in principle, be quantitatively calculated directly from metabolite concentrations and physicochemical parameters, by employing kinetic models, for example [40]. However, this quantitative approach, in general, cannot be applied for large networks due to the large number of unknown parameters, since detailed kinetic information is not available for all reactions that can occur in a cell. Quantitative models also have a computational cost increasing exponentially with network size [42]; taken together, these aspects make this approach not suitable for large networks analysis.

Multiple tools and databases have been made available in recent years to reconstruct and study an organism's metabolism. KEGG [43], BioCyc [44], and Reactome [45] are among the notable examples of manually curated databases which harbor high-quality metabolic information. As example, KEGG was used to reconstruct the metabolism and then prioritize molecular targets in the disease-causing pathogen *Mycoplasma genitalium*, for which 67 proteins were disclosed as potential targets to the study of novel compounds or as vaccine candidates [46]. A similar strategy was used against the rapidly growing mycobacterium, *Mycobacterium abscessus*, identifying 40 candidate proteins in a subtractive strategy by filtering candidates with lower potential as targets [47]. In another work, the BioCyc vocabulary was employed for the reconstruction of the metabolism of *Bacillus subtilis* [48]. Transcriptomic data can also be projected onto the metabolic model, as it was performed in the study of hypervirulent strains of *Clostridium difficile*. For these strains, the BioCyc vocabulary was used to disclose differentially expressed pathways [49].

Different strategies are used to reconstruct metabolic networks, and these can be usually divided into two broad classes, as defined by Machado et al. [50]: (1) *Bottom-up*, where a draft model is generated automatically from the genomic information for an organism, usually by finding orthology relationships against annotated genes with known associations to biochemical reactions and later subjected to extensive manual improvement; and (2) *Top-down*, where a universal model is generated and curated manually, which is then used as a template for building an organism-specific model, also employing orthology predictions to identify the reactions from the template conserved in the organism of interest; this can be made automatically once the universal model is obtained. In the following sections, we further detail how these different models can be employed to the study of bacterial metabolism, with a particular focus towards static representations of metabolism using graph models, as well as dynamic simulations using constraint-based approaches.

10.2.1.1 Static Representations of Metabolism Using Graph-Based Approaches

Graphs are mathematical structures used to model relationships between elements. In a metabolic context, these elements can be biochemical reactions themselves or its components: metabolites (substrates/products of these reactions) and enzymes. These can be represented as nodes in the graph, and there exist links (or edges) connecting these objects when they participate in a common pathway, for example. In general, a graph $\mathbf{G} = (\mathbf{V}, \mathbf{E})$ is composed of a set of finite nodes $\mathbf{V}$, where $\mathbf{V} = \{v_1, v_2, \ldots, v_n\}$, and of (directed or undirected) edges $\mathbf{E}$ connecting elements in $\mathbf{V}$. Different types of graphs can be used to represent a metabolic network, and these have been the focus of extensive reviews [51–53]. The choice of a graph representation depends on the objective of the reconstruction, for example, what biological questions will be emphasized in the modeling, and what particular metabolic aspect, if any, will be emphasized in the model. For instance, simple graphs such as compound, reaction, or enzyme graphs can be employed to focus on these specific metabolic components, and in these networks, nodes represent metabolites, biochemical reactions, and proteins, respectively. To illustrate how elements are linked in such models, in a reaction graph, an edge connects two nodes (representing two distinct reactions) when at least one compound produced by one of the reactions is used as a substrate in the other reaction.

A graph-based representation of metabolism allows the use of topological measures of network analysis and the study of its properties from a global standpoint. In the context of molecular targets, multiple topological measures have been used to identify central pathogen molecules in a metabolic reconstruction. These centrality measures use the topology of the resulting network to identify important elements therein. A simple measure, the node degree, counts the number of edges that point towards and away from a node. The relationship between node degree and protein essentiality has led to the centrality–lethality "rule" in early studies using yeast protein interaction networks [54]. While degree centrality is considered a local metric that considers only the immediate neighborhood of a node to define importance, betweenness centrality is a global centrality measure employed in many different applications [55]. In this metric, the more a node is used as an intermediate in paths reaching different parts of the network, the higher its centrality value. All these metrics, as well as others, have been used as a proxy of protein importance in networks aiming pathogens target prioritization [9, 47, 56–60]. Another relevant topological property is choke-point analysis. Choke-points (CP) are defined as reactions that uniquely consume or produce a given compound, and were first used to disclose potential targets in *Plasmodium falciparum*, the malaria-causing parasite [61]. The intuition behind their identification is that if a CP is inhibited (e.g., by a drug targeting the enzyme(s) catalyzing the CP reaction), there will be no alternative reactions to produce (or consume) these metabolites, leading to either starvation of the unique product or accumulation of potentially toxic compounds in the cell, as by definition there will be no replacement reactions to fulfill the functional activity. Other works have also employed this metric to the prioritization of targets [9, 62–67].

Although graph-based approaches represent a simple and effective way to model a metabolic network and study its properties through a topological perspective, it offers limited insight regarding the way that this network functions since these graphs are a static representation of metabolism. More sophisticated approaches are available that allows the simulation of metabolic dynamics by modeling metabolism as a set of mathematical equations. In the next sections, we detail how these approaches are formalized and used to simulate pathogen metabolism and disclose important proteins therein.

10.2.1.2 COnstraint-Based Reconstruction and Analysis (COBRA) Methods

In addition to the static representation described previously, the complete metabolic repertoire of an organism can be organized in a Genome-Scale Network Reconstruction (GENRE), which holds information about the established biochemical knowledge from an organism. A GENRE is the starting point from where a dynamic metabolic model can be derived, and these are termed Genome-Scale Models (GEMs or GSMs) [68]. GSMs go beyond the established knowledge of the metabolism of an organism contained in the GENRE by incorporating assumptions and hypotheses needed to enable computational simulations and metabolic analyses. The production and update of such a model at a genomic scale are based on an annotated genome describing the enzyme repertoire and a literature-based biochemical and genetical analysis [39]. The Biochemical, Genetic and Genomic knowledge base (BiGG) [69, 70] is nowadays one of the important repositories of genomic reconstructions. To construct a GSM, one should consider a cell subjected to a specific environment. In this context, it is considered (i) the inputs and outputs of the network, (ii) a mathematical representation of network stoichiometry, and (iii) the reaction fluxes are subjected to constraints, such as reaction directionality. GSMs are suitable for in silico analysis and are at the heart of hypothesis-driven research in metabolism based on COBRA methods. One of the most popular methods is Flux Balance Analysis (FBA) [71–73] which requires the modeler a choice of a cell objective (objective function) driving the metabolism fate. Usually, the objective function is a proxy for cellular growth, when it is also called a biomass function, and thus contains terms related to the synthesis of nucleic acids, proteins, carbohydrates, and fat, in proportions that match in vitro experiments, the simplest of which assesses the fraction of dry weight mass of each component. Using these methods, it is possible to determine the complete metabolic phenotype that a cell exhibits under a specific environmental condition.

Because of the versatility and scalability of COBRA methods, it has been adopted by several groups, and many COBRA-derived methods also appeared in the literature, which now counts more than a hundred methods, and many of them have a ready to use software to support their application [74]. Therefore, the activities of a cell are subjected to a variety of constraints that can be diverse in nature, such as follows:

(i) Physicochemical constraints: Conservation of mass and charge, availability of enzymes and substrates, and thermodynamic constraints are inviolable. They are always respected.
(ii) Topological constraints: Biochemical reactions must respect localizations of all elements in the same cell compartment.
(iii) Environmental constraints impose time-dependent external conditions such as pH, temperature, and nutrients availability.
(iv) Regulatory constraints: Metabolism is directly subjected to self-imposed genetic regulation mechanisms. Regulation selects which enzymes will be active under a specific environmental condition. This acts in favor of cell fitness, eliminating suboptimal metabolic states.

An important modeling step is to represent stoichiometry and restrictions mathematically. First, we have charge and mass conservation requiring that no charge or matter is produced or consumed in a biochemical transformation. For each metabolite, the production rate should be equal to the consumption rate. This balance of fluxes in a steady state can be represented by

$$S \cdot v = 0 \tag{10.1}$$

where v represents the flux vector and S is the stoichiometric matrix, whose lines represent metabolites, columns represent reactions, and a matrix element a_{ij} represents the stoichiometric coefficient of a metabolite i in reaction j.

Constraints are imposed on reactions flux values, reflecting the biological behavior of minimum and maximum flux rates for biochemical reactions. Thermodynamics is also reflected in these constraints since reversible reactions fluxes can span both positive and negative portions of the solution space, while irreversible reactions are associated with strictly positive flux values. These restrictions, and some other known flux limits for uptake rates, reduce the system's solution space, ensuring that expressible metabolic phenotypes are thermodynamically and biologically feasible. Both its steady-state mathematical representation (Eq. 10.1) and flux constraints defining the solution space are essential to enable the application of COBRA methods

10.2.1.3 Flux Balance Analysis (FBA)

The basic assumption of the FBA method is that cell metabolism is directed toward some kind of "objective." In normal environmental conditions, such an objective can be cell growth, as previously cited. Another possibility used in the microbial engineering area is to consider the production of a metabolite of interest as the objective. This objective can be mathematically described as an objective function Z, which is essential for the formulation of FBA as a linear optimization problem [42, 73, 75].

Equation 1 defines an indeterminate linear system, that is, a system whose number of equations to satisfy is smaller than the number of system's variables. This implies that it is not possible to find a unique solution for the system. FBA method proposes that, under this circumstance, it is possible to find at least one solution which optimizes the objective function Z, which is given by

$$Z = c \cdot v \tag{10.2}$$

where c is a vector of coefficients (weights) for fluxes in vector v. In this case, the objective function Z is associated with cell growth and coefficients in c are associated with weights of elements in cell biomass production. The biomass function stoichiometry expresses the requirements of cell growth [73].

It follows that FBA states a linear optimization problem, where it has

(i) The indeterminate linear system $S \bullet v = 0$.
(ii) The constraints on the flux solution space $v_{min} \leq v \leq v_{max}$.
(iii) The objective function to be optimized $Z = c \bullet v$.

A solution v of such an optimization problem is both thermodynamically and biologically feasible. However, the method does not ensure that the solution found is unique. In principle, there can be alternative solutions satisfying the same set of restrictions. In the case where the objective function is associated with biomass production, the linear optimizations problem maximizes Z, which is represented in the system as a stoichiometrically balanced pseudo-reaction. Its coefficients are the rates in which metabolites are converted into biomass constituents as nucleic acids, proteins, and lipids [73].

Biomass is not the only possible objective function. *Z* can also represent a specific metabolite production that one wants to optimize, that is, if the goal is to obtain a metabolic phenotype associated with its maximization or minimization [72].

Since FBA formulates a linear optimization problem whose solution is in principle not unique, it is possible to explore the solution space to evaluate the obtained metabolic phenotype using Flux Variability Analysis (FVA) to perform such exploratory endeavor. FVA establishes minimum and maximum limits of fluxes for each reaction while preserving the already obtained optimization for the objective function [76]. The method is shown to have applications in many subjects, including exploration of alternative optima, investigating network flexibility and network redundancy, and optimization for antibiotic production [76].

There are several available applications dedicated to performing FBA and FVA in GSMs, most popular are the COBRAToolbox [77] and COBRApy [78].

10.2.2 Transcriptional Regulatory Network (TRN) Modeling

Cellular organisms encode all the genetic information necessary for their survival in the DNA, and from this molecule, the flow of genetic information is possible through the processes of transcription and translation. Indeed, the cells can quickly synthesize, when necessary, large amounts of a given protein, since several copies of RNA can be produced from a single gene from the DNA and, in turn, each RNA molecule complementary to that gene can head the synthesis of many proteins. However, each gene is transcribed and translated at different rates, allowing the cell to exercise effective control, both in abundance and in the temporal pattern of gene expression. This is crucial for the proper organism response to environmental changes, availability of nutrients, and/or physicochemical variations. Thus, gene expression is a strictly controlled process through the cell's transcriptional regulatory network (TRN) [79]. In the following sections, we discuss in detail the methods for identifying transcription factors (TF) and how reconstructing transcriptional regulatory networks (TRNs). We also describe the properties of TRNs and the most used algorithms for their integration with GSMs.

10.2.2.1 Identification of Transcription Factor (TF) Repertoire

In general, the ability to respond and adapt to environmental changes is defined by the cell's repertoire of DNA-binding transcription factors (TFs) and their interactions with the *cis*-regulatory regions of their target genes, in the form of a transcriptional regulatory network [80, 81]. TFs bind to the promoter regions of specific genes to, either positively or negatively, regulate the transcriptional expression of target genes.

Due to their crucial role in coordinating the gene expression kinetics of a genome, TFs have been exhaustively studied, including mutational analysis, and sequence comparisons. In this regard, the identification of the TF repertoire from a given genome allows, on a global scale, to elucidate regulatory networks. In this context, the bacteria with the best-studied TFs, and for which the transcriptional regulatory networks have been elucidated are: *E. coli* K12 [82, 83], *Bacillus subtilis* [84, 85], *Corynebacterium glutamicum* [86, 87], and the pathogen *S. aureus* [88].

For example, in the pathogen *S. aureus*, the repertoire of TFs was identified by diverse bioinformatic tools. In brief, the open reading frames that encode predicted protein sequences of 12 *S. aureus* strains were scanned by using the domain assignations associated to DNA-binding regions in the Superfamily database [89] and Pfam protein family database [90]. Additionally, family-specific Hidden Markov Models (HMM) were used to construct the bacterial models: *E. coli* K-12, *B. subtilis*, and *C. glutamicum* were also used. These HMMs were constructed, considering the DNA-binding domain sequence (around 60 amino acids) of every protein from multiple families. *S. aureus* USA300 proteome sequences were scanned with these HMMs, and proteins with less than 60% coverage in the DNA-binding region against their corresponding HMM were excluded. Finally, a dataset that includes the regulatory proteins identified by Superfamily, Pfam and HMMs from bacterial models were considered as TFs. This analysis revealed the sheer complexity of the *S. aureus* regulatory network and offered unique insights into many yet uncharacterized TFs in this important human pathogen.

10.2.2.2 Methods to Reconstruct Transcriptional Regulatory Networks (TRNs)

The basic components of TRNs are the interactions between TFs and the promoters of their target genes to activate or repress gene transcription [91]. In the classic modeling of TRNs, the intracellular signals that regulate the activities of TFs, as well as the environmental factors that trigger cellular responses, are all excluded. Unlike metabolic networks, TRNs are not as conserved between different species, unless between strains closely related phylogenetically or that share a common lifestyle. Gene expression data associated with information encoded in a genome and literature allow the application of diverse approaches to determine the set of interactions between TFs and their respective target genes, that is, to reconstruct TRNs.

The reconstruction of TRNs provides the basic framework to understand the genome-wide perspective, the organization of gene regulation in all the organisms. In this regard, a common approach involves a comparative genomics analysis, with (at least) two methodologies: template network-based methods and TF-binding site (TFBS) data-based methods via prediction of *cis*-regulatory elements (*cis*-RE), including propagation from known regulons and *ab initio* regulon inference.

Template-based methods [92] consider the use of well-characterized networks, as described in *E. coli K-12* or *B. subtilis* strain 168, as a starting point for the reconstruction. In general, these approaches use the notion of sequence comparison to identify orthologs using the reciprocal best-hit method, on the genome of interest; however, it is useful in closely related organisms [93, 94]. In a posterior step, DNA-binding sites can be inferred on the "reconstructed" regulated network, considering the *cis*-RE. We must consider that this approach is based on the conservation of regulatory elements across different organisms. Therefore, diverse considerations are necessary, such as to collect all information concerning the DNA-binding site in a selected model organism. Then, these data are used to construct positional weight matrices (PWMs) or to achieve pattern searches. In general, these approaches have been useful to predict TRNs in diverse proteobacteria organisms [95]; however, some limitations are evident, such as the fact that organisms with no closely phylogenetic relationship with the reference genome can originate meaningless interactions. Indeed, diverse studies have suggested that the conservation to identify conservation at orthologs level to be used to reconstruct regulons is limited by the phylogenetic distance [96], as a consequence of the diverse growth environments associated with the organisms in which members of the regulatory networks are present [97]. In summary, the reconstruction based on a genome-scale approach depends on the availability of a template network.

In contrast, the emergence of a large number of experiments utilizing high-throughput technologies, including microarrays and RNA sequencing (RNA-seq), have been performed to analyze the global expression in all the organisms, allowing the identification of co-expressed genes in a specific condition. In this respect, a systematic analysis to evaluate all this information is through Gene Co-expression Networks (GCNs), where the network $G = (V, E)$ is composed of a set of nodes (V) that represent the genes and a set of edges (E) that indicate significant co-expression relationships [55, 98, 99]. In general, these networks maintain the structural properties of real networks, such as scale-free topology, that is, that there are some highly connected nodes, namely hubs, and a large number of nodes with a small number of connections [100, 101]. To do this, different algorithms have been developed to reconstruct GCNs; in particular, the Weighted Gene Co-expression Network Analysis (WGCNA) that considers the co-expression patterns between two genes and the overlapping of neighbor genes [102]. Thus, highly correlated genes are clustered into large modules based on similarities in their expression profiles. In a posterior step, these modules are evaluated to identify those enriched genes that share similar biological functions [103, 104] . WGCNA also compares different GCNs to identify conserved modules between species or cell types [105–107]. GCNs have been used to identify genes with similar expression patterns in a set of samples, allowing the prediction of gene functions at the genome level, the functional discovery of unknown genes, and their associations with diseases [108–110].

10.2.2.3 Topological Properties of Transcriptional Regulatory Networks

Genes and their protein products are strongly related to each other through a complex network of interactions [111]. These interactions are important to contend against diverse environmental challenges, and their studies have focused on the large scale modeling of genomic and proteomic data. In this regard, network theory has been considered to obtain insights into biological systems, such as in protein–protein interaction networks or gene regulatory networks (GRNs). Particularly by GRNs, it is possible to identify common principles of organization, such as the existence of a hierarchical–modular organization and the existence of global regulators or hub nodes [112, 113]. In general, GRNs have been modeled as directed graphs because only a small set of regulator genes may activate or repress other genes. Therefore, the number of transcription factors identified in bacteria follows a correlation with the genome size, suggesting that the GRN follows a similar organization, as observed in the bacterial models *E. coli* or *B. subtilis.* In this regard, the GRNs have been characterized by topological metrics such as node degree, clustering coefficient, centralities, hubs, and communities [114]. In general terms, the degree of a node (K) is defined as the number of interactions that it has with other nodes. In directed networks, input (Kin) and output degrees ($Kout$) are defined as the number of arrows that enter and leave from a node, respectively, which corresponds to the number of regulatory genes that affect the expression of a particular gene. Thus, many biological networks present a degree distribution of the form approximately a power-law $P(k) = Ak\text{-}\gamma$ distribution, where A is a constant that warrant that the $P(k)$ values are less than one, and the degree exponent γ is often between 2 and 3 [81]. The clustering coefficient Ci of the node vi is calculated as $Ci = 2Ei/(ki)(ki\text{-}1)$, where Ei is the number of edges between the neighbors of vi, which indicates the probability that two nodes with a common neighbor in a graph are interconnected. In addition, the shortest path length has been defined as the minimal number of edges needed to reach a node from the other node through a path along the edges of the network, whereas centrality assigns every vV of a given graph G a value $C(v)R$ [115].

Finally, the connectivity in a network refers to the associations between each pair of nodes, where the connected component is a set of nodes that are linked to each other by paths and give us information about how much are connected the elements in a network and their module structure. Concerning this subject, the challenge in curating or predicting GRNs from high-throughput data make the estimate of the total number of expected interactions for the complete networks, a key factor allowing algorithms to set boundaries when assessing the possibility of connections between all possible gene pairs [116].

10.2.3 Integrating Genome-Scale Models (GSMs)

Extracellular signals affect metabolic, regulatory, and signaling processes to trigger cellular responses. Although such mechanisms are highly integrated into the cell, network analysis and reconstruction methods focused on the processes independently, without considering the interaction that exists between them [117]. From high-throughput technologies, substantial omics data coming from different sources became available in public databases, quantifying several cellular components from the messenger RNA (mRNA) levels to amounts of metabolites. As an example of the omics sciences, we can mention (i) genomics that provides information about the DNA and characteristics of the organism's genome, (ii) transcriptomics that quantifies the levels of gene expression of a complete set of transcripts for a specific physiological condition, (iii) proteomics which consists of the large-scale characterization of the set of proteins expressed in a cell or tissue, and (iv) metabolomics, which is the comprehensive and quantitative analysis of the set of all low-molecular mass metabolites present in a biological system. In this context, computational methods are needed to reduce the dimensionality of omics data and promote their integration, improving the knowledge of the underlying biological processes. Thus, a central challenge for the development of Systems Biology is not limited to reconstructing biological networks but also to integrate different types of high-performance data derived from several GSMs, in order to improve the predictive value of computational models [118].

The reconstructions of metabolic networks represent an advantageous platform for the integration of omics data because they are computational structures with manual curation that allow the description of Gene–Protein–Reaction (GPR) relationships [119, 120]. In this sense, FBA analysis has been used successfully as an analytical platform for integrating omics data into metabolic models. In particular, gene expression data are used to further restrict the possible solution space, improving the predictive capacity of the model [74].

In the last two decades, the integration between genome-scale metabolic models (GSMMs) and TRNs has become an important area of research in the System Biology of microorganisms [121]. Therefore, several algorithms were developed to incorporate expression data in GSMMs, specifically FBA-driven algorithms that use the levels of mRNA (obtained experimentally) to model the reaction network, inactivating them or restricting their activity levels [122].

However, despite the abundance of data and the variety of available methods, integration with expression data face several challenges, such as inherent noise from biological experiments, variation between platforms used to quantify the transcriptome, transcript detection bias, and a still uncertain relationship between gene expression and metabolic flux [123]. Next, we will describe the most used methods for integrating GSMMs and TRNs, showing their structures and the main differences among them.

As discussed in Sect. 2.1.3, the classic formulation of the FBA does not consider the restrictions imposed by the regulation of gene expression. The work by Covert et al. [124] showed that several cellular phenotypes could not be described just by the distribution of metabolic flux. Then, this limitation motivated the development of a pioneering method for the integration of GSMMs with gene expression data called Regulatory FBA (rFBA) [125]. The rFBA uses the formalism of Boolean logic to define additional restrictions specifying which genes in the network are active (ON) or inactive (OFF), under a specific condition. Thus, the restrictions defined by transcriptional regulatory events can be described using the value 1 to indicate that the transcriptional unit (TU) is transcribed and the value 0 that it is not. The equations that describe the expression of a TU employ Boolean logic modifiers like AND, OR, and NOT.

In general, the rFBA model simplifies the relationship between regulation and metabolism in a binary process, where genes and metabolic flux of reaction can only assume two states: ON or OFF. This approach was successfully applied in the reconstruction of the first integrated GSM for *E. coli* (iMC1010V2) [121]. A strategy similar to rFBA is used in both the Steady-State Regulated FBA (SR-FBA) described by Shlomi et al. [126] and in the integrated FBA (iFBA) developed by Covert et al. [125].

However, a difficulty for the data integration with rFBA is the absence of an automated algorithm to determine the Boolean rules that correlate the regulator to its respective target at the genomic-scale of organisms. Another limitation is the qualitative interaction of Boolean logic, that is, the genes are active or inactive, intermediate values are not allowed. Finally, the rFBA requires an extensive search in the literature and, therefore, availability of huge data regarding information of gene regulation of the organisms, hampering the use of these methods in less-studied species.

In this context, the demand for a method that automates or facilitates the reconstruction of integrated GSMs was critical. Thus, Chandrasekaran and Price [127] developed a method called Probabilistic Regulation Of Metabolism (PROM) without the disadvantages of those based on Boolean rules. For the construction of the integrated model, PROM requires GSMMs, TRNs, and significant data on gene expression under various environmental conditions or genetic disorders of the organism under study. PROM applies probabilities to represent the states of genes and the interactions between TF and each of its targets. For that, PROM binarized the gene expression data according to an user-supplied threshold to assess the expression probability of a target gene, given the expression of its TF. So, the representation of the probability of gene A is active when its regulator B is inactive is given by $P(A = 1|B = 0)$, and similarly, $P(A = 1|B = 1)$ represents the probability of gene A being active when regulator B is also active. Once calculated the probabilities, PROM incorporates them into the FBA by restricting the upper limit of the reaction to $P \times V_{max}$, where P is the probability of the gene being active and V_{max} is the maximum flux through this reaction. Although PROM describes the data in a binary way to calculate the probability of a gene activity, the actual integration of the expression data in the FBA model is not limited to the ON/OFF states of the reactions. Instead, PROM restricts the maximum flux through relevant reactions, according to the activity of the TFs [128].

In the context of target selection in pathogens, the PROM's framework was applied in an unprecedented simulation experiment in *MTB*PROM2.0, an expanded genome-scale regulatory-metabolic model constructed for *M. tuberculosis* (MTB) [129]. In this integrated model, the TRN represented the most substantial expansion of its knowledge, which was based on TF-binding measured using genome-scale ChIP-seq data generated for 190 TFs (89% of an estimated 214 MTB TFs). In this project, the updated regulatory-metabolic model *MTB*PROM2.0 was used to predict the effect of TF knockouts on MTB growth rates across multiple media conditions. Precisely, the PROM's framework was applied to infer the strength of each regulatory interaction for each TF knockout simulation, where the conditional probability that each target metabolic gene would be expressed in the absence of each TF is calculated using experimental microarray gene expression data. In regard to finding potential new drug targets for more effective MTB treatments, the most relevant results using the model *MTB*PROM2.0 were the successfully predicted synergistic interactions between overexpression of the TF *whiB4* and the activity of two anti-TB drugs, ethionamide and isoniazid.

On the other hand, one of the requirements of most integration methods is that data should be discretized *a priori* into active or inactive sets, which requires the definition of subjective thresholds for the levels of expression. In 2011, Jensen and Papin [130] developed a method to map expression data from two or more experimental conditions into a GSMM without arbitrary expression thresholding, called Metabolic Adjustment by Differential Expression (MADE). The MADE's framework uses statistically significant changes in levels of gene expression to determine sequences of low or highly expressed reactions. The MADE's confidence in the statistically significant changes through various experimental conditions to restrict the reaction activity, avoids arbitrariness around the determination of an adequate threshold in previous methods. For target prioritization, MADE was used to integrate the most current GSMM version of the ESKAPE pathogen *A. baumannii* ATCC 19606 (iLP844) with transcriptomic data from the same strain in response to treatment with colistin (polymyxin E) [131], as mentioned above, one of the 'last resort' antibiotics against MDR pathogens. Condition-specific drug targets were obtained by predicting essential genes, unique to cells treated with colistin without homologues in humans, to avoid possible side effects. As a result of this integrated model, it was possible to prioritize specific targets to be taken into consideration when developing therapies in combination with colistin in an LPS-strain.

Most publicly available metabolic reconstructions are at the genomic scale, in the sense that they seek to include all reactions found in the genome annotation with or without experimental evidence. However, most reactions integrated into the GSMMs are not active under specific conditions. In this context, the development of methods capable of adjusting the reconstructions at the genomic scale in the context-specific networks contributed to the *in silico* modeling of a particular biological condition (i.e., growing in a biofilm, as discussed in Sect. 3.1). The Becker and Palsson's work [132] presented the method called Gene Inactivity Moderated by Metabolism and Expression (GIMME) that uses quantitative data of gene expression and one or more metabolic objectives to reconstruct a context-specific network, which is more consistent with the available data. The accuracy of context-specific models has been improved over the years by restricting them to other types of omics data. However, few methods have been developed to integrate metabolomics data with other omics data in GSMMs. As an example, the work by Schmidt et al. [133] described the Gene Inactivation Moderated by Metabolism, Metabolomics and Expression (GIM^{3}E) algorithm that enables the construction of context-specific models based on an objective function, transcriptomics, and cellular metabolomics data. By using the GIM^{3}E framework the authors were able to integrate additional omics datasets with the GSMM of the pathogenic bacteria belonging to the species *Salmonella enterica* serovar Typhimurium during growth in rich media and concomitantly to investigate the virulence phenotype.

Another recent software that allows the analysis of both qualitative regulatory networks and genome-scale metabolic and integrating regulatory analysis natively into all available functions is the FlexFlux [134]. Through this open-source software, the models at the genomic scale can be analyzed separately or together and, in the last case, the states of the TRNs are used to restrict the FBA. FlexFlux is able to perform both regulatory steady-state analysis (RSA) and metabolic network analyses using FBA. In the integration of the metabolic model with the regulatory pathways the FlexFlux uses the RSA analysis, by which the goal is to achieve a single steady-state constraint for each component of the regulatory network from initial values. In brief, the main features of FlexFlux are (i) it accepts regulatory networks in qualitative format, including the simplest one represented by Boolean networks; (ii) it allows the translation of discrete qualitative states in continuous user-defined intervals, which makes it possible to restrict the reaction's flux with different intervals according to the qualitative state, different from a single value as applied in the methods described previously; (iii) it uses the regulatory analysis in steady-state RSA to restrict the metabolic flows of the reactions; and (iv) it was the first flow analysis program to accept the SBML-qual file to describe the regulatory network. The main elements of the SBML-qual file are the *QualitativeSpecies* that represent the entities of the model as the molecular components of the network and the *Transitions* with rules that define the state of certain species in each iteration [135]. For validation of the FlexFlux method, the authors successfully used the mechanism described by Jacob and Monod in 1961 [136], for the *lac* operon that triggers the diauxic shift curve corresponding to the sequential consumption of two substrates (glucose versus lactose) when both are available in the medium. While the prediction using only the FBA without considering catabolic repression does not reproduce the behavior observed in vitro, FlexFlux can perform the simulation along with RSA and FBA analyzes, with biomass maximization for the *E. coli* GSMM [137] and a qualitative model of the *lac* operon [138] (translated to the SBML-qual format) in a medium containing lactose and/or D-glucose.

Finally, the algorithm described by Banos et al., called CoRegFlux [139], is one of the most recent methods that allows integrating TRNs with GSMMs based on restrictions. The CoRegFlux framework uses a statistical method of reverse engineering to infer target genes from a given set of regulators at the genomic scale and then, estimate the effect of TFs on their respective targets, quantifying their effects as activations or repressions. The values obtained represent the transcriptional states of the cell and constitute the basis for further integration with the metabolic models. As a case study chosen by authors for its validation was the complex biological phenomenon of diauxic shift for *Saccharomyces cerevisiae*, involving major changes in transcriptional and metabolic elements of yeast.

While this chapter does not describe a complete review to date of the methods integrating the GSMs, given various accessible options when addressing on target selection in pathogens, it is important to be aware of both what program is most suitable for modeling the problem and the data availability for the pathogen under study. The integration of models at the genomic scale already imposes important obstacles to be overcome; hence, it is crucial to verify the data quality. Additionally, for modeling integrated GSMs, it is important to curate the integrated model with literature information and, if feasible, validate it with experimental approaches, so to get an improved version of a model with strong predictive value.

10.2.4 Structural Information at the Genomic Scale

The research of protein structures and functions at the genomic scale led to numerous and important discoveries in life science. Therefore, the analysis of patterns of folds into the three cellular domains helped to identify diverse and relevant information, such as common and unique protein folds, to (for instance) identify potential drug targets; and the estimation of no more than 10,000 of basic unique protein shapes [140]. In the following sections, we will describe the most relevant results associated with the identification of drug targets based on protein structure information.

10.2.4.1 Modeling the Structurome

The International Year of Crystallography was an event proclaimed in 2014 by the United Nations to celebrate the centenary of the discovery of X-ray crystallography and to emphasize the importance of crystallography in human life. Crystals have fascinated humans for millennia, still, the understanding of their nature and the utilization of their properties for something else than creating expensive jewelry had to wait until the twentieth century. Although Wilhelm Röntgen discovered X-rays in 1895 [141], it took 17 years before Max von Laue shone them on a crystal of copper sulfate pentahydrate, suspecting that the wavelength of X-rays might be comparable with the interatomic distances [142]. A year after, Lawrence Bragg and his father William Bragg developed a smart mathematical explanation of the images generated by Laue, in the form of the famous Bragg's Law, allowing to determine crystal structures [143]. In 1934, John Bernal and Dorothy Crowfoot reported the first diffraction pattern of a protein crystal [144], but it was not until the 1950s and 1960s when the first protein structures were solved after Kendrew solved the structure of Myoglobin. Soon the idea to use the information derived from protein structures to guide the design of new therapeutic agents was established.

Structural biology started to have an impact in the field of drug discovery in the early seventies. The antihypertensive captopril, approved in 1981 [145], was the first drug derived from a structure-based approach, followed by dorzolamide, used for the treatment of glaucoma, in 1995 [146]. Nowadays, structural biology is fully integrated into the drug discovery pipeline by both academic and pharmaceutical companies. In the last decades, nuclear magnetic resonance spectroscopy (NMR) and cryo-electron microscopy (cryoEM) are also becoming powerful complementary techniques to obtain the 3D structure of a protein. Nevertheless, although the number of experimentally determined structures deposited in the Protein Data Bank (PDB) [147] increased by nearly 35% to more than 161,000 in the past 3 years, the number of sequences in databases, such as UniProtKB [148], grows dramatically more rapidly. The number of sequences in UniProtKB has now reached almost 180 million. Clearly, *in silico* prediction of protein structures has become essential to narrow this gap.

The need for accurate models can frequently be met by homology-based modeling. Homology based modeling is based on the observation that protein structures evolve slower than associated sequences and in consequence similar sequences adopt almost identical structures, and distantly related sequences still fold into similar structures. In the last years, due to the increase in computational power the availability of good estimators of protein structure quality, it has become possible to model not the whole *Structurome* but a good percentage of all the proteins of a given genome. The pipeline usually consists, as a first step, of the search for good templates in the PDB, usually using a psi-blast search against a template library. Since proteins usually have different structurally independent domains, to separate them by using databases as Pfam is recommended before searching for templates. It is important to note that since in the PDB there is redundancy, sequences can be grouped at a 95% sequence identity threshold using for example CD-hit [149]. Then, every target structure can be built with a plethora of comparative modeling softwares available in the literature. One of the most popular ones is the MODELLER software [150]. Depending on the modeling software more than one structure can be generated for each template, and quality measurements can be used to select the best one, we have previously shown that generating 10 different models and using the GA341 [151], score above 0.7, and QMEAN, a score between −2 and 2 [152], good models can be obtained. It is important to remark that the quality of the models should always be presented, not only of the whole structure but also for the different structural motifs, particularly when selecting a target to conduct druggability estimations and drug development campaigns.

10.2.4.2 Druggability and the Pocketome

One of the most important criteria that define an attractive molecule to be used as a drug target is its druggability, also called bindability. Druggability is a concept that describes the ability of a given protein to bind a drug-like compound, which in turn modulates its function in the desired way [153]. From a purely structural point of view, it can be referred to as the likelihood that a small molecule binds a protein with a high affinity (<1 micromolar). First genomic reports of protein druggability revealed that only 10–14% of the human genes were considered part of the "druggable genome" a subset of genes encoding proteins that can be potentially modulated using experimental small molecule compounds [153, 154]. In contrast, the percentage of genes encoding druggable proteins increases by up to 50–75% in bacterial genomes [8, 9, 67, 155]. These findings are encouraging for antibiotic discovery. Druggable proteins are expected to have pockets with suitable physicochemical properties that allow the binding of drug-like compounds. Then, the collection of the known and predicted binding pockets for the proteome structures of a given organism can be called as its organism pocketome [156].

It is not easy to summarize the diversity of approaches that have been proposed for the identification of potential and druggable pockets. Some are based on a purely geometric analysis of the protein surface [157–160], while others are based on energy calculations [161–163]. Another way to classify the different approaches is to consider detection algorithms. The diversity of available algorithms for the pocket prediction can be classified into grid-based and grid-free approximations. Grid-based algorithms [156, 158, 160, 162, 164–166], stand on covering the protein with a 3D grid and then search for the grid points that are not superimposed by any atom of the protein and that satisfy some condition.

Grid-free approximations are based on search probes or spheres as well as methods that use the concept of Voronoi diagram. The approximations of spheres (probes) are based on the positioning of these on the protein surface and the identification of clusters of spheres with certain representative properties of the pockets. One of these algorithms, fpocket [159], has been widely used for pocket detection and druggability prediction of proteins. This open-source algorithm predicts protein pockets based on the concept of α-spheres. An α-sphere is a sphere that contacts four atoms on its boundary and contains no internal atom. In a protein, very small spheres are located inside, while larger spheres are located in the outer region. In this way, the algorithm defines a maximum and minimum size of spheres to avoid identifying either very exposed to the solvent or regions inaccessible to the solvent, respectively. After the identification of α-spheres throughout the structure, they are labeled according to the type of atoms that make it up, being apolar when it contains at least three atoms with low electronegativity, and polar when it has at least two polar atoms. Therefore, the search of protein pockets is based on the search for clusters of α-spheres in the protein structure. MetaPocket is a consensus-based method that in its recent update MetaPocket 2.0 [167] aggregates results of eight different pocket detection algorithms (among them aforementioned LIGSITEcs, Q-SiteFinder, Fpocket, and ConCavity).

Pockets can be described by physicochemical information determined from a set of descriptors, that tend to correlate the binding ability of a drug-like compound. These physicochemical parameters describe the pockets using information from atoms and residues, in the case of Fpocket this can be done by using the α-spheres properties. Some of them are the number of α-spheres, the hydrophobic density of the pocket, the proportion of apolar spheres, the apolarity score (equivalent to the sum of the apolar amino acids over the total of the same wrapped in a pocket dice), and the density of spheres (defined as the average of the distances between all spheres). The predictive potential of the descriptors is based on the fact that they were estimated by using protein pockets of which their potential drug yield is known. Thus, each structure is assigned a score (called a druggability score or DS) between 0 and 1. Based on our analysis of DS distribution for all pockets that host a drug-like compound in the PDB [9, 168, 169], pockets can be classified in four categories: (i) non-druggable (ND; $DS \leq 0.2$), (ii) poorly druggable (PD; $0.2 < DS \leq 0.5$), (iii) druggable (D; $0.5 < DS \leq 0.7$), and (iv) highly druggable (HD; $DS > 0.7$).

It is important to remark that displaying a druggable pocket is a necessary, but not sufficient condition for a target to be druggable since binding to the pocket must also modify the biological activity of the protein in the desired sense. Furthermore, evaluating the relevance of a given pocket generally demands a manual inspection, as it might involve pockets other than the active site (i.e., allosteric and protein–protein interaction sites). In this context, Target-Pathogen [170], a web server developed to prioritize drug targets for antibiotics discovery, offers easy inspection of pockets together with relevant information about active site residues (obtained from the Catalytic Site atlas, CSA), PFAM-relevant residues or in the context of protein–protein complexes. Figure 10.2 shows a druggable pocket as presented in Target-Pathogen.

Once druggable proteins are selected, other layers of multi-omics data (e.g., from transcriptomics, metabolomics, system biology) must be cohesively integrated to select a set of proteins with attractive features to be used as drug targets. Once an appropriate target is selected, structural biologists, bioinformaticians, and medicinal chemists need to continue working hand in hand to deliver selective compounds with drug-like properties.

The increasing genome-scale data availability is continuously changing the paradigm of drug discovery. New technologies contribute to the evaluation of gene function, essentiality, and suitability of proteins for drug development. However, traditional target prioritization approaches, such as searching the literature and trying

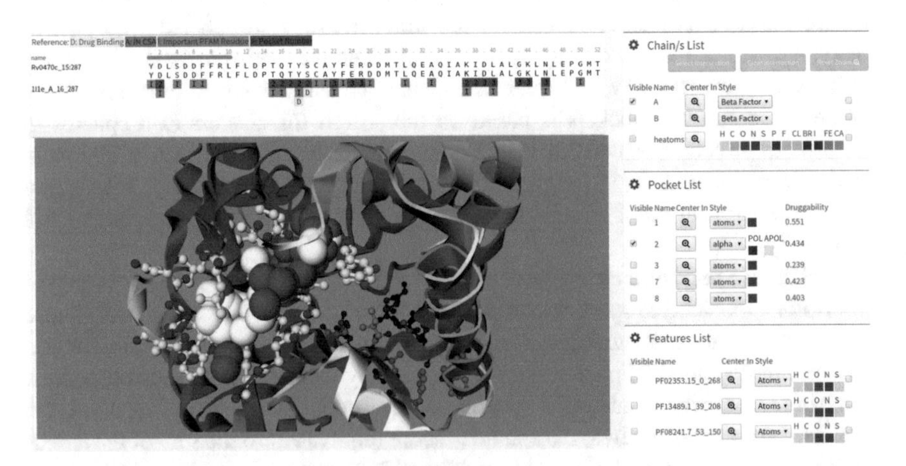

Fig. 10.2 Protein structures visualization in Target-Pathogen. The different visualization tabs that are available when searching for proteins in TP are shown. The table above shows the alignment to the corresponding crystal structure or template model. Other tabs present structure-related data, including the interactive pocket visualization module. The visualization module allows (i) to select which pocket to show (ticking the corresponding pocket Select field), (ii) display present HETATMS, assigned CSA, or PFAM-relevant residues, (iii) display the protein in different styles. In the druggable pocket of the example shown below, we depict polar alpha spheres of pocket '1' in black, while its apolar α spheres in white. The HETATMS found in the crystal structure is shown as balls and sticks in different colors. Figure from [170]

to mentally integrate diverse criteria, can quickly become overwhelming. In this sense, to fully exploit the potential of such data, the data must be effectively integrated and easy to interrogate within specialized databases. In the following section, we will mention some web servers for prioritizing targets.

10.2.5 Web Servers for Target Selection in Pathogens

Despite the various applications, databases, softwares and tools for analysis of GSMs (recently revised in [55, 171–174]), particularly genome-scale metabolic models (GSMMs), few web servers specialized for using these models to prioritize drug targets in pathogens are available to date. In general, in these web servers, the GSMMs are used only to predict gene or reaction essentiality [174]. For example, FindTargetsWEB (accessible at http://pseudomonas.procc.fiocruz.br/FindTargetsWEB), a user-friendly web application, analyzes gene essentiality using flux balance analysis (FBA), flux variability analysis (FVA) and takes as input an extended Systems Biology Markup Language (SBML) file of the GSMM of the pathogen understudy to prioritize targets [175]. In this web server, a GSM of a multidrug-resistant *P. aeruginosa* strain belonging to a clone endemic in Brazil (*P. aeruginosa* ST277) was applied as a model. For this pathogen among the prioritized targets, several had been previously reported in the literature as targets for antimicrobial development and many others have already approved drugs. FindTargetsWEB is capable of processing GSMMs of Gram-negative bacteria, Gram-positive bacteria, bacteria not classified as either Gram-positive or Gram-negative, and even integrated host–pathogen GSMMs.

Even though several user-friendly web servers allow the online use of genome-scale metabolic models without function exclusively for target selection in pathogens, they generally support tools to explore the metabolic models for *in silico* knockout experiments to predict the gene essentiality undergrowth in the selected media. In this regard, KBase [176], a collaborative, open environment for systems biology of plants, microbes, and their communities, via an interface to build and analyze metabolic models is capable of such simulation by using flux balance analysis to simulate multiple growth phenotypes (accessible at http://kbase.us/simulate-growth-on-phenotype-data-method). Also in this context, it is important to mention the MetExplore platform [177] maintained since 2010 and freely available at https://metexplore.toulouse.inra.fr/metexplore2. MetExplore offers an architecture that allows users to perform FBA and FVA (see Sect. 2.1.3) along with knockout analyzes.

Regarding databases with a focus on a single protein analysis, or on specific protein characteristics, there are few existing ones and tools are designed for target selection in a set of relevant pathogens. Potential Drug Target Database (PDTD) [178] integrated with TarFisDock (accessible at http://www.dddc.ac.cn/tarfisdock/) [179] are useful web-based resources to select potential targets using a reverse docking program. A similar tool is PharmMapper [180] (accessible at https://bio.tools/pharmmapper) server but it uses a pharmacophore mapping approach, an alternative method apart from molecular docking. The Therapeutic Targets Database [181] provides a large volume of data of already known therapeutic targets and can be freely available online at https://db.idrblab.org/ttd. Another database that includes data of known targets is TargetDB/TargetTrack [182] but only focuses on structural information. UniDrug-Target provides potential bacterial pathogenic-specific drug targets from given proteome sequences [183]. This server compares pathogenic strains proteomes sequences against beneficial strains and human proteomes to identify unique proteins in the former. Target Hunter (https://www.cbligand.org/Target Hunter/) [184] is a web server that assigns the targets associated with the most similar compounds of a query chemical as the predicted targets. Also, T-iDT [185] is a standalone tool used to identify essential genes by comparing a query against the Database of Essential Genes (DEG) [186] and exclude homologous human genes.

There are also resources focusing on a particular group of pathogens. TDR targets [187] is an interesting resource focused on tropical neglected diseases. TuberQ provides a druggability analysis of the *Mycobacterium tuberculosis* proteome contributing to a better selection of potential drug targets for screening campaigns, and for the analysis of targets in structure-based drug design projects [155]. TargetTB [188] is another server focused on *Mycobacterium tuberculosis*. It integrates network analysis of the protein–protein interactome, a flux balance analysis of the Reactome, experimentally derived phenotype essentiality data, sequence analyses, and a structural assessment of targetability. The antibacTR database [189] is designed to help in identifying potential antimicrobial drug targets on Gram-negative bacteria focusing on essentiality and conservation of proteins.

Combining structurome, drugome, and metabolic information, Target-Pathogen [170] is a novel resource to select and prioritize drug targets of several relevant pathogens, including *M. tuberculosis*, *M. leprae*, *K. pneumoniae*, *S. aureus*, *Schistosoma mansoni*, *Shigella dysenteriae Sd197*, *Toxoplasma gondii*, *Leishmania major*, and *Trypanosoma cruzi* among others. It provides a large amount of information for genes and proteins within the available genomes is actually present in Target-Pathogen including protein function, metabolic role, off-targeting, structural druggability, essentiality, and omic experiments. Using TP, researchers can easily prioritize proteins of interest in a quickly and intuitive manner, running simple queries (such as looking for proteins with high druggability score or associated with high centrality reactions), filtering by different data, assigning numerical weights for different features and combining these results to produce a ranked list of targets. In this sense, an aspect that distinguishes Target-Pathogen from other servers is its ability to upload users' own data. Once new data is uploaded, it can be included to obtain a personalized ranking of candidate drug targets in pathogens. Another feature where Target-Pathogen stands alone is its capacity to rank not only proteins but entire pathways. This feature allows users to prioritize promising pathways for new drug development in order to synergistically attack several proteins of the same metabolic pathway. As of February 2020, there are 23 of the most relevant microorganisms for human health, but the database can be easily updated with other relevant pathogens as the bioinformatic pipelines have been automatized. Target-Pathogen encourages users to request new genomes of interest by mailing to target@biargentina.com.ar.

10.3 Pathogen-Focused Applications

Nowadays, omic approaches have created new opportunities for antibiotic discovery, aiding in the evaluation of protein suitability to be used as a drug target. Previous *in silico* efforts have been used to select drug targets in clinically relevant pathogens such as *M. leprae* [190], *Shigella flexneri* [191], *Corynebacterium* spp, [8], *Treponema pallidum* [192], *Haemophilus ducreyi* [193], *Salmonella typhi* [194], *Streptococcus pneumoniae* [195], *Mycobacterium tuberculosis* [9, 10, 196–198], *Helicobacter pylori* [199], *Staphylococcus epidermidis* [11], *Klebsiella pneumoniae* [67], *E. coli* [200], *Salmonella enterica* subsp. *enterica* serovar Poona [201], *Brucella melitensis* [202], *Clostridium botulinum* [203, 204], *Fusobacterium nucleatum* [205], *Pseudomonas aeruginosa* [12], *Bacillus anthracis* [206], and many other pathogens. Regardless, in this section, we will describe different efforts to select and prioritize biological targets in the species *Staphylococcus aureus*, *Klebsiella pneumoniae*, and *Mycobacterium tuberculosis*.

10.3.1 Staphylococcus aureus

Staphylococcus aureus is a Gram-positive pathogen of wide clinical relevance, responsible for hospital and community-associated infections. [207]. In humans, can cause from the simplest and localized skin lesions such as impetigo and abscesses to the most serious and invasive infections such as pneumonia, endocarditis, osteomyelitis, and meningitis, which can progress to septicemia [208]. In addition, *S. aureus* is the etiologic agent of diseases associated with toxins such as food poisoning [209]. *S. aureus* is an aggressive pathogen as a consequence of the combination of elevated antibiotic resistance and prominent virulence with the production of an array of cell surface and secreted factors, including proteins that promote adhesion to host and some to evade triggered immune responses. [210].

One of the most relevant characteristics of *S. aureus* is its feasibility for acquiring antimicrobial resistance (AMR), as the methicillin resistance phenotype (Methicillin-Resistant *Staphylococcus aureus*, MRSA). Methicillin was introduced for clinical use in 1961, and in just six months the first cases of MRSA strains in a British hospital have been reported [211]. MRSA isolates are not restricted to the hospital environments (healthcare-associated MRSA strains, HA-MRSA), also from the 1980s, MRSA strains began to be identified in healthy individuals in the population, including Australian Indian populations (community-associated MRSA strains, CA-MRSA) [212]. Currently, MRSA is one of the main agents of healthcare-related infections (HAIs) around the world, increasing the cost of treatment and patient mortality when compared to infections associated with strains susceptible to methicillin (Methicillin-Susceptible *Staphylococcus aureus*, MSSA).

The population structure of *S. aureus* indicates that five clonal strains originated the majority of nosocomial MRSA isolated in the world and belong to the multi-locus sequence types ST5, ST8, ST22, ST30, and ST45 [213]. Probably, the various MRSA clones emerged through the horizontal transfer of the Staphylococcal Cassette Chromosome *mec* (SCC*mec*), the mobile genetic element that confers resistance to methicillin, in these five clonal strains at different times. The results obtained from studying the population structure of ST5 provided strong evidence that there is a limited geographical spread of individual clones of the predominant MRSA strains [214]. However, an exception is a hybrid MRSA clonal group ST239 which has generated multiple variants disseminated globally with well-documented hospital epidemics on all continents. The chromosome of this MRSA hybrid has parental contributions estimated at 20% of ST30 and 80% of ST8, being the first reports of *S. aureus* infections belonging to ST239 date from the late 1970s [215–219]. In Brazil, a multiresistant *S. aureus* clone (Brazilian epidemic clone, BEC) of the lineage ST239 is widely disseminated in hospitals and samples with profiles of high resistance to methicillin and several other antibiotics were isolated in geographically distant regions of the country, thus corroborating the great capacity of dissemination of this clonal complex [219].

The multidrug resistance of *S. aureus* strains, including antibiotics of 'last resort' such as vancomycin, limits the clinical practice. Beside, isolates with high adhesion capacity through the formation of biofilm promote chronic infections. The biofilm can be defined as a sessile microbial community in which cells are attached to an abiotic surface and/or to other cells and surrounded by an amorphous protective matrix [220]. In the hospital conditions, the biofilm represents an important microenvironment for the *S. aureus* fitness that allows the cell aggregation in implanted medical devices associated with persistent and chronic infections. Even worse, *S. aureus* can establish secondary infections mediated by biofilm, which are potentially more serious and resulting from the displacement of bacteria from the primary site of the infection [221, 222]. Depending on the matrix composition, the staphylococcal biofilm can be classified as *ica*-dependent or *ica*-independent. The first one is mediated by the intercellular adhesion polysaccharide PIA, whereas the mechanisms involved in the formation of the *ica*-independent biofilm are multifactorial and not well understood [223]. In general, the *ica*-independent biofilm consists of several structural cellular components that are modulated by a multifaceted and overlapping regulatory network, which results in a strain-dependent biofilm composition [224]. Together, the characteristics presented here for the several *S. aureus* clones argue for the urgent necessity to discern what mechanisms are underlying the infectious process and then how to prioritize targets for these pathogens. In this section, particularly we discuss a case study carried out by our group for a BEC strain belonging to the ST239 lineage. For which we produce a preliminary curated GSMM and integrate it with some regulatory pathways differentially expressed in biofilm conditions capable of influencing the metabolism of this pathogen.

The project (part of the doctoral thesis of Maiana de Oliveira Cerqueira e Costa) [225] began with a comparative genomics analysis with two ST239 Brazilian isolates Bmb9393 (used as a reference) and Gv69 as well as eight international strains belonging to ST239, ST8, or ST30 lineages. The Brazilian ST239 strains were nosocomial MRSA, being Bmb9393 isolated from a bloodstream infection in the Rio de Janeiro, and Gv69 from a skin infection of a patient with pneumonia in Teresina, Piauí. Our group sequenced and determined the complete genome of both strains, with the Bmb9393 consisting of a circular chromosome of 2.9 Mpb and a circular plasmid of 2,908 bp [226] and Gv69 of a 3.0-Mpb chromosome [227]. After the open reading frame (ORF) predictions and automatic annotation with the SABIA platform [228], an important manual curation process was carried out to improve the quality of the annotation of the complete genomes of Bmb9393 (GenBank: CP005288 for the chromosome and CP005289 for the plasmid) and Gv69 (GenBank: CP009681). Other results regarding the comparative study of regions of genomic plasticity (RGPs) are described by Botelho et al. [219].

Despite the great synteny among the genomes of *S. aureus* under this study, the comparative analysis provided relevant insights regarding the mechanisms of virulence and resistance of the ST239 strain. However, biofilm production, a phenotype associated with pathogenesis, is not possible to unravel merely at the genomic level. That is because there are multiple determinants whose gene expression levels are in a strain-dependent manner, defined by either the individual and/or integrated action of transcriptional regulators. Thus, to understand the superior ability of Bmb9393 to accumulate *ica*-independent biofilm to adhere and invade human airway cells, it was necessary to reconstruct the TRN of this strain and also to get gene expression data (i.e., planktonic versus biofilm conditions) for further analyzes. It is important to note that biofilm production is a multifactorial process influenced by the nutritional gradients throughout the layers and the bacteria require immediate adaptation to abrupt changes in the environment to accumulate this protective matrix [229]. Thus, the formation of biofilm in *S. aureus* is an interesting choice for an integrative modeling approach, in which genes with altered expression levels in the biofilm condition are integrated into a metabolic model of the organism model.

Thus, we started with the TRN reconstruction of the Bmb9393 strain applying a hybrid methodology based on reference TRNs and the prediction of *cis*-RE (see Sect. 2.2.2). The identification of regulatory interactions between TFs and their respective target genes, that is, the determination of the regulons present in Bmb9393, were obtained through the following steps: (i) comparative propagation of the TRN of the strain of *S. aureus* N315 [230], (ii) search for TFBSs of *S. aureus* and the model species for Gram-positive bacteria *B. subtilis* with TFs orthologous of Bmb9393, to expand those already propagated and also discover new regulons, and (iii) inference of regulons through the literature. After all these treads, the final TRN of Bmb9393 contains 1,075 interactions between 64 TFs and 805 target genes/operons (non-redundant set) (Fig. 10.3). On the other hand, the TRN of *S. aureus* strain N315 reconstructed by Ravcheev et al. [230] yielded a network with fewer elements, corresponding to just 663 interactions between 49 TFs and 547 target genes/operons (non-redundant set).

After the reconstruction of the Bmb9393 TRN, the next step was to identify which regulons were associated with the biofilm formation. The process consisted of mining all the genes differentially expressed in biofilm versus planktonic conditions

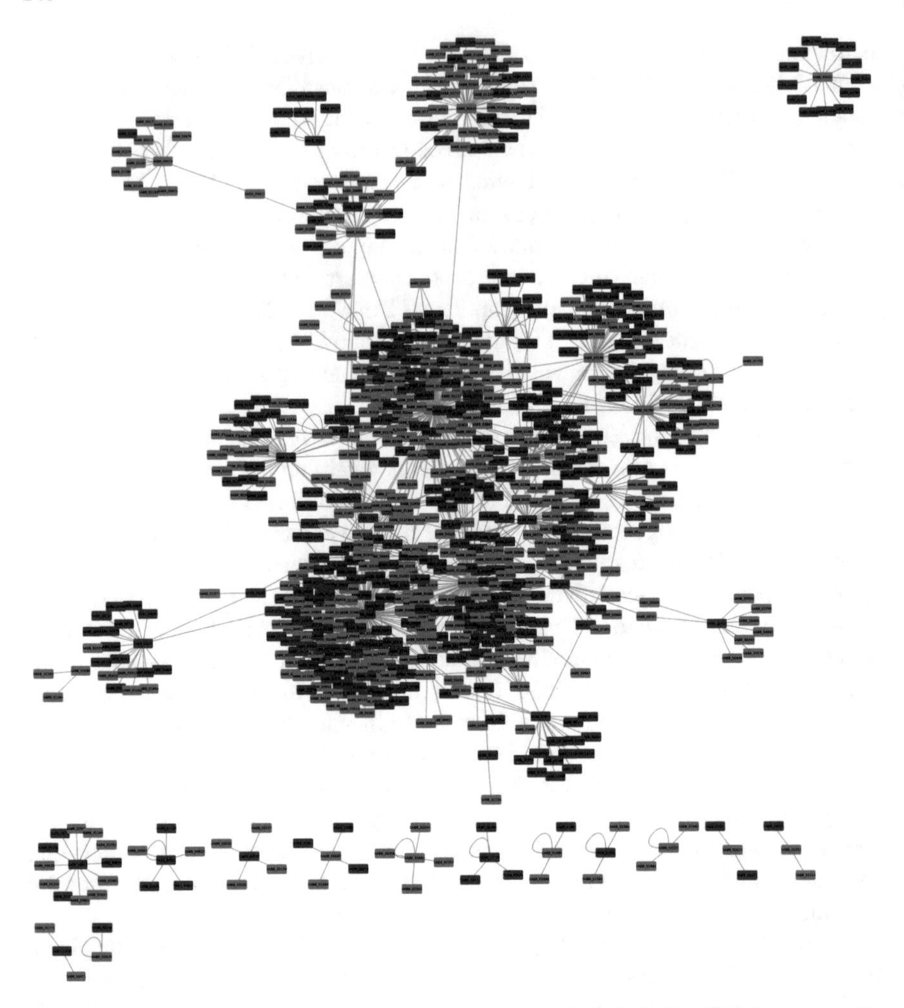

Fig. 10.3 Graphic representation of the overlap of the TRN of the Bmb9393 strain to the genes with altered expression in the biofilm condition. The red nodes depict those with altered biofilm expression and blue nodes represent those unique to Bmb9393's TRN. Figure from Costa, MOC (Tese 2018) [225]

from a lot of experiments using microarray, both available from the literature and specific databases such as SATMD [231]. This procedure yielded a non-redundant set of 992 genes altered in the biofilm condition with orthologous in the Bmb9393 genome, which were identified using the Proteinortho program [232] and Perl scripts. Subsequently, the non-redundant set of 992 genes was mapped with the TRN of the Bmb9393 strain, to recover all the regulons that had their expression altered in at least one of their respective target genes in the biofilm condition (Fig. 10.3). Out of the set of 992 altered genes in the biofilm condition, 384 are target genes of several identified regulons in the Bmb9393 TRN, of which 112 are regulated by more than one TF and/or sigma factor. The 384 genes are regulated by 48 TFs and different sigma factors, that is, the growth pattern in the biofilm alters the expression of target genes belonging to 48 out of 64 total regulons described in the reconstructed TRN for Bmb9393. Finally, through this data mining procedure, we were able to identify (out of these 48 regulons) 36 regulons that had the altered expression for all their target genes, namely *codY*, *saeR*, *sarA*, sigma factor *sigB*, *braR*, *csoR*, *hrcA*, *ctsR*, *icaR*, *bglR*, *cggR*, *fruR*, *glvR*, *mtlR*, *murR*, *arcR*, *argR*, *cymR*, *glnR*, *hisR*, *birA*, *fapR*, *comK*, *sigH*, *nreC*, *srrA*, *agrA*, *ccpE*, *czrA*, *gbsR*, *hssR*, *purR*, *rbsR*, *rex*, *rsp*, and *sarR*.

As a metabolic model for the Bmb9393 strain had already been published in the work by Bosi et al. [233], we used it as a starting point to perform a curation and refinement process. This was necessary since the reconstruction of metabolic models must be an iterative process to get the refined and accurate versions of the model, as new information and tools become available. Thus, we subjected the available SBML of the Bmb9393's GSMM to a rigorous manual curation and refinement process. After this process, we performed the mapping of GPRs in this model, which was essential for further integration with the regulatory data as well as for the *in silico* gene knockouts. Also, we carry out FBA analyzes on three models, namely the one published by Bosi et al. [233] and two of our models, the curated SBML-file of the Bmb9393's GSMM and the integrated version with the regulatory pathways, using biomass as an objective function in all cases.

The last step involved the integration of the curated Bmb9393's GSMM with the regulatory pathways associated with the biofilm, in this case with the regulons whose expression in all of its target genes was altered in the biofilm condition. For that, firstly, we obtained the SBML-qual files for each regulatory pathway, separately, for using as inputs in the FlexFlux program together with the SBML files of Bmb9393. As a qualitative model, the SBML-qual represents the necessary rules for updating the values of the components of the regulatory network, instead of the quantities associated with them (see Sect. 2.3) [135]. To obtain the necessary format, we use the Cell Collective platform [234], where you can build the regulatory path of interest using information from the literature as well as data obtained from databases and then can export the model in the desired qualitative format. After that, the integration itself was carried out, through FBA analyzes referred to each SBML-qual (related to each regulatory path at a time) along with the RSA tool available in the FlexFlux platform [134].

To exemplify here, we will relate the draft of an integrated GSMM of Bmb9393 with three selected TFs for ArcR (degradation of arginine), GlnR (assimilation of nitrogen), GlvR (use of maltose) and MurR (use of N-acetylmuramic acid, MurNAc), for being three of the best regulons with altered expression under biofilm conditions. In addition to these, the regulon associated with the use of lactose was used as a method to evaluate the methodology used, since the behavior of the lac operon is well studied and the expected results are well documented in the literature (see Flex Flux in Sect. 2.3).

Next, we present the results of the growth simulation with the integrated GSMM obtained for the use of alternative sources of carbon and the degradation of arginine. From the information on the regulatory interactions of the Bmb9393 TRN and the data available in the literature, regarding the external signals and regulation of the genes participating in the selected regulons, we built, for each of them separately, a qualitative regulatory model in the format SBML-qual, under the Flex Flux framework, to be integrated into metabolism. In the next step, each of them was integrated by the FBA analysis on the metabolic model, to respond to the conditions determined in the environment represented by the constraint file (which specifies the objective function and can contain constraints to simulate, e.g., different external metabolite concentrations). Therefore, instead of modifying the exchange reaction limit (i.e., of compounds in the model itself to simulate the medium), the integrated model can incorporate regulatory restrictions to respond adequately to nutrient availability through associations gene–protein–reactions (GPRs).

The first regulon integrated into the GSMM of Bmb9393 was the LacR repressor, which consists of the seven *lacABCDFEG* genes, and is responsible for the transport and use of lactose. As an alternative source of carbon, the lactose is under the effect of the CCR phenomenon, where there is a repression of genes associated with the use of secondary carbon sources (i.e., lactose) in the presence of a preferred carbon source, which in the case of *S. aureus* is glucose [235, 236]. So, the hierarchical use of sugars enables an "economical" and efficient use of carbon and energy sources [237]. Thus, the integrated model must be able to respond correctly, without any alteration in the metabolic model itself, to the three media conditions tested: (i) glucose only, (ii) lactose only, and (iii) the presence of glucose and lactose. When performing the three simulations, with the medium conditions specified by the restrictions file, the biomass values are (i) growth with glucose $\Rightarrow$ 2.874862 mmol/g DCW/h, (ii) growth with lactose $\Rightarrow$ 2.788074 mmol/g DCW/h, and (iii) growth with both sugars present in the medium $\Rightarrow$ 2.874862 mmol/g DCW/h. Thus, the results qualitatively reproduce the regulatory phenomenon CCR described in the literature, since the biomass value, when both sugars are available, is equal to the corresponding medium only with glucose, indicating that the preferred source was consumed first.

Beside lactose, the use of maltose regulated by GlvR and the use of N-acetylmuramic acid modulated by MurR presented the same results of biomass values, when tested under the same conditions as described for the *lac* operon. Considering that maltose and N-acetylmuramic acid are alternative carbon sources and both are under the effect of the CCR, it was expected that the integrated metabolic model would behave the same as with lactose/glucose in a medium with these carbohydrates. Previously, the metabolic model Bmb9393 described by Bossi et al. [233] had responded similarly with the reduction of the biomass value when maltose was the only available carbon source. Corroborating the importance of using alternative carbon sources in biofilm, studies have associated the global regulator CcpA, the mediator of the CCR phenomenon, and the TCA cycle to virulence, especially in the formation of biofilm. Indeed, cells in the biofilm condition represent a fairly heterogeneous population physiologically, due to oxygen, nutrient, and pH gradients. This mosaic of conditions requires permanent adaptation of the cells to survive in such a complex environment. Growth in a nutrient-rich condition reduces the bacterial demand for biosynthetic intermediates provided by the TCA cycle because they are obtained from the external environment. However, when environmental conditions change and nutrients become limiting, as in the case of growth in biofilm, *S. aureus* increases the activity of the TCA cycle and catabolizes alternative carbon sources, such as maltose or N-acetylmuramic acid [238].

In addition to the genes associated with alternative sources of carbohydrate, ArcR is also under the effect of CCR through the global regulator CcpA. In *S. aureus*, ArcR is a TF activator of the arginine catabolism pathway called arginine deaminase (ADI), consisting of the enzymes arginine deaminase (ArcA), ornithine transcarbamylase (ArcB), and carbamate kinase (ArcC), pathway responsible for converting the arginine in the ornithine, ammonia, and carbon dioxide, with concomitant ATP production [239]. The genes encoding the enzymes of the ADI pathway are organized in the operon *arc*, along with ArcD, the arginine–ornithine transporter that catalyzes arginine uptake and the concomitant export of ornithine from the cell. While ArcR is not essential for aerobic growth, it supports growth under anaerobic conditions when arginine is the only available carbon source. This behavior was confirmed by the FBA analysis on the curated and integrated model for the ArcR regulon. Thus, using the restriction file simulating an anaerobic medium, without glucose and only with arginine as a carbon source, the resulting biomass value of the model was 2.705626 mmol/g DCW/h. This result reproduces an important growth under biofilm condition, as described in the literature [240, 241]. It is important to note that this result indicates an improvement over the model of Bmb9393 by Bosi et al. [233], which was unable to grow using arginine as a carbon source under the same anaerobiosis condition.

In *S. aureus*, the transcription of the operon *arc* is dependent on the ArcR activator and the ArgR repressor, in addition to the regulation of CcpA through the CCR, its expression is induced under anaerobic growth conditions [242]. In addition to its function in obtaining energy, ADI is one of the two main pathways that produce ammonia, being used by many lactic acid bacteria in maintaining pH homeostasis and survival during low pH conditions, by neutralizing acids resulting from bacterial fermentation [243].

The results obtained with the draft integrated genome-scale model of Bmb9393 represent important advances for this strain. We highlighted the build of the first metabolic model integrated with the relevant regulatory pathways for the formation of biofilm in *S. aureus*, which can satisfactorily respond to the tested environment conditions, such as some alternative carbon sources, contrary to the metabolic model alone. As perspectives in this project, we will carry out transcriptome experiments under anaerobic conditions, and on the alternative carbon sources described here, that is, maltose, lactose, and N-acetylmuramic acid, to validate the integrated model and simulate the medium found in deeper layers of the biofilm. In the future, with a metabolic model integrated with regulatory data from *S. aureus* with greater predictive value, we hope to indicate some TFs candidate targets for further alternative therapies, such as silencing CRISPR-Cas associated with phage therapy.

10.3.2 *Klebsiella pneumoniae*

K. pneumoniae is a Gram-negative bacterium belonging to the Enterobacteriaceae family that occupies diverse ecological niches ranging from soil to water [244]. From a human health perspective, it represents one of the most important opportunistic pathogens [245–247]. The *K. pneumoniae* opportunistic nosocomial and healthcare-associated infections (HAIs) commonly are manifested by pneumonia, urinary tract, and wound infections, any of which can progress to bacteremia [248]. Whereas outside the hospital setting, *K. pneumoniae* can cause severe community-acquired infections (CAIs), which include endophthalmitis, pneumonia, necrotizing fasciitis, non-hepatic abscess, meningitis and pyogenic liver abscess in the absence of biliary tract disease [249–251].

K. pneumoniae is intrinsically resistant to aminopenicillin used to treat Gram-negative infections (because of the production of the SHV β-lactamase encoded by the chromosomal core gene *blaSHV*). Also, resistance to all drug classes used to treat this species has been observed clinically most associated with horizontally acquired antimicrobial resistance (AMR) genes [252–254]. Interestingly, AMR genes are non-randomly distributed in the *K. pneumoniae* population, and the number of acquired AMR genes per strain follows a bimodal distribution, observing most strains carrying either zero acquired AMR genes or ~10 acquired AMR genes (encoding resistance to multiple drugs) [248].

Although, among the AMR repertoire of *K. pneumoniae*, the production of *K. pneumoniae* carbapenemase (KPC), the main Ambler's class A carbapenemase found in this species, is particularly worrisome since it confers resistance to all beta-lactams [255]. Indeed, infections caused by carbapenem-resistant *K. pneumoniae* (CRKp) represent a high burden of disease worldwide, especially in countries like Argentina, Brazil, China, Colombia, Greece, Israel, Italy, Poland, and the USA where KPC-2-producing *K. pneumoniae* are endemic [256, 257].

There are hundreds of distinct phylogenetic lineages or clones within the *K. pneumoniae* population, but some of them are highlighted as global problem clones since they are multidrug-resistant (MDR) (e.g., clonal group 15 (CG15), CG20, CG29, CG37, CG147, CG101, CG258, and CG307) or hypervirulent clones (e.g., CG23, CG25, CG65, CG66, CG86, and CG380) [248]. Regarding the hypervirulent infection, a diversity of virulence-associated determinants have already been described, some of them are (i) capsule-locus genes (*magA*, K1-specific *wzy* capsule repeat unit polymerase, and *wcaG*, a component of K-locus 1 (KL1) as well as KL6, KL16, KL54, KL58, KL63, and KL113), (ii) clone-specific markers (*kfu*—ferric iron uptake—and *alls*—allantoin metabolism—are conserved in CG23 and *kvgAS* is conserved in CG86, CG65, and CG25), and (iii) hypermucoidy phenotype associated to the presence of *rmpA* genes (revised in [248]).

However, the most concerning problem now is the convergence of resistance and virulence in the one single *K. pneumoniae* strain. Pertinent findings revealed that MDR clones pose the greatest risk, as they are more likely to acquire virulence genes by horizontal gene transfer than hypervirulent clones to acquire resistance genes. This scenario is particularly critical from the perspective of hospital infection control, as these MDR clones are responsible for frequent hospital outbreaks [258].

In Brazil, strains of *K. pneumoniae* MDR causing hospital infections in adult patients hospitalized at the intensive care units in 2018 were most frequently resistant to both broad-spectrum cephalosporins and carbapenems (last report of the Brazilian Health Surveillance Agency, 2018, accessible at https://bit.ly/2MRfYvE). Thus, in these cases, polymyxins have become last-resort antimicrobials for the treatment of serious infections caused by these MDR/CRKp strains. Unfortunately, to make things even worse, an important increase in the resistance rates to polymyxins has been observed among carbapenem-resistant *K. pneumoniae*. Braun and collaborators have observed an increase in polymyxin B resistance rates from 0 to 30.6% among *K. pneumoniae* isolates recovered from blood cultures between 2009 and 2015 in a tertiary Brazilian hospital [259]. Because of this, search for new strategies to counter these infections is ongoing and include the use of novel approaches such as polymyxin combination therapy, immunotherapy based on monoclonal mAbs or even nanoparticles in combination with antibiotics [260].

In this section, we report our recently published case study of a multidimensional data integration strategy to prioritize drug targets in MDR/CRKp [67]. In this work, we used several layers of 'omics' data and 'Metadata' related to the species *Klebsiella pneumoniae* (using strain Kp13 as a genome reference) to prioritize proteins with characteristics of attractive potential targets for the development of new antimicrobials. Finally, genomic, transcriptomic information, structural, regulatory, metabolic, and functional data were integrated into the Target-Pathogen (accessible at http://target.sbg.qb.fcen.uba.ar), which allows the prioritization of candidate targets based on a scoring function. Particularly, we incorporated information about polymyxin resistance in our analyses, in order to enrich for targets that would also be useful against the MDR/CRKp strains (Fig. 10.4).

This study started with the genomic analysis of *Klebsiella pneumoniae* Kp13 (referred to as Kp13 throughout the text), responsible for a hospital outbreak in southern Brazil in 2009, being CRKp and resistant to many antibiotics, including polymyxin B. Our group has determined its complete genome, which comprises one 5.3 Mbp circular chromosome and six plasmids (totaling 0.43 Mbp), and we have manually annotated its predicted coding sequences (CDS), composed of 5,736 predicted peptides (BioProject/NCBI PRJNA78291) [261]. Firstly, the Kp13 proteome was used as a query in BLASTp against the predicted human proteome (from version GRCh38.p10) to identify non-host homologous targets (identity $\geq$ 40% with a human protein were filtered out). Similarly, orthologous proteins were filtered by comparison to the proteins of the gut flora sequenced by the Human Microbiome Project [262].

Following off-target proteins removal, the remaining Kp13 proteins were classified according to their druggability (see definition in Sect. 2.4.2). Using a methodology previously described by our group for druggability prediction [155], based on the open-source pocket detection algorithm fpocket [169], several physicochemical descriptors to estimate the pocket druggability were used on a large genomic scale of Kp13. All pockets that are found to host a drug-like compound in the PDB, pockets

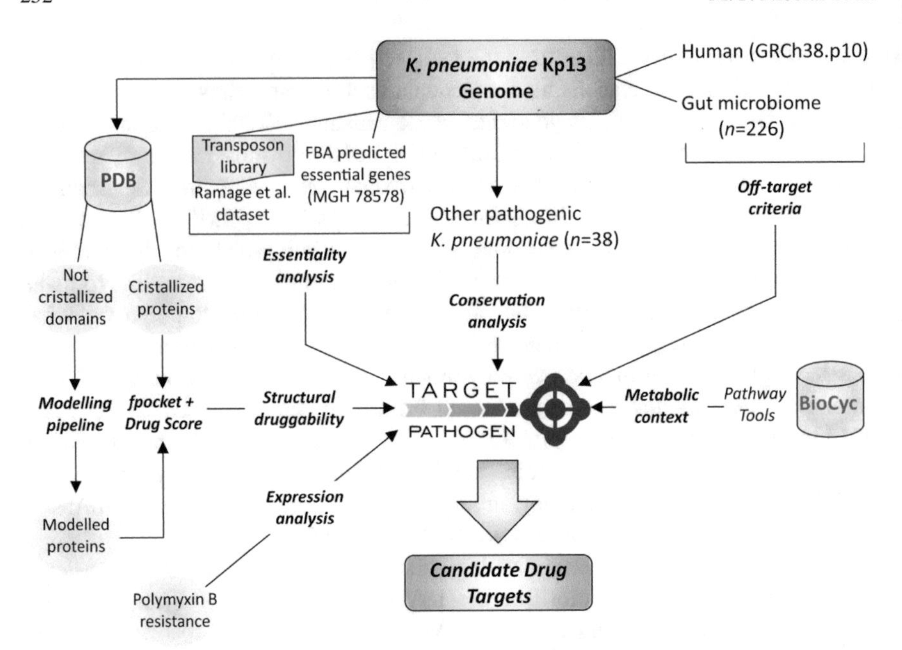

Fig. 10.4 A general outline of the prioritization pipeline. Analysis in silico of druggability, subcellular location, gene essentiality, conservation in pathogens, off-target in humans, and gut microbiota, gene expression, and metabolic context. Figure from [67]

were classified into four categories: non-druggable ($0.0 \leq DS < 0.2$), poorly druggable ($0.2 \leq DS < 0.5$), druggable ($0.5 \leq DS < 0.7$) and highly druggable ($0.7 \leq DS \leq 1.0$). All proteins for which we obtained structural models were subject to this classification (refer to http://target.sbg.qb.fcen.uba.ar/patho/public/docs/TargetMethodology.pdf for further details). By applying these filters, we obtained 2,950 candidate druggable proteins (out of 3,194 total proteins) with no close homologs in the human genome.

Also, the proteome of Kp13 was used to search for groups of orthologs among different *K. pneumoniae* proteomes (with protein identity $\geq$ 60% and coverage $\geq$ 70%). The conservation of a protein in multiple *K. pneumoniae* genomes implies that a drug binding such target could be used to control multiple strains of this bacterium, including from different sequence types (STs). Thus, while we used Kp13 as a reference organism, our results can be broadly expanded to *K. pneumoniae* bacteria disseminated in various geographical regions.

We investigated the essentiality criteria because to achieve this human pathogen, an ideal target drug should be essential for the pathogen in all the host microenvironments it encounters. Thus, to consider whether a gene was essential, we used an available large-scale study (arrayed transposon mutant library) identifying essential growth genes in *K. pneumoniae* described Ramage et al. [263]. Furthermore, we also analyzed an experimentally validated *in silico* genome-scale metabolic reconstruction available for *K. pneumoniae* MGH 78578 described by Liao et al. [263, 264].

For this integration analysis, we performed a whole-genome-based reconstruction of the Kp13 metabolic network (Kp13-MN) using Pathway Tools algorithm and incorporating evidence from a previously curated *K. pneumoniae* metabolic network [264], followed by manual inspection and curation of the resulting Kp13 network. Once constructed, this network was analyzed from a graph-theoretic point of view as a reaction graph, allowing the calculation of topological metrics that relate to node importance. Using this strategy, we obtained a Kp13-MN containing a total of 1,969 reactions compose, with 1,847 being enzyme-catalyzed and forming part of 321 predicted metabolic pathways. We also identified choke-points (CPs) in the Kp13-MN, that is, reactions that uniquely consume (input CPs) or produce (output CPs) a given compound (see definition in Sect. 2.1.1). A total of 145 reactions were strictly classified as input CPs, while 154 reactions were strictly output CPs. On the other hand, 149 reactions were classified as CPs on both producing and consuming sides of the reaction. Mapping these CPs reactions to proteins resulted in a total of 841 proteins, corresponding 34% of these to essential proteins. The projection of Kp13-MN onto a reaction graph allowed the calculation of topological metrics, such as betweenness centrality. Figure 10.5 depicts the resultant Kp13-MN graph, with node sizes proportional to this metric. The presence of few high-centrality nodes indicates that these hubs may be of special importance to the cohesiveness of the network.

For target prioritization, firstly, we applied Eq. (10.3) that defines the importance of a protein as a target according to essentiality, conservation, and metabolic context criteria, which allows the identification of general targets against *K. pneumoniae*. We applied the Eq. (10.3) for the 2,950 candidate druggable proteins with no close homologs in the human genome. Thus, for each protein, we defined its score based on the following function:

$$SF = \left(E_{\text{mgh}} + E_{\text{deg}}\right)/2 + C_v + C_y + chk \tag{10.3}$$

where the first term of the equation incorporates essentiality analysis as described, with E_{mgh} and E_{kpn} assumed to be 1 if the protein has a hit with an essential gene; otherwise, these terms are zeroed. Cv is the proportion of hits of the protein in different pathogenic *K. pneumoniae*. Cy is the ratio between the node betweenness centrality of the associated reaction and the node with the highest centrality in Kp13-MN and chk defines if the protein is associated with a chokepoint reaction ($chk = 1$, otherwise $chk = 0$). Based on the Eq. (10.3), we got the first ranking of *K. pneumoniae* proteins, a set of candidates which we called 'general targets'.

The 15 highest ranking proteins are involved in fatty acids, lipopolysaccharide (LPS), peptidoglycan, pyrimidine deoxyribonucleotides, and purine nucleotides biosynthesis. Proteins involved in fatty acid biosynthesis components pathways (e.g., 3-oxoacyl-[acyl-carrier-protein] synthase 1 and 3 and enoyl-[acyl-carrier-protein] reductase [NADH]) are druggable, essential, conserved, and majorly related to important reactions from the metabolic point of view and principally are choke-points. This pathway allows homeostasis of the bacterial membrane, in particular, the enoyl-[acyl-carrier-protein] reductase [NADH] (FabI) has been targeted for development of new

Fig. 10.5 Metabolic network of *K. pneumoniae* Kp13 represented as a reaction graph. Nodes depict reactions in the network, and there exists an edge between two nodes when the product of a reaction is used as the substrate in the following reaction. Node size is proportional to betweenness centrality, and MetaCyc accessions (http://metacyc.org) for hub reactions are shown. Figure from [67]

antibacterial agents [27, 265]. Synthesis of lipopolysaccharide (LPS), an essential component of the Gram-negative outer membrane, also appeared top-ranked, with cytoplasmic enzymes LpxA, LpxC, and LpxD involved in the initial steps of lipid A production through the Raetz pathway [266]. In the last two decades, numerous LpxC inhibitors have been developed as bactericidal agents against several enterobacteria including *K. pneumoniae*, and recently, a novel inhibitor (LPC-069) was demonstrated to be efficient against extremely drug-resistant strains with no known adverse effects in mice [267]. Accordingly, our results showed that the gene encoding LpxC protein is conserved in all studied pathogenic strains of *K. pneumoniae* and does not present close homologs within the human proteome.

Once targets that complied with rules associated with gene essentiality, metabolic importance, and broad *K. pneumoniae* conservation were identified, we included in our scoring function the Eq. (10.4) that incorporates polymyxin B (PB) resistance:

$$SF = \left(\left(E_{\mathrm{mgh}} + E_{\mathrm{deg}}\right)/2 + C_v + C_y + chk\right)/4 + P_b \tag{10.4}$$

where P_b is 1 if the protein is overexpressed in PB presence.

PB is an antibiotic considered as 'last resort' in the treatment of infections caused by carbapenems resistant pathogens. We reasoned that Kp13 proteins that are overexpressed when exposed to the antibiotic may have a role in counteracting the deleterious effects of the drug in the bacterial cell, as was shown for other pathogens [268]. Further, these proteins need not be necessarily resistance-related or involved in alterations to outer membrane components. Notably, polymyxins have been shown to induce rapid killing at concentrations considerably lower than that required for cytoplasmic membrane permeabilization or depolarization, which suggests that other bactericidal effects may be involved [269]. We have previously shown that the gene expression response elicited by PB treatment in Kp13 affects a myriad of transcriptional regulators such as two-component systems, which in turn impact the expression of a broad and diverse set of genes, along with various metabolic shifts in *K. pneumoniae* [270].

As the last step in this *K. pneumoniae* target prioritization pipeline, we sought to identify candidate proteins for which developed drugs would present enhanced selectivity toward bacterial pathogens, thus minimizing the impact to the commensal gut microbiota. This was achieved by modifying the scoring function in order to include a term that penalizes the score of a protein in up to 50% with the increasing presence of orthologs in gut commensal species (n = 226, full list is provided in [67]), following the Eq. (10.5):

$$SF = \left(\left(E_{\mathrm{mgh}} + E_{\mathrm{deg}}\right)/2 + C_v + C_y + chk\right)/4 + G_M \tag{10.5}$$

where GM is the number of gut microbiome organisms that have at least one homologous protein in the Kp13 genome, normalized by the total number of analyzed organisms. For each sequence present in the Kp13 proteome, we analyzed the number of organisms that present at least one significant hit (E-value $\leq 10-5$; identity $\geq$40%).

We have intentionally not set any a priori weights on each of the terms that compose the scoring functions in order to avoid incurring in possible representation biases, as we posited that all considered variables play an important role in defining a suitable target.

The top 15 highest ranking proteins as candidate targets shortlisted from Eqs. (10.3), (10.4), and (10.5) are provided in [67]. Here, we show the result of their interrelations (Fig. 10.6), which comprise 29 unique proteins with characteristics that are desirable from a druggability perspective, and their follow-up could be promising given the need for novel drugs developed for controlling infections caused by resistant bacteria. In this sense, proteins that are high-ranking all display interesting features that could be exploited in future drug development works. For instance, the integrated analysis of the 29 unique candidates reveals the emergence of a core metabolic subset shared between many of them, comprising biosynthesis of amino acids, fatty acids,

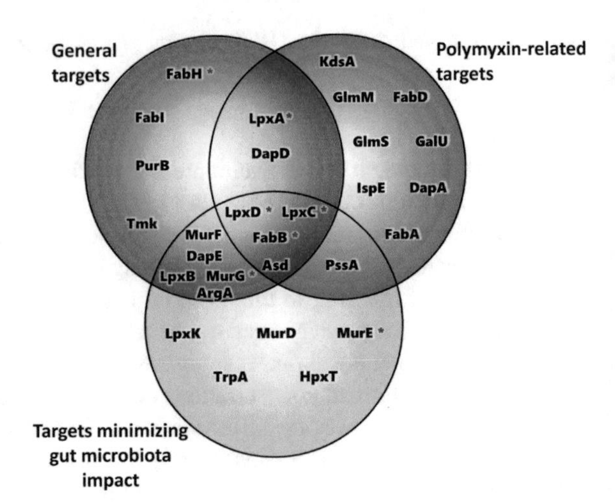

Fig. 10.6 Venn diagram showing the number of unique and shared targets identified using the three different ranking strategies for drug targeting. Targets that have been experimentally tested with inhibitors in *K. pneumoniae* are marked with an asterisk: LpxA [278, 279], LpxD [278], FabB/FabH [280], LpxC [281], and MurG/MurE [282]. Figure from [67]

and cell wall components. These represent fundamental metabolic aspects of the bacterial cellular machinery, and the specific proteins identified here represent a good starting point for further experimental exploration as well as confirmation of their relevance in ongoing efforts against some of these.

Two of these unique 29 most promising candidate proteins that can be highlighted here, because they are attractive either in monotherapy or in drug combination therapy, are IspE and PssA.

IspE is a cytoplasmic kinase of the MEP pathway that is involved in the biosynthesis of the isoprenoids used by many Gram-negative bacteria (including *E. coli*, *Salmonella enterica*, *P. aeruginosa*, and *H. influenzae*), as well as Gram-positive bacteria such as *Clostridium difficile* and *Bacillus subtilis*, *M. tuberculosis*, and even few apicoplast protozoa such as *Plasmodium falciparum* [271]. Because isoprenoids are involved in a wide variety of vital biological functions, the seven enzymes without close human homologs that participate in their metabolism (Dxs, IspC, IspD, IspE, IspF, IspG, IspH) are favorable candidate drug targets and several inhibitors have been already reported [272], mainly as antimalarial targets [273, 274]. In Gram-negative bacteria, compounds from the isoxazol-5(4 H)-one series have been evaluated (e.g., PubChem compound ID 3768522 and DrugBank DB03687) showing inhibitory activities against IspE/Ipk from *E. coli* and *Y. pestis* [275], and the study of its effect in *Klebsiella* bacteria is also appealing in light of our results.

The active form of phosphatidylserine (PtdSer) synthase (PssA) from some bacteria is a cytoplasmic membrane-associated enzyme that converts cytidine diphosphate diacylglycerol (CDP-DAG) and serine (L-Ser) to PtdSer, a negatively charged phospholipid that is rapidly decarboxylated by PtdSer decarboxylase (Psd) to generate phosphatidylethanolamine (PtdEtn), the major phospholipid of membranes [276]. PssA has been shown to play a significant role in virulence of Brucella abortus in a mouse model of infection with a Δ*pssA* mutant [277]. Interestingly, the Brucella cell envelope is normally resistant to the bactericidal action of polycationic peptides such as PB, but the Δ*pssA* mutant of *B. abortus* showed loss of PtdEtn and increased sensitivity to this drug, without any changes in its LPS structure [277]. Thus, an evaluation of a combination of PB and an inhibitor against PssA seems like a plausible approach.

Thus, this section exemplifies an integrative analysis framework for the prioritization of protein targets using as model organism *Klebsiella pneumoniae* strain Kp13, a multidrug-resistant (including polymyxin) bacterium responsible for nosocomial infections. Various layers of information were combined, including whole-structural, metabolic, genomic, and expression data as input to scoring functions that allowed a short-listing of targets with desirable characteristics from a druggability standpoint. Out of 5,736 predicted proteins that form the proteome of Kp13, we obtained structural models for 3,194 of them and predicted the presence and location of pockets that were characterized by their druggability. The reconstruction and annotation of the metabolic network of this strain allowed the identification of the metabolic complement and enzymatic activities performed by Kp13 and related bacteria, as well as important topological metrics in this network. This was used to contextualize the functional aspects of the candidate targets identified. All this information, along with other genomic features calculated for each protein, were loaded into the openly available web server TP [170] that allows easy retrieval of any of the generated data, along with parameter customization.

10.3.3 Mycobacterium tuberculosis

Tuberculosis (TB) is considered one of the first human infectious diseases. Its causative agent, *Mycobacterium tuberculosis* (Mtb), was responsible for twenty percent of all deaths in the western world between the seventeenth and nineteenth centuries, and still remains an enormous health concern worldwide. TB is estimated to be responsible for 1.3 million deaths among human immunodeficiency virus (HIV)-negative people each year [283]. In 2017, more than 10 million people fell ill with TB, with more than 95% of cases and deaths occurring in developing countries [283]. The human immune response to Mtb infection relies on phagocytosis of the bacteria by macrophages leading to granuloma formation. Inside the macrophages, bacilli face stressful conditions characterized by the presence of Reactive Nitrogen and Oxygen Species (RNOS), produced by the macrophage inducible Nitric Oxide Synthase and NADH Oxidase, hypoxia and nutrient deprivation. In these conditions, Mtb switches to a latent state and can remain alive (and dormant) for decades [284]. Latent-infected people, of which there are estimated to be ~2 billion in the world, represent an enormous reservoir of potential reactivation TB, which can result in recurrent acute infections and also spread to other people. Key to TB biology in the latent phase is its capacity to fight the host's RNOS-mediated attack. These reactive species are known to display a concentration-dependent mycobactericidal activity [285]. Based on these observations an interesting hypothesis emerges that identification of Mtb RNOS protein targets would allow us to design inhibitors against them to synergize with the macrophages attack and kill TB in the latent phase. In this section, we present a multidimensional genome-wide data integration strategy in order to prioritize drug targets in latent tuberculosis based on this idea.

The analysis began by classifying all available protein structures (including experimental obtained and homology-based models) according to their structural druggability. For this purpose, we divided the proteins into four groups. The first group corresponds to proteins from Mtb, which have been crystallized in the presence of a drug-like compound, and thus are defined as crystallized with drugs (CWD group), which are druggable *per se*. The second group includes proteins when there is at least one protein from the same PFAM family, that has been crystallized in the presence of a drug-like compound. This group is potentially druggable as referred to as Domain-Based Druggable (DBD) group. The remaining proteins correspond to all structures that have no relation to any structure hosting a drug-like compound. This group is divided into experimentally obtained structures and homology-based models. For all structures, all pockets and their corresponding Druggability Score (DS) were calculated, using fpocket (as described above for *K. pneumoniae*), and the Druggable (D) and Highly Druggable (HD) proteins are selected as the best candidates.

From a general point of view, an interesting result is that over half of the analyzed structures are druggable or highly druggable. The DBD-HD group, composed of 1,187 proteins, is the most attractive for target discovery due to the fact that both, an association criteria (assignation to DBD) and a structural criteria (DS > 0.7) match for several cases. Also, there are about 641 proteins, which we predict are highly druggable solely from a structural point of view. From a total of 1828 HD proteins, 743 are reported to be essential for *in vitro* growth [9].

To further rank the group of HD and essential proteins described above by their potential as drug targets, an analysis of the available data related to their expression level under infection mimicking conditions (Hypoxia, Starvation, RNOS stress, and infection in mice) was performed. All proteins were ranked according to the number of conditions where the protein was found to be overexpressed. The DBD group includes 48 druggable proteins, of which 20 are essential, which are also overexpressed in 3 or 4 of the above-mentioned conditions. In this group, it is worth mentioning the inclusion DevS, known to be involved in RNOS sensing, and signal transduction, harboring a druggable kinase ATP-binding pocket. Another interesting case concerns the 3-methyl-2-oxobutanoate hydroxymethyltransferase, a protein involved in both Hypoxia and Infection. Among the remaining proteins we find, for example, proteins like l,d-transpeptidase 2, the alpha–beta hydrolase, and DNApol III delta subunit. Interestingly, almost half of the analyzed essential proteins are overexpressed in some of the conditions mimicking infection. This fact may reflect that *M. tuberculosis* protein expression is highly regulated allowing the pathogen to adapt to a variety of different external stimuli.

As mentioned before, one key hypothesis to fight latent TB is to identify which proteins are already targeted by the RNOS attack of macrophages, and therefore try to synergistically inhibit them with drugs. However, little is known concerning the potential targets of RNOS in Mtb; in other words, which enzymes would be partially or totally inhibited by these reactive species [286]. Interestingly, RNOS protein reactivity can be nailed down to the presence of reactive Cys, Tyr, or transition metal centers in proteins [287]. Therefore, besides the expression analysis, structure information combined with chemical reactivity knowledge was used to predict possible sensitivity towards RNOS. With this analysis, 800 potentially sensitive to RNOS proteins were obtained. Combination of RNOS sensitivity with the previous criteria shows that there are about 200 druggable, essential, and sensitive to RNOS proteins which are also over-expressed in different infection models. This shortlist of 200 proteins emerges as the most attractive targets. Within this group, for example, the Inositol-3-phosphate synthase (I3PS, ino1) has a druggable pocket harboring one oxidable cysteine and two oxidable tyrosines and overlaps the NAD-binding site.

As a last step in the prioritization procedure, the Pathway tools software was used to build a comprehensive *M. tuberculosis* metabolic network (Mtb-MN). As described for *K. pneumoniae*, one of our goals was also to identify choke-points because genes associated with these reactions are supposed to be relevant for Mtb. Another relevant parameter is the centrality of the reactions, meaning the many pathways converge or emerge from them. In this MtB-MN, pathways were scored according to their potential to be used as a target in latent tuberculosis drug discovery projects. With this tool, instead of prioritizing only enzymes, we could prioritize the entire metabolic pathways. To do this, all pathways that do not have at least one druggable protein were ruled out and a scoring function was defined in order to assign a score to each pathway in the following Eq. (10.6):

$$SF = \frac{1}{2} * \left((NR + C_y + C_{hk} + E_s + C)/5 + (S + H + S_t + I)/4 \right) \qquad (10.6)$$

where *NR* is the total number of reactions in each pathway normalized by the number of reactions of the highest pathway. *Cy* is the average of the ratio between the centrality of each node in the pathway and the node with the highest centrality in Mtb-MN. *Chk* is the proportion of choke-point reactions in each pathway. *E* reflects the number of essential genes present in each pathway. *C* ("completeness") is the proportion of reactions to which a gene could be assigned during metabolic reconstruction. *S*, *H*, *St*, *I* are the proportions of genes associated with RNOS stress, hypoxia, starvation, and infection models, respectively (see above). This score sums up data on the relevance of individual genes to characterize the whole pathway. A value close to 1 would mean that 'all genes'in the pathway are essential, relevant for stress, and so on.

This analysis revealed several high scoring "druggable" pathways. One of them was the mycothiol biosynthesis pathway depicted in Fig. 10.7. Mycothiol is crucial for the intracellular redox balance and plays a key role in Mtb survival within macrophages [288]. All the involved enzymes are essential, and two of the four crystallized proteins (Ino1 and MshB), have a DS greater than 0.7. The gene encoding Ino1 is also over-expressed in RNOS stress, hypoxia and starvation while *mshB* is over-expressed in RNOS stress conditions (Fig. 10.7).

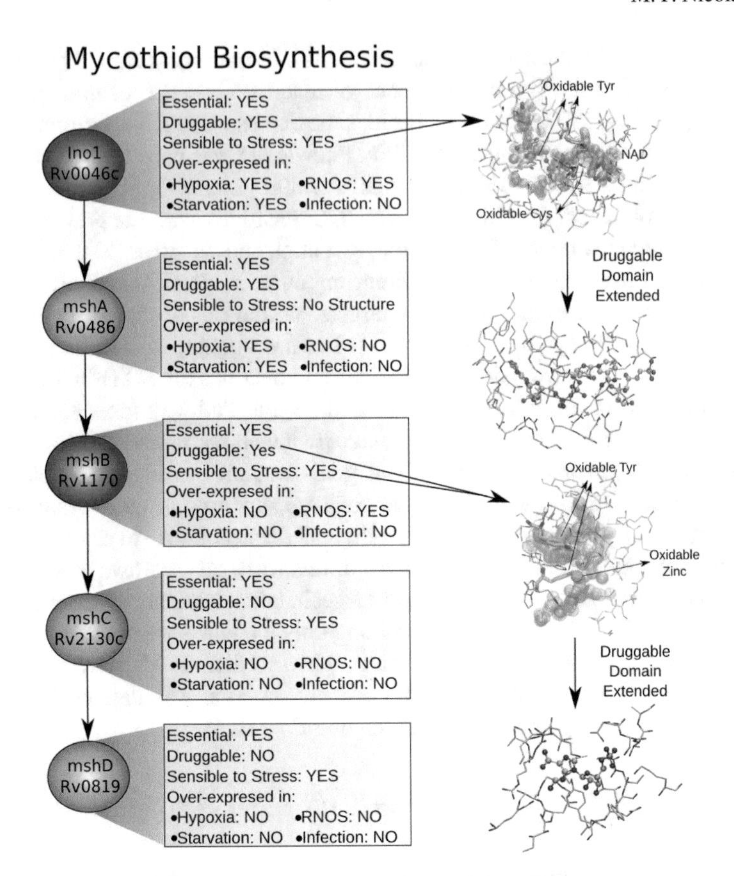

Fig. 10.7 Representation of the Mycothiol biosynthesis pathway. Two outstanding protein pockets depicted with their corresponding DBD partner pockets are shown. The most important proteins in this pathway Ino1 and MshB are shown in red. A representation of the pocket and of the known drugged pocket of the PFAM Family (DBD) are shown next to their respective protein

Another top-ranking pathway is the one responsible for histidine biosynthesis, which has been suggested as harboring potential drug targets due to its absence in mammals [289]. The pathway is composed of eight essential proteins. Five of them (Rv1600, Rv1603, Rv1605, Rv1606, and Rv2121c) harbor druggable pockets.

A further top-ranked path is the well-characterized mycolate biosynthesis pathway. This pathway is targeted by first-line tuberculosis drugs such as isoniazid and ethambutol [290]. This pathway compromises 22 proteins. The importance of this pathway lies in the number of genes that are druggable (60%) and essential (83%). Besides, almost 90% of the reactions that participate in this pathway were found to be choke points.

Mtb-MN analysis also reveals the relevance of sulfur metabolism, essential for the survival and virulence of Mtb. Moreover, most genes are absent in humans. Finally, another pathway highlighted by the Mtb-MN analysis was “alanine degradation IV” performed by a unique gene (Rv2780). Although not reported as essential, this l-alanine dehydrogenase was the first antigen reported to be present in pathogenic Mtb, but not in *M. bovis* [291]. Furthermore, it was suggested that the lack of a functional alanine dehydrogenase impairs BCG replication in the human host [292]. This protein is druggable and appears to have a crucial function in RNOS stress since it was overexpressed in all the studied conditions.

In summary, in this section, we showed that our integrated analysis of Mtb metabolism along with expression, essentiality, and druggability data has allowed the identification of important proteins and pathways harboring promising therapeutic targets.

10.4 Conclusions and Perspectives

With the revolution of the genome era, that had great advances in the early 2000s and became accessible worldwide in the last 10 years, the omic revolution started. Usually, the word omics to refer to the field of study in biological sciences that ends with -omics, such as genomics, transcriptomics, proteomics, or metabolomics and the ending -ome is used to address the objects of study of such fields, such as the genome, proteome, transcriptome, or metabolome, respectively. The amount of data generated in each -ome, and the connections between them, can only be integrated and managed with the help of computational methods such as the ones reviewed in this chapter. The availability of these results through web applications that can be easily used and customized allows for a more widespread adoption of these techniques. One of the main aims of integrating these data is to analyze a specific pathogen and be able to identify key targets, in general proteins that are relevant for drug discovery campaigns against that pathogen. In the present chapter, we have described current efforts that allow target prioritization applying integrated multi-omics approaches. In this integrative approach, we have shown how to include and integrate diverse layers of 'multi-omics' data, that allow the generation of genome-scale models including genetic, structural/functional and cellular information, metabolic context, and expression data to prioritize relevant proteins that can be attractive targets for the development of new antimicrobials, both in a general way and in specific applications to *Staphylococcus aureus*, *Klebsiella pneumoniae* and *Mycobacterium tuberculosis*.

It is clear that we are facing a paradigm shift in drug development projects. Most of the previous efforts were focused on the specific knowledge of a group of scientists that, analyzing the literature and performing state-of-the-art experiments were able to pinpoint specific targets and then initiate the search for new possible inhibitors, but today this is being replaced by approaches that include the integration of multi-omic data. Also important, the development of new drugs on the basis of the 'single target–single drug' concept, despite its initial favorable outcome, is now facing reduced success rates and prolonged development times. Taking this into account, the adoption of GSMs models, evaluating drug effects as the result of multiple interactions in a biological network, has yielded unprecedented opportunities to understand the functioning of biological systems, to identify the molecular mechanisms of drug action beyond the previous thinking of inhibiting a specific enzyme activity. We have described that network modeling is a key tool, not only to identify a target but to assess the effect of its inhibition in a cellular context.

Despite the relative success of these new approaches, we believe that many challenges in the near future need to be addressed to accelerate drug discovery. The most important ones according to our consideration are: (a) To be able to generate data that can be integrated in a quantitative manner, curation of the data and therefore integration is still a bottleneck in most systems biology approaches; (b) To develop better algorithms that are able to reconstruct and identify connections between genes and their regulation besides already known ones; (c) To develop GSM models of pathogens that can be challenged towards cellular outcomes and not specific pathway down or upregulation; (d) To be able to analyze at the structurome level the inhibition of a drug and not at a single target; and (e) To couple target identification with drug development tools to quickly identify possible lead compounds.

An important problem in drug discovery for many pathogens is that when searching for targets there is no "One size fits all," which in other words means that different pathogens may present different optimal targets. Good targets are not involved in one specific cellular process such as cell growth, on the contrary, depending on the pathogen and the context of the infection, that is, how it affects the host, target proteins could be involved in many different cellular processes. Based on this observation, it is very difficult to develop an artificial intelligence algorithm that can be trained to identify targets that when modulated would allow the host's body to get rid of the infection. Therefore, most applications, despite their unique usefulness, still have to be used by a researcher that assesses the relevance of the highlighted targets. Again, system biology approaches will be able to help in this regard.

Acknowledgments This work was supported by fellowships from CNPq (process no. 306894/2019-0) and grant by CAPES (process no. 88887.368759/2019-00) to M.F.N. E.P-R was supported by Dirección General de Asuntos del Personal Académico-Universidad Nacional Autónoma de México (IN-209620) and Programa Iberoamericano de Ciencia y Tecnologìa para el Desarrollo (P918PTE0261).

References

1. Levy SB, Marshall B (2004) Antibacterial resistance worldwide: causes, challenges and responses. Nat Med 10:S122–S129
2. New report calls for urgent action to avert antimicrobial resistance crisis, 11 Jun 2019. http://www.who.int/news-room/detail/29-04-2019-new-report-calls-for-urgent-action-to-avert-antimicrobial-resistance-crisis
3. Wenzel RP (2004) The antibiotic pipeline—challenges, costs, and values. N Engl J Med 523–526. http://dx.doi.org/10.1056/nejmp048093

4. Blundell TL, Sibanda BL, Montalvão RW, Brewerton S, Chelliah V, Worth CL et al (2006) Structural biology and bioinformatics in drug design: opportunities and challenges for target identification and lead discovery. Philos Trans R Soc Lond B Biol Sci 361:413–423
5. Pires DP, Cleto S, Sillankorva S, Azeredo J, Lu TK (2016) Genetically engineered phages: a review of advances over the last decade. Microbiol Mol Biol Rev 80:523–543
6. Ekins S, Freundlich JS (2013) Computational models for tuberculosis drug discovery. Methods Mol Biol 993:245–262
7. Galizzi J-P, Lockhart BP, Bril A (2013) Applying systems biology in drug discovery and development. Drug Metabol Drug Interact 28:67–78
8. Radusky LG, Hassan S, Lanzarotti E, Tiwari S, Jamal S, Ali J et al (2015) An integrated structural proteomics approach along the druggable genome of Corynebacterium pseudotuberculosis species for putative druggable targets. BMC Genom 16(Suppl 5):S9
9. Defelipe LA, Do Porto DF, Pereira Ramos PI, Nicolás MF, Sosa E, Radusky L et al (2016) A whole genome bioinformatic approach to determine potential latent phase specific targets in Mycobacterium tuberculosis. Tuberculosis. 97:181–192
10. Kaur D, Kutum R, Dash D, Brahmachari SK (2017) Data intensive genome level analysis for identifying novel, non-toxic drug targets for multi drug resistant mycobacterium tuberculosis. Sci Rep. 7:46595
11. Wadood A, Ghufran M, Khan A, Azam SS, Uddin R, Waqas M, et al (2017) The methicillin-resistant S. epidermidis strain RP62A genome mining for potential novel drug targets identification. Gene Rep 8:88–93
12. Uddin R, Jamil F (2018) Prioritization of potential drug targets against P. aeruginosa by core proteomic analysis using computational subtractive genomics and Protein-Protein interaction network. Comput Biol Chem 115–122. http://dx.doi.org/10.1016/j.compbiolchem.2018.02.017
13. Bergen PJ, Forrest A, Bulitta JB, Tsuji BT, Sidjabat HE, Paterson DL, et al (2011) Clinically relevant plasma concentrations of colistin in combination with imipenem enhance pharmacodynamic activity against multidrug-resistant pseudomonas aeruginosa at multiple inocula. Antimicrob Agents Chemother 5134–5142. http://dx.doi.org/10.1128/aac.05028-11
14. Bergen PJ, Bulman ZP, Landersdorfer CB, Smith N, Lenhard JR, Bulitta JB et al (2015) Optimizing polymyxin combinations against resistant gram-negative bacteria. Infect Dis Ther 4:391–415
15. Wang Q, Chang C-S, Pennini M, Pelletier M, Rajan S, Zha J et al (2016) Target-agnostic identification of functional monoclonal antibodies against klebsiella pneumoniae multimeric MrkA fimbrial subunit. J Infect Dis 213:1800–1808
16. Szijártó V, Guachalla LM, Visram ZC, Hartl K, Varga C, Mirkina I et al (2015) Bactericidal monoclonal antibodies specific to the lipopolysaccharide O antigen from multidrug-resistant Escherichia coli clone ST131-O25b:H4 elicit protection in mice. Antimicrob Agents Chemother 59:3109–3116
17. Szijártó V, Nagy E, Nagy G (2018) Directly bactericidal anti-escherichia coli antibody. Trends Microbiol 642–644
18. Storek KM, Auerbach MR, Shi H, Garcia NK, Sun D, Nickerson NN et al (2018) Monoclonal antibody targeting the β-barrel assembly machine of Escherichia coli is bactericidal. Proc Natl Acad Sci U S A 115:3692–3697
19. Steiner B, Swart AL, Hilbi H (2019) Perturbation of legionella cell infection by RNA interference. Methods Mol Biol 1921:221–238
20. de la Fuente-Núñez C, Lu TK (2017) CRISPR-Cas9 technology: applications in genome engineering, development of sequence-specific antimicrobials, and future prospects. Integr Biol 9:109–122
21. Ligon BL (2004) Penicillin: its discovery and early development. Semin Pediatr Infect Dis 15:52–57
22. Brown ED, Wright GD (2016) Antibacterial drug discovery in the resistance era. Nature 336–343. http://dx.doi.org/10.1038/nature17042
23. Wright GD (2012) Antibiotics: a new hope. Chem Biol 19:3–10

24. Kolter R, van Wezel GP (2016) Goodbye to brute force in antibiotic discovery? Nat Microbiol 1:15020
25. Reymond J-L, van Deursen R, Blum LC, Ruddigkeit L (2010) Chemical space as a source for new drugs. Med Chem Comm 30. http://dx.doi.org/10.1039/c0md00020e
26. Selzer PM, Brutsche S, Wiesner P, Schmid P, Müllner H (2000) Target-based drug discovery for the development of novel antiinfectives. Int J Med Microbiol 191–201. http://dx.doi.org/10.1016/s1438-4221(00)80090-9
27. Payne DJ, Gwynn MN, Holmes DJ, Pompliano DL (2007) Drugs for bad bugs: confronting the challenges of antibacterial discovery. Nat Rev Drug Discov 6:29–40
28. Projan SJ (2003) Why is big Pharma getting out of antibacterial drug discovery? Curr Opin Microbiol 6:427–430
29. Hackbarth CJ, Chen DZ, Lewis JG, Clark K, Mangold JB, Cramer JA et al (2002) N-alkyl urea hydroxamic acids as a new class of peptide deformylase inhibitors with antibacterial activity. Antimicrob Agents Chemother 46:2752–2764
30. Flores A, Quesada E (2013) Entry inhibitors directed towards glycoprotein gp120: an overview on a promising target for HIV-1 therapy. Curr Med Chem 20:751–771
31. Farha MA, Leung A, Sewell EW, D'Elia MA, Allison SE, Ejim L et al (2013) Inhibition of WTA synthesis blocks the cooperative action of PBPs and sensitizes MRSA to β-lactams. ACS Chem Biol 8:226–233
32. Starkey M, Lepine F, Maura D, Bandyopadhaya A, Lesic B, He J, et al (2014) Identification of anti-virulence compounds that disrupt quorum-sensing regulated acute and persistent pathogenicity. PLoS Pathogens e1004321. http://dx.doi.org/10.1371/journal.ppat.1004321
33. Feist AM, Palsson BØ (2008) The growing scope of applications of genome-scale metabolic reconstructions using Escherichia coli. Nat Biotechnol 26:659–667
34. Kim TY, Sohn SB, Kim YB, Kim WJ, Lee SY (2012) Recent advances in reconstruction and applications of genome-scale metabolic models. Curr Opin Biotechnol 23:617–623
35. Kim HU, Kim TY, Lee SY (2010) Genome-scale metabolic network analysis and drug targeting of multi-drug resistant pathogen Acinetobacter baumannii AYE. Mol BioSyst 6:339–348
36. Lewis NE, Schramm G, Bordbar A, Schellenberger J, Andersen MP, Cheng JK et al (2010) Large-scale in silico modeling of metabolic interactions between cell types in the human brain. Nat Biotechnol 28:1279–1285
37. Kim HU, Kim SY, Jeong H, Kim TY, Kim JJ, Choy HE et al (2011) Integrative genome-scale metabolic analysis of Vibrio vulnificus for drug targeting and discovery. Mol Syst Biol 7:460
38. Bidkhori G, Benfeitas R, Elmas E, Kararoudi MN, Arif M, Uhlen M et al (2018) Metabolic network-based identification and prioritization of anticancer targets based on expression data in hepatocellular carcinoma. Front Physiol 9:916
39. Thiele I, Palsson BØ (2010) A protocol for generating a high-quality genome-scale metabolic reconstruction. Nat Protoc 5:93–121
40. Varma A, Palsson BO (1994) Metabolic Flux Balancing: Basic Concepts, Scientific and Practical Use. Nat Biotechnol 12:994–998
41. Seif Y, Kavvas E, Lachance J-C, Yurkovich JT, Nuccio S-P, Fang X et al (2018) Genome-scale metabolic reconstructions of multiple Salmonella strains reveal serovar-specific metabolic traits. Nat Commun 9:3771
42. Price ND, Papin JA, Schilling CH, Palsson BO (2003) Genome-scale microbial in silico models: the constraints-based approach. Trends Biotechnol 21:162–169
43. Kanehisa M, Furumichi M, Tanabe M, Sato Y, Morishima K (2017) KEGG: new perspectives on genomes, pathways, diseases and drugs. Nucleic Acids Res 45:D353–D361
44. Karp PD, Billington R, Caspi R, Fulcher CA, Latendresse M, Kothari A et al (2019) The BioCyc collection of microbial genomes and metabolic pathways. Brief Bioinform 20:1085–1093
45. Jassal B, Matthews L, Viteri G, Gong C, Lorente P, Fabregat A et al (2020) The reactome pathway knowledgebase. Nucleic Acids Res 48:D498–D503

46. Butt AM, Tahir S, Nasrullah I, Idrees M, Lu J, Tong Y (2012) Mycoplasma genitalium: a comparative genomics study of metabolic pathways for the identification of drug and vaccine targets. Infect Genet Evol 12:53–62
47. Shanmugham B, Pan A (2013) Identification and characterization of potential therapeutic candidates in emerging human pathogen Mycobacterium abscessus: a novel hierarchical in silico approach. PLoS ONE 8:e59126
48. Belda E, Sekowska A, Le Fèvre F, Morgat A, Mornico D, Ouzounis C, et al (2013) An updated metabolic view of the Bacillus subtilis 168 genome. Microbiology 757–770. http://dx.doi.org/10.1099/mic.0.064691-0
49. Scaria J, Mao C, Chen J-W, McDonough SP, Sobral B, Chang Y-F (2013) Differential stress transcriptome landscape of historic and recently emerged hypervirulent strains of Clostridium difficile strains determined using RNA-seq. PLoS ONE 8:e78489
50. Machado D, Andrejev S, Tramontano M, Patil KR (2018) Fast automated reconstruction of genome-scale metabolic models for microbial species and communities. Nucleic Acids Res 46:7542–7553
51. Lacroix V, Cottret L, Thebault P, Sagot MF (2008) An introduction to metabolic networks and their structural analysis. IEEE/ACM Trans Comput Biol Bioinform 594–617. http://dx.doi.org/10.1109/tcbb.2008.79
52. Montañez R, Medina MA, Solé RV, Rodríguez-Caso C (2010) When metabolism meets topology: reconciling metabolite and reaction networks. BioEssays 32:246–256
53. Cottret L, Jourdan F (2010) Graph methods for the investigation of metabolic networks in parasitology. Parasitology 137:1393–1407
54. Jeong H, Mason SP, Barabási AL, Oltvai ZN (2001) Lethality and centrality in protein networks. Nature 41–42. http://dx.doi.org/10.1038/35075138
55. Ramos PIP, Arge LWP, Lima NCB, Fukutani KF, de Queiroz ATL (2019) Leveraging user-friendly network approaches to extract knowledge from high-throughput omics datasets. Front Genet 10:1120
56. Giuliani S, Silva ACE, Borba JVVB, Ramos PIP, Paveley RA, Muratov EN, et al (2018) Computationally-guided drug repurposing enables the discovery of kinase targets and inhibitors as new schistosomicidal agents. PLoS Comput Biol (no prelo)
57. Zhang M, Su S, Bhatnagar RK, Hassett DJ, Lu LJ (2012) Prediction and analysis of the protein interactome in Pseudomonas aeruginosa to enable network-based drug target selection. PLoS One 7:e41202
58. Watowich AF, Muskus S, Current C (2012) Advances in computational strategies for drug discovery in leishmaniasis. Curr Top Trop Med http://dx.doi.org/10.5772/28292
59. Ochoa R, Martínez-Pabón MC, Arismendi-Echeverri MA, Rendón-Osorio WL, Muskus-López CE (2017) In silico search of inhibitors of Streptococcus mutans for the control of dental plaque. Arch Oral Biol 83:68–75
60. Gupta SK, Gross R, Dandekar T (2016) An antibiotic target ranking and prioritization pipeline combining sequence, structure and network-based approaches exemplified for Serratia marcescens. Gene 591:268–278
61. Yeh I, Hanekamp T, Tsoka S, Karp PD, Altman RB (2004) Computational analysis of Plasmodium falciparum metabolism: organizing genomic information to facilitate drug discovery. Genome Res 14:917–924
62. Singh S, Malik BK, Sharma DK (2007) Choke point analysis of metabolic pathways in E. histolytica: a computational approach for drug target identification. Bioinformation 68–72 http://dx.doi.org/10.6026/97320630002068
63. Rahman SA, Schomburg D (2006) Observing local and global properties of metabolic pathways: "load points" and "choke points" in the metabolic networks. Bioinformatics 22:1767–1774
64. Jadhav A, Ezhilarasan V, Prakash Sharma O, Pan A (2013) Clostridium-DT(DB): a comprehensive database for potential drug targets of Clostridium difficile. Comput Biol Med 43:362–367

65. Sharma A, Pan A (2012) Identification of potential drug targets in Yersinia pestis using metabolic pathway analysis: MurE ligase as a case study. Eur J Med Chem 57:185–195
66. Gupta M, Prasad Y, Sharma SK, Jain CK (2017) Identification of phosphoribosyl-AMP cyclohydrolase, as drug target and its inhibitors in Brucella melitensis bv. 1 16 M using metabolic pathway analysis. J Biomol Struct Dyn 35:287–299
67. Ramos PIP, Fernández Do Porto D, Lanzarotti E, Sosa EJ, Burguener G, Pardo AM, et al (2018) An integrative, multi-omics approach towards the prioritization of Klebsiella pneumoniae drug targets. Sci Rep 8:10755
68. Heavner BD, Smallbone K, Barker B, Mendes P, Walker LP (2012) Yeast 5—an expanded reconstruction of the Saccharomyces cerevisiae metabolic network. BMC Syst Biol 55. http://dx.doi.org/10.1186/1752-0509-6-55
69. Schellenberger J, Park JO, Conrad TM, Palsson BØ (2010) BiGG: a biochemical genetic and genomic knowledgebase of large scale metabolic reconstructions. BMC Bioinform 11:213
70. Norsigian CJ, Pusarla N, McConn JL, Yurkovich JT, Dräger A, Palsson BO et al (2020) BiGG models 2020: multi-strain genome-scale models and expansion across the phylogenetic tree. Nucleic Acids Res 48:D402–D406
71. Raman K, Chandra N (2009) Flux balance analysis of biological systems: applications and challenges. Brief Bioinform 10:435–449
72. Lee JM, Gianchandani EP, Papin JA (2006) Flux balance analysis in the era of metabolomics. Brief Bioinform 7:140–150
73. Orth JD, Thiele I, Palsson BØ (2010) What is flux balance analysis? Nat Biotechnol 28:245–248
74. Lewis NE, Nagarajan H, Palsson BO (2012) Constraining the metabolic genotype-phenotype relationship using a phylogeny of in silico methods. Nat Rev Microbiol 10:291–305
75. Edwards JS, Palsson BO (2000) The Escherichia coli MG1655 in silico metabolic genotype: its definition, characteristics, and capabilities. Proc Natl Acad Sci U S A 97:5528–5533
76. Gudmundsson S, Thiele I (2010) Computationally efficient flux variability analysis. BMC Bioinform 11:489
77. Schellenberger J, Que R, Fleming RMT, Thiele I, Orth JD, Feist AM, et al (2011) Quantitative prediction of cellular metabolism with constraint-based models: the COBRA Toolbox v2.0. Nat Protoc 6:1290–1307
78. Ebrahim A, Lerman JA, Palsson BO, Hyduke DR (2013) COBRApy: constraints-based reconstruction and analysis for python. BMC Syst Biol 7:74
79. Karlebach G, Shamir R (2008) Modelling and analysis of gene regulatory networks. Nat Rev Mol Cell Biol 9:770–780
80. Browning DF, Busby SJ (2004) The regulation of bacterial transcription initiation. Nat Rev Microbiol 2:57–65
81. Barabási A-L, Oltvai ZN (2004) Network biology: understanding the cell's functional organization. Nat Rev Genet 5:101–113
82. Flores-Bautista E, Cronick CL, Fersaca AR, Martinez-Nuñez MA, Perez-Rueda E (2018) Functional prediction of hypothetical transcription factors of escherichia coli K-12 based on expression data. Comput Struct Biotechnol J 16:157–166
83. Gama-Castro S, Salgado H, Santos-Zavaleta A, Ledezma-Tejeida D, Muñiz-Rascado L, García-Sotelo JS, et al (2016) RegulonDB version 9.0: high-level integration of gene regulation, coexpression, motif clustering and beyond. Nucleic Acids Res 44:D133–D143
84. Moreno-Campuzano S, Janga SC, Pérez-Rueda E (2006) Identification and analysis of DNA-binding transcription factors in Bacillus subtilis and other Firmicutes–a genomic approach. BMC Genom 7:147
85. Kobayashi H, Akitomi J, Fujii N, Kobayashi K, Altaf-Ul-Amin M, Kurokawa K et al (2007) The entire organization of transcription units on the Bacillus subtilis genome. BMC Genom 8:197
86. Brune I, Brinkrolf K, Kalinowski J, Pühler A, Tauch A (2005) The individual and common repertoire of DNA-binding transcriptional regulators of Corynebacterium glutamicum, Corynebacterium efficiens, Corynebacterium diphtheriae and Corynebacterium jeikeium deduced from the complete genome sequences. BMC Genom 6:86

87. Brinkrolf K, Brune I, Tauch A (2006) Transcriptional regulation of catabolic pathways for aromatic compounds in Corynebacterium glutamicum. Genet Mol Res 5:773–789
88. Ibarra JA, Pérez-Rueda E, Carroll RK, Shaw LN (2013) Global analysis of transcriptional regulators in Staphylococcus aureus. BMC Genom 14:126
89. Pandurangan AP, Stahlhacke J, Oates ME, Smithers B, Gough J (2019) The SUPERFAMILY 2.0 database: a significant proteome update and a new webserver. Nucleic Acids Res 47:D490–D494
90. El-Gebali S, Mistry J, Bateman A, Eddy SR, Luciani A, Potter SC et al (2019) The Pfam protein families database in 2019. Nucleic Acids Res 47:D427–D432
91. Imam S, Schäuble S, Brooks AN, Baliga NS, Price ND (2015) Data-driven integration of genome-scale regulatory and metabolic network models. Front Microbiol 6:409
92. Madan Babu M, Teichmann SA, Aravind L (2006) Evolutionary dynamics of prokaryotic transcriptional regulatory networks. J Mol Biol 358:614–633
93. Medeiros Filho F, do Nascimento APB, Dos Santos MT, Carvalho-Assef APD, da Silva FAB (2019) Gene regulatory network inference and analysis of multidrug-resistant Pseudomonas aeruginosa. Mem Inst Oswaldo Cruz 114:e190105
94. Tsoy OV, Ravcheev DA, Čuklina J, Gelfand MS (2016) Nitrogen fixation and molecular oxygen: comparative genomic reconstruction of transcription regulation in alphaproteobacteria. Front Microbiol 7:1343
95. Pérez AG, Angarica VE, Vasconcelos ATR, Collado-Vides J. Tractor_DB (2007) (version 2.0): a database of regulatory interactions in gamma-proteobacterial genomes. Nucleic Acids Res 35:D132–D136
96. Lozada-Chávez I, Janga SC, Collado-Vides J (2006) Bacterial regulatory networks are extremely flexible in evolution. Nucleic Acids Res 34:3434–3445
97. Santos-Zavaleta A, Pérez-Rueda E, Sánchez-Pérez M, Velázquez-Ramírez DA, Collado-Vides J (2019) Tracing the phylogenetic history of the crl regulon through the bacteria and archaea genomes. BMC Genom 20:299
98. Stuart JM, Segal E, Koller D, Kim SK (2003) A gene-coexpression network for global discovery of conserved genetic modules. Science 302:249–255
99. Law KL, Su GY, Lin N, Lin HY, Jang WT, Chi CS (1989) Acute peritoneal dialysis in low birth weight infants. Zhonghua Yi Xue Za Zhi 43:119–124
100. van Noort V, Snel B, Huynen MA (2004) The yeast coexpression network has a small-world, scale-free architecture and can be explained by a simple model. EMBO Rep 5:280–284
101. Tsaparas P, Mariño-Ramírez L, Bodenreider O, Koonin EV, Jordan IK (2006) Global similarity and local divergence in human and mouse gene co-expression networks. BMC Evol Biol 6:70
102. Zhang B, Horvath S (2005) A general framework for weighted gene co-expression network analysis. Stat Appl Genet Mol Biol 4(17)
103. Mueller AJ, Canty-Laird EG, Clegg PD, Tew SR (2017) Cross-species gene modules emerge from a systems biology approach to osteoarthritis. NPJ Syst Biol Appl 3:13
104. van Dam S, Võsa U, van der Graaf A, Franke L, de Magalhães JP (2018) Gene co-expression analysis for functional classification and gene-disease predictions. Brief Bioinform 19:575–592
105. Hosseinkhan N, Mousavian Z, Masoudi-Nejad A (2018) Comparison of gene co-expression networks in Pseudomonas aeruginosa and Staphylococcus aureus reveals conservation in some aspects of virulence. Gene 639:1–10
106. Bakhtiarizadeh MR, Hosseinpour B, Shahhoseini M, Korte A, Gifani P (2018) Weighted gene co-expression network analysis of endometriosis and identification of functional modules associated with its main hallmarks. Front Genet 9:453
107. Yang Y, Han L, Yuan Y, Li J, Hei N, Liang H (2014) Gene co-expression network analysis reveals common system-level properties of prognostic genes across cancer types. Nat Commun 5:3231
108. Amar D, Safer H, Shamir R (2013) Dissection of regulatory networks that are altered in disease via differential co-expression. PLoS Comput Biol 9:e1002955

109. Emilsson V, Thorleifsson G, Zhang B, Leonardson AS, Zink F, Zhu J et al (2008) Genetics of gene expression and its effect on disease. Nature 452:423–428
110. Carlson MRJ, Zhang B, Fang Z, Mischel PS, Horvath S, Nelson SF (2006) Gene connectivity, function, and sequence conservation: predictions from modular yeast co-expression networks. BMC Genom 7:40
111. Hartwell LH, Hopfield JJ, Leibler S, Murray AW (1999) From molecular to modular cell biology. Nature 402:C47–C52
112. Jeong H, Tombor B, Albert R, Oltvai ZN, Barabási AL (2000) The large-scale organization of metabolic networks. Nature 407:651–654
113. Ravasz E, Somera AL, Mongru DA, Oltvai ZN, Barabási AL (2002) Hierarchical organization of modularity in metabolic networks. Science 297:1551–1555
114. Potapov AP (2008) Signal Transduction and Gene Regulation Networks. In: Junker BH, Schreiber F (eds) Analysis of biological networks. Wiley, Hoboken, NJ, USA, pp 181–206
115. Junker BH, Schreiber F (2011) Analysis of biological networks. books.google.com, https://books.google.com/books?hl=en&lr=&id=YeXLbClh1SIC&oi=fnd&pg=PT4&dq=Junker+BH+Schreiber+F+Analysis+of+biological+networks+1st+ed+Wiley+Interscience+2008+ISBN+978-0-470-04144-4&ots=0El1K8ARJZ&sig=y2bn9u3wWL3bNNdxk7ZTyOgVa3M
116. Greenfield A, Madar A, Ostrer H, Bonneau R (2010) DREAM4: combining genetic and dynamic information to identify biological networks and dynamical models. PLoS One 5:e13397
117. Lee JM, Gianchandani EP, Eddy JA, Papin JA (2008) Dynamic analysis of integrated signaling, metabolic, and regulatory networks. PLoS Comput Biol 4:e1000086
118. Price ND, Shmulevich I (2007) Biochemical and statistical network models for systems biology. Curr Opin Biotechnol 18:365–370
119. Palsson B (2002) In silico biology through "omics". Nat Biotechnol 20:649–650
120. Chavali AK, D'Auria KM, Hewlett EL, Pearson RD, Papin JA (2012) A metabolic network approach for the identification and prioritization of antimicrobial drug targets. Trends Microbiol 20:113–123
121. Covert MW, Knight EM, Reed JL, Herrgard MJ, Palsson BO (2004) Integrating high-throughput and computational data elucidates bacterial networks. Nature 429:92–96
122. Covert MW, Palsson BØ (2002) Transcriptional regulation in constraints-based metabolic models of Escherichia coli. J Biol Chem 277:28058–28064
123. Zhang W, Li F, Nie L (2010) Integrating multiple "omics" analysis for microbial biology: application and methodologies. Microbiology 156:287–301
124. Covert MW, Schilling CH, Palsson B (2001) Regulation of gene expression in flux balance models of metabolism. J Theor Biol 213:73–88
125. Covert MW, Xiao N, Chen TJ, Karr JR (2008) Integrating metabolic, transcriptional regulatory and signal transduction models in Escherichia coli. Bioinformatics 24:2044–2050
126. Shlomi T, Eisenberg Y, Sharan R, Ruppin E (2007) A genome-scale computational study of the interplay between transcriptional regulation and metabolism. Mol Syst Biol 3:101
127. Chandrasekaran S, Price ND (2010) Probabilistic integrative modeling of genome-scale metabolic and regulatory networks in Escherichia coli and Mycobacterium tuberculosis. Proc Natl Acad Sci U S A 107:17845–17850
128. Blazier AS, Papin JA (2012) Integration of expression data in genome-scale metabolic network reconstructions. Front Physiol 3:299
129. Ma S, Minch KJ, Rustad TR, Hobbs S, Zhou S-L, Sherman DR et al (2015) Integrated modeling of gene regulatory and metabolic networks in mycobacterium tuberculosis. PLoS Comput Biol 11:e1004543
130. Jensen PA, Papin JA (2011) Functional integration of a metabolic network model and expression data without arbitrary thresholding. Bioinformatics 27:541–547
131. Presta L, Bosi E, Mansouri L, Dijkshoorn L, Fani R, Fondi M (2017) Constraint-based modeling identifies new putative targets to fight colistin-resistant A. baumannii infections. Sci Rep 7:3706

132. Becker SA, Palsson BO (2008) Context-specific metabolic networks are consistent with experiments. PLoS Comput Biol 4:e1000082
133. Schmidt BJ, Ebrahim A, Metz TO, Adkins JN, Palsson BØ, Hyduke DR (2013) GIM3E: condition-specific models of cellular metabolism developed from metabolomics and expression data. Bioinformatics 29:2900–2908
134. Marmiesse L, Peyraud R, Cottret L (2015) FlexFlux: combining metabolic flux and regulatory network analyses. BMC Syst Biol 9:93
135. Chaouiya C, Bérenguier D, Keating SM, Naldi A, van Iersel MP, Rodriguez N et al (2013) SBML qualitative models: a model representation format and infrastructure to foster interactions between qualitative modelling formalisms and tools. BMC Syst Biol 7:135
136. Jacob F, Monod J (1961) Genetic regulatory mechanisms in the synthesis of proteins. J Mol Biol 3:318–356
137. Feist AM, Henry CS, Reed JL, Krummenacker M, Joyce AR, Karp PD et al (2007) A genome-scale metabolic reconstruction for Escherichia coli K-12 MG1655 that accounts for 1260 ORFs and thermodynamic information. Mol Syst Biol 3:121
138. Busby S, Ebright RH (1999) Transcription activation by catabolite activator protein (CAP). J Mol Biol 293:199–213
139. Banos DT, Trébulle P, Elati M (2017) Integrating transcriptional activity in genome-scale models of metabolism. BMC Syst Biol 11:134
140. Koonin EV, Wolf YI, Karev GP (2002) The structure of the protein universe and genome evolution. Nature 420:218–223
141. Röntgen WC (1896) On a new kind of rays. Science 3:227–231
142. Friedrich W, Knipping P, Laue M (1913) Interferenzerscheinungen bei Röntgenstrahlen. Annalen der Physik 971–988 http://dx.doi.org/10.1002/andp.19133461004
143. Bragg WL (1913) The diffraction of short electromagnetic waves by a crystal. Proc Camb Philos Soc, 17:43–57 (1913). Communicated by Professor Sir Thomson JJ (1966) Read 11 November 1912. X-ray and neutron diffraction 109–125. http://dx.doi.org/10.1016/b978-0-08-011999-1.50015-8
144. Bernal JD, Crowfoot D (1934) X-Ray photographs of crystalline pepsin. Nature 794–795. http://dx.doi.org/10.1038/133794b0
145. El-Mehairy MM, Shaker A, Ramadan M, Hamza S, Tadros SS (1981) Control of essential hypertension with captopril, an angiotensin converting enzyme inhibitor. Br J Clin Pharmacol 469–475. http://dx.doi.org/10.1111/j.1365-2125.1981.tb01152.x
146. Renaud J (2020) The evolving role of structural biology in drug discovery. Struct Biol Drug Discov 1–22. http://dx.doi.org/10.1002/9781118681121.ch1
147. Berman HM, Battistuz T, Bhat TN, Bluhm WF, Bourne PE, Burkhardt K, et al (2002) The protein data bank. Acta Crystallogr Sect D Biol Crystallogr 899–907. http://dx.doi.org/10.1107/s0907444902003451
148. UniProt Consortium T (2018) UniProt: the universal protein knowledgebase. Nucleic Acids Res 46:2699
149. Li W, Godzik A (2006) Cd-hit: a fast program for clustering and comparing large sets of protein or nucleotide sequences. Bioinformatics 22:1658–1659
150. Eswar N, Eramian D, Webb B, Shen M-Y, Sali A (2008) Protein structure modeling with MODELLER. Methods Mol Biol 145–159. http://dx.doi.org/10.1007/978-1-60327-058-8_8
151. Melo F, Sali A (2007) Fold assessment for comparative protein structure modeling. Protein Sci 2412–2426. http://dx.doi.org/10.1110/ps.072895107
152. Benkert P, Tosatto SCE, Schomburg D (2008) QMEAN: a comprehensive scoring function for model quality assessment. Proteins: Struct, Funct, Bioinform 261–277. http://dx.doi.org/10.1002/prot.21715
153. Hopkins AL, Groom CR (2002) The druggable genome. Nat Rev Drug Discov 727–730. http://dx.doi.org/10.1038/nrd892
154. Cheng AC, Coleman RG, Smyth KT, Cao Q, Soulard P, Caffrey DR, et al (2007) Structure-based maximal affinity model predicts small-molecule druggability. Nat Biotechnol 71–75. http://dx.doi.org/10.1038/nbt1273

155. Radusky L, Defelipe LA, Lanzarotti E, Luque J, Barril X, Marti MA, et al (2014) TuberQ: a Mycobacterium tuberculosis protein druggability database. Database 2014:bau035
156. An J, Totrov M, Abagyan R (2005) Pocketome via comprehensive identification and classification of ligand binding envelopes. Mol Cell Proteomics 4:752–761
157. Hendlich M, Rippmann F, Barnickel G (1997) LIGSITE: automatic and efficient detection of potential small molecule-binding sites in proteins. J Mol Graph Model 15:359–363, 389
158. Weisel M, Proschak E, Schneider G (2007) PocketPicker: analysis of ligand binding-sites with shape descriptors. Chem Cent J 1:7
159. Le Guilloux V, Schmidtke P, Tuffery P (2009) Fpocket: an open source platform for ligand pocket detection. BMC Bioinform 10:168
160. Coleman RG, Sharp KA (2006) Travel depth, a new shape descriptor for macromolecules: application to ligand binding. J Mol Biol 362:441–458
161. Goodford P (2005) The basic principles of GRID. Methods Princ Med Chem 1–25. http://dx.doi.org/10.1002/3527607676.ch1
162. Laurie ATR, Jackson RM (2005) Q-SiteFinder: an energy-based method for the prediction of protein-ligand binding sites. Bioinformatics 21:1908–1916
163. Hernandez M, Ghersi D, Sanchez R (2009) SITEHOUND-web: a server for ligand binding site identification in protein structures. Nucleic Acids Res 37:W413–W416
164. Jiménez J, Doerr S, Martínez-Rosell G, Rose AS, De Fabritiis G (2017) DeepSite: protein-binding site predictor using 3D-convolutional neural networks. Bioinformatics 33:3036–3042
165. Levitt DG, Banaszak LJ (1992) POCKET: a computer graphics method for identifying and displaying protein cavities and their surrounding amino acids. J Mol Graph 10:229–234
166. Venkatachalam CM, Jiang X, Oldfield T, Waldman M (2003) LigandFit: a novel method for the shape-directed rapid docking of ligands to protein active sites. J Mol Graph Model 21:289–307
167. Zhang Z, Li Y, Lin B, Schroeder M, Huang B (2011) Identification of cavities on protein surface using multiple computational approaches for drug binding site prediction. Bioinformatics 27:2083–2088
168. Schmidtke P, Le Guilloux V, Maupetit J, Tufféry P (2010) fpocket: online tools for protein ensemble pocket detection and tracking. Nucleic Acids Res 38:W582–W589
169. Schmidtke P, Barril X (2010) Understanding and predicting druggability. A high-throughput method for detection of drug binding sites. J Med Chem 53:5858–5867
170. Sosa EJ, Burguener G, Lanzarotti E, Defelipe L, Radusky L, Pardo AM et al (2018) Target-pathogen: a structural bioinformatic approach to prioritize drug targets in pathogens. Nucleic Acids Res 46:D413–D418
171. Karp PD, Ivanova N, Krummenacker M, Kyrpides N, Latendresse M, Midford P et al (2019) A comparison of microbial genome web portals. Front Microbiol 10:208
172. Jensen PA (2018) Coupling fluxes, enzymes, and regulation in genome-scale metabolic models. Methods Mol Biol 1716:337–351
173. Mienda BS, Salihu R, Adamu A, Idris S (2018) Genome-scale metabolic models as platforms for identification of novel genes as antimicrobial drug targets. Future Microbiol 13:455–467
174. Gu C, Kim GB, Kim WJ, Kim HU, Lee SY (2019) Current status and applications of genome-scale metabolic models. Genome Biol 20:121
175. Merigueti TC, Carneiro MW, Carvalho-Assef APD, Silva-Jr FP, da Silva FAB (2019) FindTargetsWEB: a user-friendly tool for Identification of potential therapeutic targets in metabolic networks of bacteria. Front Genet 10:633
176. Arkin AP, Cottingham RW, Henry CS, Harris NL, Stevens RL, Maslov S et al (2018) KBase: the united states department of energy systems biology knowledgebase. Nat Biotechnol 36:566–569
177. Cottret L, Frainay C, Chazalviel M, Cabanettes F, Gloaguen Y, Camenen E, et al (2018) MetExplore: collaborative edition and exploration of metabolic networks. Nucleic Acids Res 46:W495–W502
178. Gao Z, Li H, Zhang H, Liu X, Kang L, Luo X et al (2008) PDTD: a web-accessible protein database for drug target identification. BMC Bioinform 9:104

179. Li H, Gao Z, Kang L, Zhang H, Yang K, Yu K et al (2006) TarFisDock: a web server for identifying drug targets with docking approach. Nucleic Acids Res 34:W219–W224
180. Wang X, Shen Y, Wang S, Li S, Zhang W, Liu X, et al (2017) PharmMapper 2017 update: a web server for potential drug target identification with a comprehensive target pharmacophore database. Nucl Acids Res W356–W360. http://dx.doi.org/10.1093/nar/gkx374
181. Yang H, Qin C, Li YH, Tao L, Zhou J, Yu CY, et al (2016) Therapeutic target database update 2016: enriched resource for bench to clinical drug target and targeted pathway information. Nucl Acids Res D1069–D1074. http://dx.doi.org/10.1093/nar/gkv1230
182. Chen L, Oughtred R, Berman HM, Westbrook J (2004) TargetDB: a target registration database for structural genomics projects. Bioinformatics 20:2860–2862
183. Chanumolu SK, Rout C, Chauhan RS (2012) UniDrug-target: a computational tool to identify unique drug targets in pathogenic bacteria. PLoS One 7:e32833
184. Wang L, Ma C, Wipf P, Liu H, Su W, Xie X-Q (2013) TargetHunter: an in silico target identification tool for predicting therapeutic potential of small organic molecules based on chemogenomic database. AAPS J 395–406. http://dx.doi.org/10.1208/s12248-012-9449-z
185. Singh NK, Selvam SM, Chakravarthy P (2006) T-iDT: tool for identification of drug target in bacteria and validation by Mycobacterium tuberculosis. Silico Biol 6:485–493
186. Luo H, Lin Y, Gao F, Zhang C-T, Zhang R (2014) DEG 10, an update of the database of essential genes that includes both protein-coding genes and noncoding genomic elements. Nucleic Acids Res 42:D574–D580
187. Magariños MP, Carmona SJ, Crowther GJ, Ralph SA, Roos DS, Shanmugam D et al (2012) TDR Targets: a chemogenomics resource for neglected diseases. Nucleic Acids Res 40:D1118–D1127
188. Raman K, Yeturu K, Chandra N (2008) targetTB: a target identification pipeline for Mycobacterium tuberculosis through an interactome, reactome and genome-scale structural analysis. BMC Syst Biol 2:109
189. Panjkovich A, Gibert I, Daura X (2014) antibacTR: dynamic antibacterial-drug-target ranking integrating comparative genomics, structural analysis and experimental annotation. BMC Genom 15:36
190. Shanmugam A, Natarajan J (2010) Computational genome analyses of metabolic enzymes in Mycobacterium leprae for drug target identification. Bioinformation 4:392–395
191. Oany AR, Mia M, Pervin T, Hasan MN, Hirashima A (2018) Identification of potential drug targets and inhibitor of the pathogenic bacteria Shigella flexneri 2a through the subtractive genomic approach. Silico Pharmacol 6:11
192. Kumar Jaiswal A, Tiwari S, Jamal SB, Barh D, Azevedo V, Soares SC (2017) An in silico identification of common putative vaccine candidates against treponema pallidum: a reverse vaccinology and subtractive genomics based approach. Int J Mol Sci 18. http://dx.doi.org/10.3390/ijms18020402
193. de Sarom A, Kumar Jaiswal A, Tiwari S, de Castro Oliveira L, Barh D, Azevedo V, et al (2018) Putative vaccine candidates and drug targets identified by reverse vaccinology and subtractive genomics approaches to control Haemophilus ducreyi, the causative agent of chancroid. J R Soc Interface 15. http://dx.doi.org/10.1098/rsif.2018.0032
194. Mukherjee S, Gangopadhay K, Mukherjee SB (2019) Identification of potential new vaccine candidates in Salmonella typhi using reverse vaccinology and subtractive genomics-based approach. bioRxiv. biorxiv.org. https://www.biorxiv.org/content/10.1101/521518v1.abstract
195. Song J-H, Ko KS (2008) Detection of essential genes in Streptococcus pneumoniae using bioinformatics and allelic replacement mutagenesis. Methods Mol Biol 416:401–408
196. Hasan S, Daugelat S, Rao PSS, Schreiber M (2006) Prioritizing genomic drug targets in pathogens: application to Mycobacterium tuberculosis. PLoS Comput Biol 2:e61
197. Cloete R, Oppon E, Murungi E, Schubert W-D, Christoffels A (2016) Resistance related metabolic pathways for drug target identification in Mycobacterium tuberculosis. BMC Bioinform 17:75
198. Lee D-Y, Chung BKS, Yusufi FNK, Selvarasu S (2011) In silico genome-scale modeling and analysis for identifying anti-tubercular drug targets. Drug Dev Res 121–129. http://dx.doi.org/10.1002/ddr.20408

199. Neelapu NRR, Mutha NVR (2015) Identification of potential drug targets in Helicobacter pylori strain HPAG1 by in silico genome analysis. Infect Disord-Drug. ingentaconnect.com, https://www.ingentaconnect.com/content/ben/iddt/2015/00000015/00000002/art00006
200. Mondal SI, Ferdous S, Jewel NA, Akter A, Mahmud Z, Islam MM et al (2015) Identification of potential drug targets by subtractive genome analysis of Escherichia coli O157:H7: an in silico approach. Adv Appl Bioinform Chem 8:49–63
201. Hossain T, Kamruzzaman M, Choudhury TZ, Mahmood HN, Nabi AHMN, Hosen MI (2017) Application of the subtractive genomics and molecular docking analysis for the identification of novel putative drug targets against salmonella enterica subsp. enterica serovar Poona. Biomed Res Int 2017:3783714
202. Pradeepkiran JA, Kumar KK, Kumar YN, Bhaskar M (2015) Modeling, molecular dynamics, and docking assessment of transcription factor rho: a potential drug target in Brucella melitensis 16M. Drug Des Devel Ther 9:1897–1912
203. Muhammad SA, Ahmed S, Ali A, Huang H, Wu X, Yang XF et al (2014) Prioritizing drug targets in Clostridium botulinum with a computational systems biology approach. Genomics 104:24–35
204. Bhardwaj T, Somvanshi P (2017) Pan-genome analysis of Clostridium botulinum reveals unique targets for drug development. Gene 623:48–62
205. Kumar A, Thotakura PL, Tiwary BK, Krishna R (2016) Target identification in Fusobacterium nucleatum by subtractive genomics approach and enrichment analysis of host-pathogen protein-protein interactions. BMC Microbiol 16:84
206. Rahman A, Noore S, Hasan A, Ullah R, Rahman H, Hossain A et al (2014) Identification of potential drug targets by subtractive genome analysis of Bacillus anthracis A0248: An in silico approach. Comput Biol Chem 52:66–72
207. David MZ, Daum RS (2010) Community-associated methicillin-resistant Staphylococcus aureus: epidemiology and clinical consequences of an emerging epidemic. Clin Microbiol Rev 23:616–687
208. McCarthy AJ, Lindsay JA (2010) Genetic variation in Staphylococcus aureus surface and immune evasion genes is lineage associated: implications for vaccine design and host-pathogen interactions. BMC Microbiol 10:173
209. Le Loir Y, Baron F, Gautier M (2003) Staphylococcus aureus and food poisoning. Genet Mol Res 2:63–76
210. Plata K, Rosato AE, Wegrzyn G (2009) Staphylococcus aureus as an infectious agent: overview of biochemistry and molecular genetics of its pathogenicity. Acta Biochim Pol 56:597–612
211. Eriksen KR (1961) "Celbenin"-resistant staphylococci. Ugeskr Laeger 123:384–386
212. Lee AS, de Lencastre H, Garau J, Kluytmans J, Malhotra-Kumar S, Peschel A et al (2018) Methicillin-resistant Staphylococcus aureus. Nat Rev Dis Primers 4:18033
213. Enright MC, Robinson DA, Randle G, Feil EJ, Grundmann H, Spratt BG (2002) The evolutionary history of methicillin-resistant Staphylococcus aureus (MRSA). Proc Natl Acad Sci U S A 99:7687–7692
214. Nübel U, Roumagnac P, Feldkamp M, Song J-H, Ko KS, Huang Y-C et al (2008) Frequent emergence and limited geographic dispersal of methicillin-resistant Staphylococcus aureus. Proc Natl Acad Sci U S A 105:14130–14135
215. Pavillard R, Harvey K, Douglas D, Hewstone A, Andrew J, Collopy B et al (1982) Epidemic of hospital-acquired infection due to methicillin-resistant Staphylococcus aureus in major Victorian hospitals. Med J Aust 1:451–454
216. Aires De Sousa M, Miragaia M, Sanches IS, Avila S, Adamson I, Casagrande ST et al (2001) Three-year assessment of methicillin-resistant Staphylococcus aureus clones in Latin America from 1996 to 1998. J Clin Microbiol 39:2197–2205
217. Aires de Sousa M, Sanches IS, Ferro ML, Vaz MJ, Saraiva Z, Tendeiro T, et al (1998) Intercontinental spread of a multidrug-resistant methicillin-resistant Staphylococcus aureus clone. J Clin Microbiol 36:2590–2596

218. Szczepanik A, Kozioł-Montewka M, Al-Doori Z, Morrison D, Kaczor D (2007) Spread of a single multiresistant methicillin-resistant Staphylococcus aureus clone carrying a variant of staphylococcal cassette chromosome mec type III isolated in a university hospital. Eur J Clin Microbiol Infect Dis 26:29–35
219. Botelho AMN, Cerqueira E Costa MO, Moustafa AM, Beltrame CO, Ferreira FA, Côrtes MF, et al (2019) Local diversification of methicillin- resistant staphylococcus aureus ST239 in South America after its rapid worldwide dissemination. Front Microbiol 10:82
220. Donlan RM, Costerton JW (2002) Biofilms: survival mechanisms of clinically relevant microorganisms. Clin Microbiol Rev 15:167–193
221. Otto M (2018) Staphylococcal biofilms. Microbiol Spectr 6. http://dx.doi.org/10.1128/microbiolspec.GPP3-0023-2018
222. Otto M (2013) Staphylococcal infections: mechanisms of biofilm maturation and detachment as critical determinants of pathogenicity. Annu Rev Med 64:175–188
223. Otto M (2019) Staphylococcal biofilms. In: Fischetti VA, Novick RP, Ferretti JJ, Portnoy DA, Braunstein M, Rood JI (eds) Gram-positive pathogens. ASM Press, Washington, DC, USA, pp 699–711
224. McCarthy H, Rudkin JK, Black NS, Gallagher L, O'Neill E, O'Gara JP (2015) Methicillin resistance and the biofilm phenotype in Staphylococcus aureus. Front Cell Infect Microbiol 5:1
225. Costa MOC (2018) Modelo metabólico em escala genômica integrado com as vias regulatórias associadas ao biofilme de Staphylococcus aureus ST239-SCCmecIII (Bmb9393). Petrópolis. Thesis [Ph.D in Computational Modeling]—Laboratório Nacional de Computação Científica
226. Costa MOC, Beltrame CO, Ferreira FA, Botelho AMN, Lima NCB, Souza RC, et al (2013) Complete genome sequence of a variant of the methicillin-resistant staphylococcus aureus ST239 lineage, strain BMB9393, displaying superior ability to accumulate ica-independent biofilm. Genome Announc 1. http://dx.doi.org/10.1128/genomeA.00576-13
227. Botelho AMN, Costa MOC, Beltrame CO, Ferreira FA, Côrtes MF, Bandeira PT et al (2016) Complete genome sequence of an agr-dysfunctional variant of the ST239 lineage of the methicillin-resistant Staphylococcus aureus strain GV69 from Brazil. Stand Genomic Sci 11:34
228. Almeida LGP, Paixão R, Souza RC, da Costa GC, Barrientos FJA, dos Santos MT et al (2004) A system for automated bacterial (genome) integrated annotation–SABIA. Bioinformatics 20:2832–2833
229. Lister JL, Horswill AR (2014) Staphylococcus aureus biofilms: recent developments in biofilm dispersal. Front Cell Infect Microbiol 4:178
230. Ravcheev DA, Best AA, Tintle N, Dejongh M, Osterman AL, Novichkov PS et al (2011) Inference of the transcriptional regulatory network in Staphylococcus aureus by integration of experimental and genomics-based evidence. J Bacteriol 193:3228–3240
231. Nagarajan V, Elasri MO (2007) SAMMD: Staphylococcus aureus microarray meta-database. BMC Genom 8:351
232. Lechner M, Findeiss S, Steiner L, Marz M, Stadler PF, Prohaska SJ (2011) Proteinortho: detection of (co-)orthologs in large-scale analysis. BMC Bioinform 12:124
233. Bosi E, Monk JM, Aziz RK, Fondi M, Nizet V, Palsson BØ (2016) Comparative genome-scale modelling of Staphylococcus aureus strains identifies strain-specific metabolic capabilities linked to pathogenicity. Proc Natl Acad Sci U S A 113:E3801–E3809
234. Helikar T, Kowal B, McClenathan S, Bruckner M, Rowley T, Madrahimov A et al (2012) The cell collective: toward an open and collaborative approach to systems biology. BMC Syst Biol 6:96
235. Görke B, Stülke J (2008) Carbon catabolite repression in bacteria: many ways to make the most out of nutrients. Nat Rev Microbiol 6:613–624
236. Oskouian B, Stewart GC (1990) Repression and catabolite repression of the lactose operon of Staphylococcus aureus. J Bacteriol 172:3804–3812
237. Brückner R, Titgemeyer F (2002) Carbon catabolite repression in bacteria: choice of the carbon source and autoregulatory limitation of sugar utilization. FEMS Microbiol Lett 209:141–148

238. Somerville GA, Proctor RA (2009) At the crossroads of bacterial metabolism and virulence factor synthesis in Staphylococci. Microbiol Mol Biol Rev 73:233–248
239. Lindgren JK, Thomas VC, Olson ME, Chaudhari SS, Nuxoll AS, Schaeffer CR et al (2014) Arginine deiminase in Staphylococcus epidermidis functions to augment biofilm maturation through pH homeostasis. J Bacteriol 196:2277–2289
240. Beenken KE, Dunman PM, McAleese F, Macapagal D, Murphy E, Projan SJ et al (2004) Global gene expression in Staphylococcus aureus biofilms. J Bacteriol 186:4665–4684
241. Resch A, Rosenstein R, Nerz C, Götz F (2005) Differential gene expression profiling of Staphylococcus aureus cultivated under biofilm and planktonic conditions. Appl Environ Microbiol 71:2663–2676
242. Makhlin J, Kofman T, Borovok I, Kohler C, Engelmann S, Cohen G et al (2007) Staphylococcus aureus ArcR controls expression of the arginine deiminase operon. J Bacteriol 189:5976–5986
243. Fernández M, Zúñiga M (2006) Amino acid catabolic pathways of lactic acid bacteria. Crit Rev Microbiol 32:155–183
244. Podschun R, Pietsch S, Höller C, Ullmann U (2001) Incidence of Klebsiella species in surface waters and their expression of virulence factors. Appl Environ Microbiol 67:3325–3337
245. Podschun R, Ullmann U (1998) Klebsiella spp. as nosocomial pathogens: epidemiology, taxonomy, typing methods, and pathogenicity factors. Clin Microbiol Rev 11:589–603
246. Rendueles O (2020) Deciphering the role of the capsule of Klebsiella pneumoniae during pathogenesis: a cautionary tale. Mol Microbiol http://dx.doi.org/10.1111/mmi.14474
247. Martin RM, Bachman MA (2018) Colonization, infection, and the accessory genome of Klebsiella pneumoniae. Front Cell Infect Microbiol 8:4
248. Wyres KL, Lam MMC, Holt KE (2020) Population genomics of Klebsiella pneumoniae. Nat Rev Microbiol http://dx.doi.org/10.1038/s41579-019-0315-1
249. Ko W-C, Paterson DL, Sagnimeni AJ, Hansen DS, Von Gottberg A, Mohapatra S et al (2002) Community-acquired Klebsiella pneumoniae bacteremia: global differences in clinical patterns. Emerg Infect Dis 8:160–166
250. Paczosa MK, Mecsas J (2016) Klebsiella pneumoniae: going on the offense with a strong defense. Microbiol Mol Biol Rev 80:629–661
251. Meatherall BL, Gregson D, Ross T, Pitout JDD, Laupland KB (2009) Incidence, risk factors, and outcomes of Klebsiella pneumoniae bacteremia. Am J Med 122:866–873
252. Holt KE, Wertheim H, Zadoks RN, Baker S, Whitehouse CA, Dance D et al (2015) Genomic analysis of diversity, population structure, virulence, and antimicrobial resistance in Klebsiella pneumoniae, an urgent threat to public health. Proc Natl Acad Sci U S A 112:E3574–E3581
253. Navon-Venezia S, Kondratyeva K, Carattoli A (2017) Klebsiella pneumoniae: a major worldwide source and shuttle for antibiotic resistance. FEMS Microbiol Rev 41:252–275
254. Wyres KL, Holt KE (2016) Klebsiella pneumoniae population genomics and antimicrobial-resistant clones. Trends Microbiol 24:944–956
255. Ambler RP (1980) The structure of beta-lactamases. Philos Trans R Soc Lond B Biol Sci 289:321–331
256. Escandón-Vargas K, Reyes S, Gutiérrez S, Villegas MV (2017) The epidemiology of carbapenemases in Latin America and the Caribbean. Expert Rev Anti Infect Ther 15:277–297
257. Lee C-R, Lee JH, Park KS, Kim YB, Jeong BC, Lee SH (2016) Global dissemination of carbapenemase-producing klebsiella pneumoniae: epidemiology, genetic context, treatment options, and detection methods. Front Microbiol 7:895
258. Wyres KL, Wick RR, Judd LM, Froumine R, Tokolyi A, Gorrie CL et al (2019) Distinct evolutionary dynamics of horizontal gene transfer in drug resistant and virulent clones of Klebsiella pneumoniae. PLoS Genet 15:e1008114
259. Braun G, Cayô R, Matos AP, de Mello Fonseca J, Gales AC (2018) Temporal evolution of polymyxin B-resistant Klebsiella pneumoniae clones recovered from blood cultures in a teaching hospital during a 7-year period. Int J Antimicrob Agents 51:522–527
260. Tiwari V, Tiwari M, Solanki V (2017) Polyvinylpyrrolidone-capped silver nanoparticle inhibits infection of carbapenem-resistant strain of acinetobacter baumannii in the human pulmonary epithelial cell. Front Immunol 8:973

261. Ramos PIP, Picão RC, de Almeida LGP, Lima NCB, Girardello R, Vivan ACP, et al (2014) Comparative analysis of the complete genome of KPC-2-producing Klebsiella pneumoniae Kp13 reveals remarkable genome plasticity and a wide repertoire of virulence and resistance mechanisms. BMC Genomics 54. http://dx.doi.org/10.1186/1471-2164-15-54
262. NIH HMP Working Group, Peterson J, Garges S, Giovanni M, McInnes P, Wang L, et al (2009) The NIH human microbiome project. Genome Res 19:2317–2323
263. Ramage B, Erolin R, Held K, Gasper J, Weiss E, Brittnacher M, et al (2017) Comprehensive arrayed transposon mutant library of Klebsiella pneumoniae outbreak strain KPNIH1. J Bacteriol 199. http://dx.doi.org/10.1128/JB.00352-17
264. Liao Y-C, Huang T-W, Chen F-C, Charusanti P, Hong JSJ, Chang H-Y et al (2011) An experimentally validated genome-scale metabolic reconstruction of Klebsiella pneumoniae MGH 78578, iYL1228. J Bacteriol 193:1710–1717
265. Payne DJ, Miller WH, Berry V, Brosky J, Burgess WJ, Chen E et al (2002) Discovery of a novel and potent class of FabI-directed antibacterial agents. Antimicrob Agents Chemother 46:3118–3124
266. Joo SH (2015) Lipid a as a drug target and therapeutic molecule. Biomol Ther 23:510–516
267. Lemaître N, Liang X, Najeeb J, Lee C-J, Titecat M, Leteurtre E, et al (2017) Curative treatment of severe gram-negative bacterial infections by a new class of antibiotics targeting LpxC. MBio 8. http://dx.doi.org/10.1128/mBio.00674-17
268. Peleg AY, Seifert H, Paterson DL (2008) Acinetobacter baumannii: emergence of a successful pathogen. Clin Microbiol Rev 21:538–582
269. Daugelavicius R, Bakiene E, Bamford DH (2000) Stages of polymyxin B interaction with the Escherichia coli cell envelope. Antimicrob Agents Chemother 44:2969–2978
270. Ramos PIP, Custódio MGF, Quispe Saji GDR, Cardoso T, da Silva GL, Braun G et al (2016) The polymyxin B-induced transcriptomic response of a clinical, multidrug-resistant Klebsiella pneumoniae involves multiple regulatory elements and intracellular targets. BMC Genom 17:737
271. Heuston S, Begley M, Gahan CGM, Hill C (2012) Isoprenoid biosynthesis in bacterial pathogens. Microbiology 158:1389–1401
272. Masini T, Hirsch AKH (2014) Development of inhibitors of the 2C-methyl-D-erythritol 4-phosphate (MEP) pathway enzymes as potential anti-infective agents. J Med Chem 57:9740–9763
273. Saggu GS, Pala ZR, Garg S, Saxena V (2016) New insight into isoprenoids biosynthesis process and future prospects for drug designing in plasmodium. Front Microbiol 7:1421
274. Kadian K, Vijay S, Gupta Y, Rawal R, Singh J, Anvikar A et al (2018) Structural modeling identifies Plasmodium vivax 4-diphosphocytidyl-2C-methyl-d-erythritol kinase (IspE) as a plausible new antimalarial drug target. Parasitol Int 67:375–385
275. Tang M, Odejinmi SI, Allette YM, Vankayalapati H, Lai K (2011) Identification of novel small molecule inhibitors of 4-diphosphocytidyl-2-C-methyl-D-erythritol (CDP-ME) kinase of Gram-negative bacteria. Bioorg Med Chem 19:5886–5895
276. Zhang Y-M, Rock CO (2008) Membrane lipid homeostasis in bacteria. Nat Rev Microbiol 6:222–233
277. Bukata L, Altabe S, de Mendoza D, Ugalde RA, Comerci DJ (2008) Phosphatidylethanolamine synthesis is required for optimal virulence of Brucella abortus. J Bacteriol 190:8197–8203
278. Postma TM, Liskamp RMJ (2016) Triple-targeting Gram-negative selective antimicrobial peptides capable of disrupting the cell membrane and lipid A biosynthesis. RSC Adv R Soc Chem 6:65418–65421
279. Design and synthesis of novel Azatidinones analogues as potential antimicrobials. J Chem, Biol Phys Sci (2017) http://dx.doi.org/10.24214/jcbps.a.7.3.68187
280. Bommineni GR, Kapilashrami K, Cummings JE, Lu Y, Knudson SE, Gu C et al (2016) Thiolactomycin-based inhibitors of bacterial β-Ketoacyl-ACP synthases with in vivo activity. J Med Chem 59:5377–5390

281. Serio AW, Kubo A, Lopez S, Gomez M, Corey VC, Andrews L, et al (2013) Structure, potency and bactericidal activity of ACHN-975, a first-in-class LpxC inhibitor. In: 53rd interscience conference on antimicrobial agents and chemotherapy. Achaogen, pp 10–13
282. Pahal V (2018) Significance of apigenin and rosmarinic acid mediated inhibition pathway of MurG, MurE and DNA adenine methylase enzymes with antibacterial potential derived from the methanolic extract of Ocimum sanctum. MOJ Drug Des Dev & Ther http://dx.doi.org/10.15406/mojddt.2018.02.00031
283. World Health Organization (2018) Global Tuberculosis Report 2018. World Health Organization
284. Lillebaek T, Dirksen A, Vynnycky E, Baess I, Thomsen VØ, Andersen ÅB (2003) Stability of DNA patterns and evidence ofmycobacterium tuberculosisreactivation occurring decades after the initial infection. J Infect Dis 1032–1039. http://dx.doi.org/10.1086/378240
285. Ascenzi P, Visca P (2008) Scavenging of reactive nitrogen species by mycobacterial truncated hemoglobins. Methods Enzymol 436:317–337
286. Rhee KY, Erdjument-Bromage H, Tempst P, Nathan CF (2005) S-nitroso proteome of mycobacterium tuberculosis: enzymes of intermediary metabolism and antioxidant defense. Proc Natl Acad Sci 467–472 http://dx.doi.org/10.1073/pnas.0406133102
287. Schopfer MP, Mondal B, Lee D-H, Sarjeant AAN, Karlin KD (2009) Heme/O2/•NO Nitric Oxide Dioxygenase (NOD) reactivity: phenolic nitration via a putative heme-peroxynitrite intermediate. J Am Chem Soc 11304–11305. http://dx.doi.org/10.1021/ja904832j
288. Buchmeier NA, Newton GL, Koledin T, Fahey RC (2003) Association of mycothiol with protection of Mycobacterium tuberculosis from toxic oxidants and antibiotics. Mol Microbiol 1723–1732 http://dx.doi.org/10.1046/j.1365-2958.2003.03416.x
289. Lunardi J, Nunes J, Bizarro C, Basso L, Santos D, Machado P (2013) Targeting the histidine pathway in mycobacterium tuberculosis. Curr Top Med Chem 2866–2884. http://dx.doi.org/10.2174/15680266113136660203
290. Barry CE, Crick DC, McNeil MR (2007) Targeting the formation of the cell wall core of M. tuberculosis. Infect Disord Drug Targets 7:182–202
291. Chan K, Knaak T, Satkamp L, Humbert O, Falkow S, Ramakrishnan L (2002) Complex pattern of Mycobacterium marinum gene expression during long-term granulomatous infection. Proc Natl Acad Sci U S A 99:3920–3925
292. Scandurra GM, Ryan AA, Pinto R, Britton WJ, Triccas JA (2006) Contribution ofL-alanine dehydrogenase toin vivopersistence and protective efficacy of the BCG vaccine. Microbiol Immunol 805–810. http://dx.doi.org/10.1111/j.1348-0421.2006.tb03856.x

Chapter 11
Modelling Oxidative Stress Pathways

Harry Beaven and Ioly Kotta-Loizou

Abstract Oxidative stress occurs as a result of an imbalance in reactive oxygen species (ROS) and antioxidant defences within the cells. It plays a key role in many physiological disease states, ranging from Alzheimer's to cancer, as well as in infectious diseases; for instance, oxidative stress is significant during bacterial infection where macrophages and neutrophils subject pathogenic bacteria to oxidising environments or upon exposure to antibiotics. Therefore, it is vital to understand the systems biology of oxidative stress in order to effectively tackle the many issues that it is related to. In this chapter, computational approaches applied for understanding oxidative stress in bacteria and eukaryotes will be detailed together with the relevant biological advances. These approaches include construction of protein–protein interaction networks, logical and flux balance modelling techniques, machine learning applications and, lastly, high-throughput genomic methods such as next-generation sequencing, which generates data to be used in the aforementioned techniques. Finally, several case studies will be presented and discussed in the context of oxidative stress.

Keywords Oxidative stress · Protein–protein interaction networks · Network analysis · Flux balance analysis · Boolean networks · Multiomics · Machine learning

11.1 Introduction

Oxidative stress is a concept that has a received much attention in recent years—it has been linked to a wide range of diseases including diabetes, cancer and Alzheimer's [1]. It can be defined as an imbalance between oxidants and antioxidants in favour of the oxidants, which leads to the perturbation of redox signalling and molecular damage [2]. These oxidants, known as reactive oxygen species (ROS), are generated under normal cellular activities such as mitochondrial respiration and intracellular

H. Beaven · I. Kotta-Loizou (✉)
Department of Life Sciences, Faculty of Natural Sciences, Imperial College London, London, UK
e-mail: i.kotta-loizou13@imperial.ac.uk

F. A. B. da Silva et al. (eds.), *Networks in Systems Biology*, Computational Biology 32,
https://doi.org/10.1007/978-3-030-51862-2_11

signalling as well as in response to bacterial infection and xenobiotics [3, 4]. ROS include radical and non-radical molecules that contain oxygen such as peroxides, superoxides and hydroxyl radicals. Traditionally, ROS are known to cause damage to biological macromolecules, i.e. DNA, proteins and lipids, resulting in altered gene expression and leading to disease. In spite of the damage that oxidative stress can cause to biological organisms, cells have evolved complex mechanisms to both protect against and make use of ROS [5]. For instance, mitochondrial ROS, originally thought of as an unfortunate but necessary by-product of respiration, have been shown to play signalling roles in autophagy [6] and the immune response [7].

In addition to oxidative stress, the state of reductive stress is evident where the imbalance between oxidants and antioxidants favours antioxidants and is currently becoming more prevalent in the literature. Reductive stress occurs when cellular conditions lead to the redox states of key molecules such as NAD^+/NADH shifting the cell to a more reducing environment [8]. Similar to oxidative stress, reductive stress has also been linked to many disease states including Alzheimer's [9].

Due to the close relationship that oxidative stress has with multiple disease states, as well as its relevance in bacterial infection, understanding the molecular mechanisms that underpin it from a systems perspective is key in the prediction, prevention and cure of these diseases. In this review, a variety of methods for understanding oxidative stress in the context of systems biology will be discussed, including protein–protein interaction networks (PPIN), Boolean networks, flux balance analysis (FBA), high-throughput techniques such as RNA sequencing (RNA-seq), and, finally, machine learning and advanced computational methods that can be applied in combination with the aforementioned techniques. Please note that this chapter does not intend to cover all aspects of the methodologies that it discusses. Instead, it aims to provide a broad overview of the techniques that a research scientist can use to probe a system. The review focuses on well-established methods with proven success as well as more novel approaches that may be used to rapidly enhance the progression of the research.

11.2 Protein–Protein Interaction Networks (PPIN)

Interactions between proteins play an important role in cell biology and are collectively known as the protein interactome. Proteins often function as necessary components in various cellular processes, including but not limited to cell signalling, catalysis and transport. Given the range of cellular processes that oxidative stress has been implicated in, mapping the protein–protein interactions that underpin both oxidative stress and the diseases caused is crucial for determining novel targets for medical intervention.

To this end, interaction networks can be constructed and analysed to extract information that may shed light on the complex relationship between oxidative stress and the progression of disease. This process can be broadly summarised in four steps: (i) collection protein–protein interactions, either through public interaction databases or novel experimentally determined interactions; (ii) interaction quality control and selection since many available interactions can be considered to have a low confidence score due to their source or the experimental method used to determine them; (iii) network construction and visualisation, using available software such as Cytoscape; (iv) network analysis, using a range of available methods such as Gene Ontology (GO).

PPINs have already successfully been used in the context of oxidative stress to illustrate that it is responsible for the underlying mechanism of both type 2 diabetes and hypertension [10]. In this study, the Unified Human Interactome database (UniHI) was used for network construction, visualisation and analysis. UniHI was specifically designed for the analysis of the human protein interactome and offers a single-access platform for human protein interaction data.

Before constructing a PPIN, it is important to define what is meant by protein–protein interaction. A working definition for the construction of PPINs is "a specific physical contact between proteins that occurs by selective molecular docking" [11]. A PPIN is made up of nodes, which represent proteins, and edges, which represent the specific physical interaction between the proteins. These edges may also be weighted and contain information such as the method by which the interaction was discovered or a confidence score for the probability of the interaction occuring.

11.2.1 Interaction Databases

While methods exist to determine novel protein–protein interactions for the system of interest, discussed in Sect. 11.2.4, many interaction databases are already available and populated with previously characterised interactions. These databases can be classified into three categories: primary databases, metadatabases and prediction databases [12]. Primary databases, such as BioGRID [13], IntAct [14] and MatrixDB [15], collect experimentally verified interactions from publications and can be further classified into general databases that curate interactions for a variety of organisms and into more specific databases, such as MatrixDB which curates interactions for extracellular matrix proteins in humans. Metadatabases consolidate interactions from multiple primary databases, with some also collecting original data from the literature. Similar to primary databases, metadatabases can be general in their coverage or organism-specific; popular examples include Mentha [16] and APID [17]. Finally, predictive databases such as STRING [18] use various computational techniques to predict protein interactions. These techniques have been comprehensively reviewed and include gene clustering and neighbourhoods, phylogenetic profiles, sequence co-evolution, Rosetta Stone and machine learning methods such as random decision forests [19]. A non-exhaustive list of 14 popular interaction databases is shown in Table 11.1.

Table 11.1 Popular protein–protein interaction databases

Database	Classification	Organism	Refs.
APID	Metadatabase	General	[17]
BIND	Primary database	General	[100]
BioGRID	Primary database	General	[13]
BioPlex	Primary database	*Homo sapiens*	[101]
DIP	Primary database	General	[102]
IID	Metadatabase	*Homo sapiens*	[103]
InnateDB	Primary database	*Homo sapiens*; *Mus musculus*; *Bos taurus*	[104]
IntAct	Primary database	General	[14]
iRefWeb	Metadatabase	General	[105]
MatrixDB	Primary database	*Homo sapiens*	[15]
Mentha	Metadatabase	General	[16]
MINT	Primary database	General	[106]
MPIDB	Metadatabase	Bacteria	[107]
STRING	Predictive database	General	[18]

11.2.2 The Problem of Redundant Interactions

Redundant interactions need to be accounted for when selecting interactions for network construction. Redundancy can arise from the use of multiple databases for network construction and a thorough review investigating this redundancy between databases was carried out by Dong and Provart, showing that in most cases the overlap between databases is less than 50% [20]. Additionally, redundancy can arise from within the same database if the same interaction has been submitted multiple times due to different methods of discovery. For example, BioGRID currently contains 1,770,514 genetic and physical interactions of which only 1,376,013 are non-redundant [13].

Network visualisation software, such as Cytoscape, has tools built-in that remove redundant interactions. However, these functions are not entirely effective when integrating data from multiple databases and may require human intervention through the use of visual inspection or programming scripts that automatically flag redundant interactions for removal.

11.2.3 Interaction Reliability

Once redundant interactions have been screened for and successfully removed, a quality control process must be carried out on the remaining interactions. The reliability of the interactions within a database will depend on both the experimental method used to detect the interaction as well as the literature curation efficiency of the database itself. Several papers have assessed the efficiency of the literature curation process and the results are varied. One review found a 2–9% error rate from datasets covering three different species curated by five different interaction databases [21]. Another review found a 1.58% error rate in model organism databases like EcoCyc and CGD [22]. Databases such as STRING, which uses text mining and computational techniques to predict interactions, provide the user with a confidence score from 0 to 1 representing the likelihood that the interaction is accurate. Depending on the scope of the study, it may be suitable to set an arbitrary threshold for a reaction confidence score.

11.2.4 Generating Novel Interactions

When dealing with a system that is novel and uncharacterised, or with interaction databases that do not build a clear picture of the interaction landscape, it may be necessary to carry out experiments to determine novel interactions. A large variety of in vitro and in vivo methods exists for characterising protein–protein interactions. These methods have been reviewed in detail and include yeast-2-hybrid (Y2H), fluorescence resonance energy transfer (FRET) and affinity-purification mass spectrometry (AP-MS) [23, 24]. While choosing the best method, it is important to consider both scientific and logistical factors [24]. Studies with a difference in scope may lend themselves towards different techniques; for instance, high-throughput techniques such as AP-MS may be suitable for a general whole proteome study. Another question that must be considered is the type of interaction that is being studied. For example, if the interaction requires a co-factor or is localised to a specific area of the cell other than the nucleus or the cytoplasm then Y2H may not be a suitable method. Finally, logistical factors such as limitations in cost, time or technical expertise have to be taken into consideration.

11.2.5 Network Construction

Once the desired protein–protein interactions have been selected and screened, the construction of an interaction network can begin. There are many available programs for both the visualisation and analysis of PPINs. Perhaps the most popular of these is Cytoscape, an open-source Java application available on Linux, Windows and Mac OS operating system [25]. Currently, Cytoscape has 361 plugins, known as apps, that carry out a number of functions such as data visualisation, network analysis, clustering, enrichment analysis and network comparison. While Cytoscape has a wide range of tools and functions that are useful for network analysis, it suffers from a high computational demand when dealing with large networks. Therefore, other programs may be more suitable. NetworkX [26] is a Python-based tool that is less computationally intensive than Cytoscape and can also be easily integrated into bioinformatics pipelines. Similar to Cytoscape, NetworkX is also continuously updated, an issue that other software face. A variety of other tools exist that can analyse constructed networks, such as NetworkBLAST which analyses PPIs across species to discover protein complexes that are conserved through evolution [27].

11.2.6 Dynamic Interaction Networks

Construction and evaluation of PPINs are useful for gaining insight into biological systems and how they work; however, it is important to understand that these networks are simply snapshots under set temporal and spatial conditions. Real living systems are much more dynamic and responsive than these PPINs give them credit for. In the context of oxidative stress, there is a variety of antioxidant defence systems that can be upregulated during stress. It is known that these defence systems can react to specific ROS; for instance, the bacterial transcription factors OxyR and SoxRS undergo conformational changes in response to hydrogen peroxide and superoxide, respectively, subsequently activating the expression of genes under their control [28]. It is, therefore, advantageous to understand these networks in a more dynamic manner.

Traditional methods that discover protein–protein interactions such as Y2H do not provide any of this dynamic information. Much like the PPINs that they help build, they are a snapshot in time. However, progress has been made, particularly with computational methods, to create a more dynamic picture of protein–protein interactions. For example, one study used FRET in vivo to successfully create a dynamic map of the chemotaxis pathway in *E. coli* with nine protein interactions dependent on chemotactic stimulation [29].

11.2.7 Network Analysis

Several methods exist for PPIN analysis. Topological analysis can be carried out to determine clustering and centrality (discussed in Sect. 11.5) and it is also relevant to Boolean networks. Annotation enrichment analysis is another way in which PPINs can be analysed, following initial annotation of the proteins in the network using information from a database such as Gene Ontology (GO) [30]. Statistical analysis can then be carried out to determine if any particular annotation is over-represented in the network relative to the number expected in the same sample size. GO analysis can be carried out to determine the molecular function, cellular component and biological process that the protein is involved with. These annotations can then be used to describe the network. Annotation enrichment analysis can also be used to determine the reliability of the PPIN [31]. GO analysis has been successfully used in the context of oxidative stress to discover a link between oxidative stress-induced serine/threonine kinase activation and frontotemporal dementia. The analysis was carried out on a novel genetic interaction map [32].

11.3 Flux Balance Analysis (FBA)

A variety of methods exist for the simulation of cellular metabolism. A popular method is FBA, a constraint-based mathematical technique for the analysis of metabolic networks. FBA relies on the principle of conservation of mass in a system to determine the optimal distribution of fluxes through reactions that maximises a pre-determined objective function. This means that very little a priori knowledge is required to model the system and is a distinct advantage of FBA when compared to mechanistic models that require in-depth kinetic parameters to successfully model. Another advantage of FBA is the computational speed even when dealing with large metabolic networks. The construction of a simple FBA model can be broken down into four steps: (i) metabolic reconstruction, (ii) construction of the stoichiometric matrix, (iii) definition of a biologically relevant objective function and (iv) optimisation of the flux distribution in the system to maximise the objective function. Other variations of FBA exist and additional constraints can be applied to the system to further reduce solution space.

FBA has been used successfully to model oxidative stress in ischaemia–reperfusion injury, occurring after the restoration of blood supply to an organ after a previous disruption, causing the generation of mitochondrial ROS and leading to cell death and oxidative damage as well as an immune response [33]. Ischaemia–reperfusion injury underpins many diseases including heart attack and stroke and the mechanism behind mitochondrial ROS generation was previously thought to be a non-specific reaction to reperfusion. In this study, Chouchani et al. [33] use an untargeted and targeted metabolomics approach in conjunction with FBA to show that generation of mitochondrial ROS is caused by a build-up of the TCA-cycle intermediate succinate. This investigation also illustrated that inhibition of succinate accumulation is a possible therapeutic target in treating a wide range of these diseases.

11.3.1 Metabolic Reconstruction

The first step in FBA analysis is a metabolic reconstruction of the system of interest, consisting of a list of stoichiometrically balanced biochemical reactions. Whole genome-scale metabolic reconstructions have been carried out across multiple organisms including *Escherichia coli* [34], *Homo sapiens* [35] and *Saccharomyces cerevisiae* [36]. Furthermore, a variety of tools exist for the genome-scale reconstruction of hundreds of microorganisms with industrial or human pathogen-related relevance and these tools have been thoroughly reviewed [37]. One example is AutoKEEGRec [38], an automated program that uses the KEGG database to construct models for any microorganism in the database. This tool runs in MATLAB and is compatible with the constraint-based reconstruction and analysis (COBRA) toolbox [39]. AutoKEEGRec is used for the construction of draft models that may be utilised as a starting point for further refinement. Many papers have been published that thoroughly review the methodologies used in the construction of whole-genome metabolic networks [40, 41].

11.3.2 Construction of the Stoichiometric Matrix

Once the list of stoichiometrically balanced reactions has been curated, the creation of a stoichiometric matrix (S matrix) can begin. In the stoichiometric matrix each row represents a metabolite and each column represents a reaction, as seen in Fig. 11.1. In each column, there is a negative value for each metabolite consumed and a positive

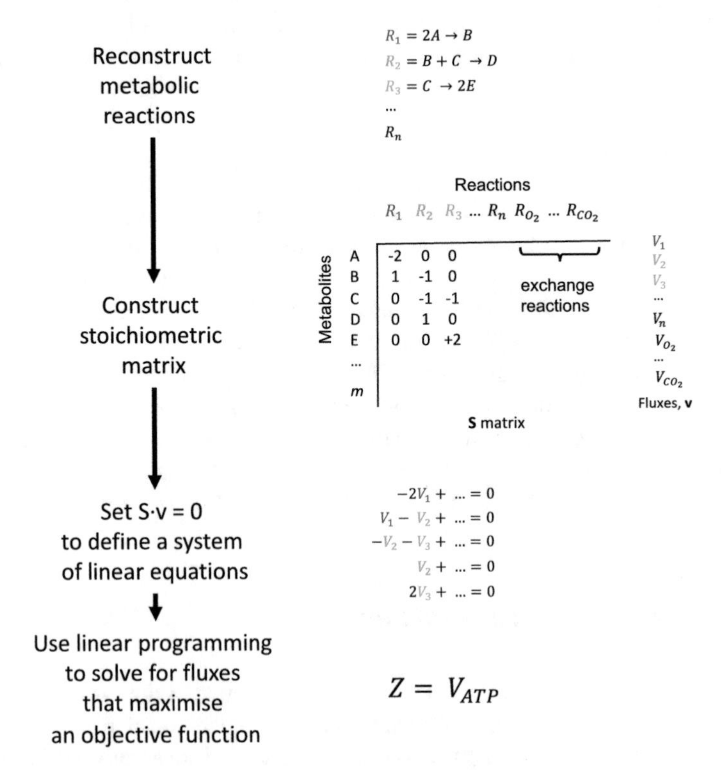

Fig. 11.1 Schematic detailing the steps for FBA. (1) Reconstruct the metabolic pathway in the form of a list of stoichiometrically balanced equations. (2) Construct the stoichiometric matrix (S matrix) to impose mass balance constraints on the system. (3) Set $S \cdot v = 0$ to define a set of linear equations representing the system at steady state. (4) Solve using linear programming to maximise or minimise an objective function Z, such as the production of ATP

value for each metabolite produced in the corresponding reaction. A value of zero indicates that the metabolite is not involved in the reaction. The stoichiometric matrix imposes the first constraint, flux (mass) balance, on the system. At steady state, this means that the total amount of a metabolite being produced is equal to the total amount of it being consumed. Exchange reactions can be incorporated to represent the movement of metabolites in and out of the cell. These reactions can include metabolites such as glucose, lactate, H_2O, CO_2 and O_2. Further constraints can be placed on the system by applying upper and lower bounds to the reaction fluxes, defining an upper and lower limit to the allowed fluxes through a particular reaction. These boundaries can also impose directionality to the reactions by setting the upper or lower limit to zero. Further constraints can also be applied to the network such as constraints based on the thermodynamic link between the flux direction in a reaction and metabolite concentration [42]. At steady state, the flux through each reaction is given by the equation:

$$S \cdot v = 0$$

where S is the stoichiometric matrix and v is a vector containing the flux through all the reactions in the system. As there are more unknown variables than there are equations, there is no single solution to the equation. The constraints applied to the system through the stoichiometric matrix, flux bounds, as well as any further constraints will reduce the solution space.

11.3.3 Defining an Objective Function

The next step in FBA analysis is to define a biologically relevant objective function; an example may be maximising for growth and a biomass reaction may be placed in the S matrix that draws precursor metabolites such as DNA, RNA, protein and lipids at stoichiometrically appropriate rates to simulate biomass production. FBA can now take place to calculate the flux distribution that results in the maximum flux through the biomass reaction. The objective function is given by:

$$Z = c \cdot v$$

where c is a vector of weights governing how much each reaction v contributes to the objective function Z.

FBA uses linear programming to identify network fluxes that are optimal relative to this defined objective function. The objective function could be defined as growth rate, ATP production or, in the case of metabolic engineering, the production of the desired product. One challenge with FBA is the choice of a biologically relevant objective function. When carrying out FBA in the context of genetic engineering, the objective function may simply be optimising for the maximum production of the desired product. However, when using FBA to model a system undergoing oxidative stress, this may not be a suitable objective function. Studies have shown that optimising for biomass production and growth gives predictions that are consistent with experimental data 86% of the time [43]. Several methods have been proposed to determine a biologically relevant objective function. One study proposes a modified version of FBA termed OM-FBA where multiomics data can be utilised to guide a suitable objective function [44]. This study successfully applied the method to sugar metabolism in *S. cerevisiae* and using transcriptomics data to guide the objective function and accurately predicted ethanol yield >80% of the time.

11.3.4 Flux Balance Analysis Tools

A number of tools exist for the analysis of FBA problems such as the leading COBRA toolbox [39]. An in-depth review of the COBRA Toolbox v.3.0 for MATLAB, as well as a tutorial for its use, is provided by Fleming et al. [39]. CellNetAnalyzer is another popular FBA analysis software for MATLAB [45]. Implementations also exist in Python such as COBRApy [46] which can carry out basic COBRA methods but does not require MATLAB to run. It also allows for interface with the COBRA Toolbox for MATLAB.

11.4 Boolean Networks

Complex systems of interactions, relying on both temporal and spatial parameters, often underpin biological processes: gaining insight into this regulation is of great benefit when trying to understand a system at an abstract level. Often the most in-depth method to model these processes is through continuous differential equations, requiring detailed knowledge of these parameters before the model is created; this is unlikely to be possible when dealing with unknown processes [47].

Boolean networks were first applied to biological systems by Stuart Kauffman [48]. Boolean networks, also known as discrete dynamical networks, provide a simplistic alternative that relies on Boolean (true/false) logic to predict the steady-state solution of a series of positive and negative interactions. In spite of their relative simplicity, Boolean networks have been shown to give highly accurate predictions of systems regulation under specific conditions [49]. A Boolean network much like a PPIN is comprised of nodes, which could represent a protein, DNA, RNA, a co-factor or any other biomolecule, and edges which represent an inhibition or activation between nodes. Boolean networks can model any type of interaction between these nodes provided positive or negative regulation is involved [50].

Boolean networks have been created in the context of oxidative stress. One study created a Boolean network that modelled the oxidative stress response and linked it to apoptosis *via* the PI3k/Akt pathway [51]. This has possible implications for cancerous cell development where apoptosis pathways can be downregulated.

11.4.1 Network Construction

The construction of a Boolean network shares many similar steps with the construction of protein–protein interaction networks. The first step involves the construction of an interaction network with positive and negative interactions between the nodes. The construction can take a "bottom-up" research approach, whereby literature and interaction databases are thoroughly reviewed to create a functional model of the system. Alternatively, or in conjunction with the "bottom-up" approach, a "top-down" data-driven approach can be taken where regulation is inferred from large datasets. This can be carried out through small-scale experiments that show differential expression of a gene under a particular condition, or through a computational approach. Computational approaches include algorithms that search large datasets from high-throughput techniques such as DNA microarrays [52]. Next, Boolean logic must be applied, and initial conditions determined so that the network can be constructed and analysed.

11.4.2 Network Analysis Tools

Once the network has been constructed and Boolean logic has been applied analysis can take place. Many programs, across multiple programming languages, exist for the analysis of Boolean networks. CellNetAnalyzer (CNA) is a MATLAB-based software, BooleanNet is a Python-based library and BoolNet is an R-based program. Finally, BoolSim a web-based program running Java allows for the analysis of simple Boolean network with no programming experience required. Boolean networks are time-discrete models, however, methods exist to convert them into continuous models such as the MATLAB package Odefy which creates a system of ODEs from a Boolean network [53]. This allows for the continuous dynamic modelling of a system without a priori knowledge of complex kinetic parameters.

11.4.3 A Simplified Example of a Boolean Network

A workflow for the creation of a simple Boolean Network has been presented in Fig. 11.2 with the subsequent analysis carried out by BooleanNet [54]. The simple Boolean network predicts that during stress conditions that result in the production of ROS, node C will be switched off. This can be experimentally confirmed in the laboratory. If experiments contradict the model then the hypothesis can be defined

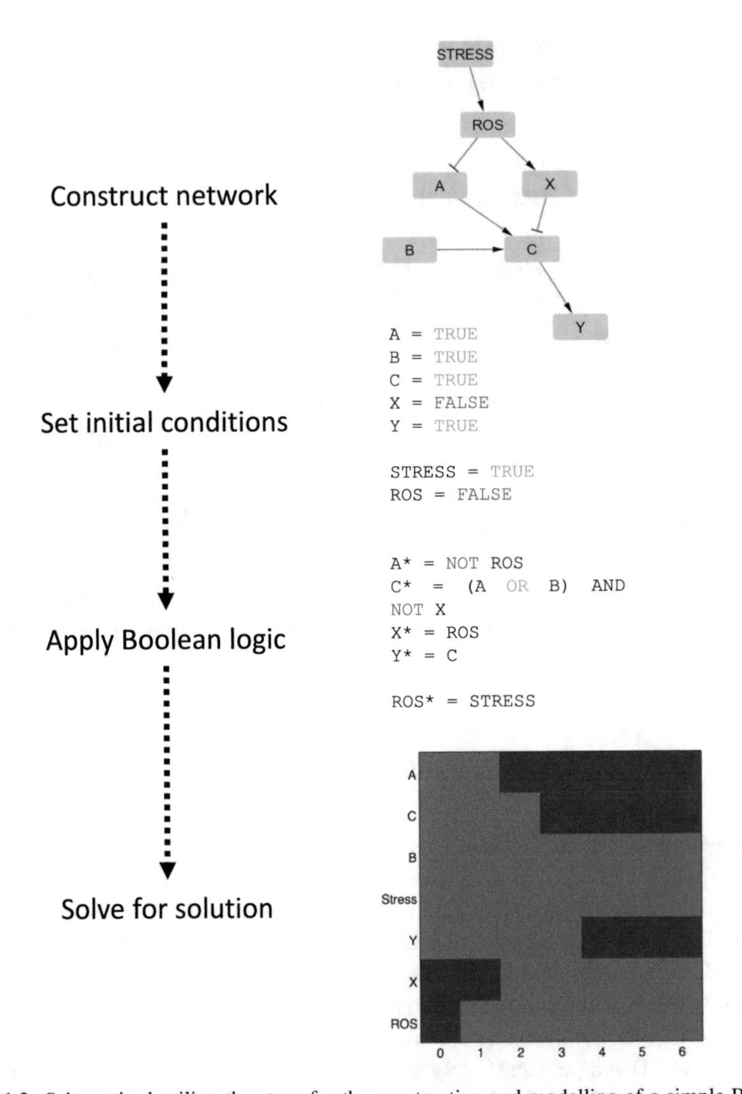

Fig. 11.2 Schematic detailing the steps for the construction and modelling of a simple Boolean network. (1) Define the network using a research and or a data driven approach. This network was constructed using Cytoscape. (2) Determine suitable initial conditions for the system. (3) Apply Boolean logic to the network. (4) Model the network using available software. This network was modelled with BooleanNet in Python 3.7

based on the results. In the case of the simple example, it may be that node C undergoes further regulation than previously thought and further interactions could be experimentally determined.

11.5 Centrality and Clustering in Biological Networks

The topological analysis of biological networks can be used to provide insight into biologically important members of the network. There are two main methods used in topological analysis; centrality and clustering.

Centrality measures are used to determine the most influential and important nodes in large networks. A number of methods exist for the analysis of 'central' nodes in both Boolean networks and PPINs and have been systematically reviewed in the context of PPINs [55]. Popular measures of centrality include closeness, betweenness, degree, eigenvector and eccentricity. Closeness centrality is given by the inverse of the sum of the shortest distances between the node in question and all other nodes in the network [56]. Betweenness centrality measures the number of times a node is included in the shortest path between any pair of nodes [56]. Betweenness centrality can, therefore, be considered a measure of how much information flows through a specific point in the network and how much that node can influence information flow in that network. Degree centrality is the idea that a central node will have the most connections to other nodes [56]. Networks with directionality such as Boolean networks can also be evaluated with an in-degree and an out-degree, where the in-degree is the number of edges coming towards a node and the out-degree is the number of edges leaving a node [57]. Eccentricity centrality is defined as the largest geodesic distance between the node of interest and any other node in the network [58]. Eigenvector centrality is given by measuring the influence of a node on the network, it is calculated by first giving each node in the network a relative score, nodes connected to high-scored nodes will then be considered to have more influence on the network [59].

Clustering analysis can be carried out to determine functional modules within the network. Clustering searches for communities of nodes where the nodes in the community are more connected to each other than the other nodes in the network. The score is given as a coefficient and can be given as a global score which refers to the clustering of the entire network, or as a local score which measures the local density of a single node [50]. Interestingly, nearly all real networks show a high degree of clustering when compared to a randomly generated network [60]. For example, clustering analysis has shown a higher degree of clustering of frontotemporal dementia associated genes than with the rest of the interactome, revealing a functional network [32].

11.6 High-Throughput and Omic-Based Screening Methods and Their Application in Systems Biology

In this section, several high-throughput techniques will be discussed. These methods allow for the collection of large datasets that can either be used in tandem with the other methods discussed in this review or as "stand-alone" data to gain an understanding of a system of interest. The omics suffix implies the comprehensive analysis of a particular set of molecules in a sample. The most popular omic methods include genomics, transcriptomics, proteomics and metabolomics. Here, some of these methods will be discussed, as well as secretomics and lipidomics due to their known association with oxidative stress. Early work with omic data focused solely on one discipline. Ultimately, it would be extremely beneficial to fully integrate these techniques into a systems biology approach. The integration of multiomics data has already been used to reveal a dynamic oxidative stress responses to manganese in neuroblastoma cells [61]. As previously discussed, methods such as OM-FBA allow for the integration of omics data into classical FBA. Another technique for this integration is through machine learning and is discussed in detail in Sect. 11.7.2.

11.6.1 Transcriptomics

Transcriptomics is the study of the transcriptome, the complete set of the RNA transcripts within a cell. Historically, methods such as northern blotting and reverse transcription-quantitative PCR (RT-qPCR) were popular when probing the transcriptome, however, these methods are relatively low-throughput and only capture a small fraction of the total transcriptome. Currently, two dominant techniques exist; microarrays which allow for the quantification for a predetermined set of RNA sequences and RNA-seq which takes advantage of high-throughput sequencing methods to analyse all the RNA in a given sample [62].

RNA-seq is a technique that can be used to investigate the transcriptome of cells under a specific physiological condition. The method has been thoroughly reviewed but can be broken down into a few steps: RNA extraction and subsequent cDNA synthesis resulting in a construction of a library, which is then sequenced with a high-throughput DNA sequencing technology such as Illumina. The reads can then be mapped to a transcriptome and quantified and statistical analysis can then be carried out to measure differential gene expression between samples of interest [63]. As a "stand-alone" method, RNA-seq has been used successfully in the context of oxidative stress to reveal a functional link between a long non-coding (lnc) RNA and cardiomyopathy. Oxidative stress was induced using a high-glucose medium in cardiomyocytes and differentially expressed lncRNAs were recorded. Subsequent studies inhibiting one of the most upregulated lncRNA (NONRATT007560.2), resulted in reduced production of ROS and apoptosis [64]. Furthermore, much like other omics-based approaches RNA-seq can be used as a data collection method and applied to techniques such as Boolean modelling and FBA [65].

One limitation with transcriptomic approaches is the issue of making assumptions between the transcriptome and the proteome, especially in eukaryotes where complex post-transcriptional processes such as RNA editing and nuclear export occur. Transcript stability also becomes a factor as well as the stability of the protein itself, with protein degradation and synthesis rates varying between both genes and cells [66]. Another compounding issue is the alteration of protein degradation rates during oxidative stress [67]. A genome-wide study of over 5000 genes in mammalian cells found that ~40% of protein levels could be explained by their corresponding mRNA level but that the level of translation is the major mechanism that controls the abundance of protein in the cell [68]. It is, therefore, advantageous to generate and integrated view of the system by combining transcriptomic approaches with proteomic approaches.

11.6.2 Proteomics and the Redox Proteome

Proteomics is the analysis of the total protein content in a cell, known as the proteome. This includes the expression, structure, function, interactions and any modifications of the protein at any stage of its existence. As previously discussed in Sect. 11.6.1, the proteome is dynamic and changes spatially and temporally as well as between cells and in response to external stimuli such as environments that result in oxidative stress. The proteome is not only defined by mRNA levels but is also subject to regulations affecting synthesis and degradation.

The redox proteome is a collective term for all proteins that undergo reversible redox reactions as well as those that are modified irreversibly during oxidative stress. Three amino acids undergo reversible redox reactions, cysteine, methionine and selenocysteine; however, many other amino acids, including the peptide backbone, undergo irreversible reactions during oxidative stress [69]. Redox signalling plays a vital role in both health and disease and a key characteristic of redox signalling is a redox reaction or covalent modification to a sensor protein which then regulates a cellular response [70]. The bacterial transcription factor OxyR is a well-characterised protein in redox signalling and a key regulator in the response to oxidative stress. The protein senses low levels of hydrogen peroxide through cysteine residues that become oxidised to form intramolecular disulphide bonds which result in a structural change in the protein regulatory domain and activate transcription of the genes necessary to defend against oxidative stress [71]. Similarly, the major oxidative stress response pathway in humans Keap1–Nrf2 is a redox-sensitive pathway: under normal cellular conditions Keap1 (Kelch ECH-associating protein 1) represses Nrf2 (nuclear factor erythroid 2-related factor 2) and promotes its degradation *via* the ubiquitin–proteasome pathway. Human Keap1 contains 27 cysteine residues, three of which (C151, C273 and C288) have been shown to play a role in altering the conformational state of Keap1 allowing nuclear translocation of Nrf2 and expression of target genes through the binding of Nrf2 to antioxidant response element (ARE) sequences in the regulatory regions of the genes [72]. Redox proteomics has been used to identify and characterise oxidised proteins in Alzheimer's disease [73]. The function of these oxidised proteins may then be used to provide explanations for the underlying mechanisms of Alzheimer's. Proteins identified in the study carried out a wide range of functions including energy-related enzymes and proteasome-related proteins.

The techniques used in proteomic studies, where mass spectrometry (MS) plays a central role as an extremely valuable tool to identify, characterise and quantify proteins and their post-translational modifications on both a high throughput and large scale, have been thoroughly reviewed [74, 75]. Spatial characterisation of the proteome is feasible by taking advantage of methods that allow for the specific purification of selected cellular compartments [76]. A dynamic understanding of proteomics can also be achieved through the use of isotopic labelling and MS to measure the synthesis and degradation rates of proteins in vivo at both a targeted level and across the entire proteome [77]. Moreover, it is now possible to quantify the entire proteome of a cell with high sensitivity [78].

11.6.3 Secretomics

Secretomics represent a sub-section of the field of proteomics which analyses proteins that are secreted by a cell, known as the secretome. Secreted proteins represent around 10% of the human genome and play vital roles in a variety of physiological processes [79]. There are two primary techniques used in secretomic workflows, based on liquid chromatography (LC) in combination with MS: in-solution digestion followed by LC-MS/MS, and SDS-gel fractionation and in-gel digestion followed by LC-MS/MS. These techniques have been thoroughly reviewed [80]. Recent studies have characterised the secretome under specific disease states to discover disease-specific biomarkers within the secretome. These studies have shown differential secretion of proteins in response to induced oxidative stress in vascular smooth muscle cells [81]. These secreted factors were termed secreted oxidative stress-induced factors (SOXF).

Computational methods also exist for the prediction of secreted proteins from transcriptomic and genomic data. This is made possible by conserved characteristics across both eukaryotes and prokaryotes, for example, classically secreted proteins in eukaryotes contain a cleavable N-terminal signal peptide of 15–30 amino acids [82]. Tools for predication fall into three categories based on their approach: random weights, sequence alignment and machine learning algorithms. These methods have been thoroughly reviewed [82].

11.6.4 Lipidomics

Lipids are a vital component of cellular membranes and lipoproteins and play crucial roles in a variety of cellular functions such as signalling and energy storage. In addition to the plasma membrane, they also have structural roles in the endoplasmic reticulum, Golgi apparatus and the nuclear membrane. Alterations in lipid homeostasis have, therefore, been closely linked to multiple disease states including obesity, cancer, diabetes, autoimmune diseases and neurodegenerative disorders [83]. Lipid peroxidation is the degradation of lipids through the attack of ROS and is a key contributor to cardiovascular disease [84]. For instance, oxidised phospholipids are known to play roles in inflammatory situations and are linked to conditions such as ischemia–reperfusion injury and atherosclerosis [85]. Lipids are diverse molecules with respect to their chemical structure and their cellular localisation [86]. This diversity presents a challenge in the analysis of lipids and is compounded by the wide variety of oxidants that can modify them [85]. The techniques for lipid analysis such as shotgun lipidomics, a highly accurate method to analyse thousands of lipid species, have been thoroughly reviewed [83, 85, 87].

11.6.5 Metabolomics—Biomarkers and Mechanisms

Metabolomics is the quantitative analysis of the metabolites in biological samples such as cells, biological fluids and tissues. Metabolomics has been used to both identify biomarkers of specific disease states and gain a deeper understanding of the mechanisms underpinning the disease.

The primary methodologies used in metabolomics are either targeted or untargeted MS-based approaches [88]. Untargeted metabolomics is a global analysis of the metabolites in a given sample. While this method offers large datasets and can be carried out without any prior knowledge, the methods of extraction affect the types of metabolites present in the sample resulting in incomplete samples. Targeted metabolomics involves the quantitative analysis of a set of predetermined metabolites. This allows for an optimised workflow for the extraction and analysis of the metabolites; however, prior knowledge of the system is required. Targeted metabolomics is commonly used to validate the results of untargeted metabolomics.

Metabolomics has successfully been used in the context of oxidative stress to identify multiple biomarkers associated with disease states [89–91]. Due to advances in technology and integration of metabolomics with other methods, such as genomics and proteomics, metabolomics can move beyond simply detecting biomarkers and be used to determine the underlying mechanisms behind the disease [88]. For example, as previously discussed, metabolomics was used in combination with FBA to determine the underlying mechanism of ROS production in ischaemia–reperfusion injury [33].

Much like a change in the transcriptome may not have a direct correlation with what is happening in the proteome, a change in the proteome may not have a direct link with the metabolome. MS-based proteomic studies are revealing that the proteome is extremely complex with most proteins undergoing post-translational modifications that regulate function [78].

11.7 Machine Learning in Systems Biology

As previously discussed, oxidative stress has been shown to play a complex role in multiple disease states. In this section, the application of machine learning techniques to previously mentioned topics will be covered. Given the extreme power of machine learning, it has become a vital tool for the progression of research in biology. Several papers review the basics of machine learning and how it can be applied to solve a biological problem [92, 93].

Two primary types of machine learning exist—supervised and unsupervised learning [93]. Supervised learning takes input data and learns the relationship between the input data and predetermined outputs to predict the outcomes of new inputs. Examples of supervised learning techniques include decision trees , support vector machines (SVM) and neural networks . Unsupervised learning algorithms infer these patterns from the data without the use of pre-trained labels. Popular methods include clustering and principle component analysis (PCA) and can be used to find patterns in high-dimensionality data such as the omics data discussed in this review.

Deep learning is a type of machine learning that takes advantage of large neural networks. It can be utilised in both a supervised and unsupervised manner. It generally requires massive and well-annotated datasets, a key disadvantage for biology as this may often not be attainable [94]. Typically, datasets of this quality have been found in genomic and image data.

11.7.1 Machine Learning and the Redox Proteome

Machine learning has been successfully applied to a wide variety of problems in biology. In the context of oxidative stress, machine learning has been used to predict redox-sensitive cysteine residues. As previously discussed, cysteine residues play a vital role in the redox proteome and cellular signalling. The group developed a support SVM-based classifier that predicted redox-sensitive cysteine residues with an accuracy of 0.68 [95]. Unlike other methods that focused specifically on catalytic cysteine residues or were heavily reliant on protein structural data, this approach allows for the prediction from primary sequence data alone. Similar work has also been carried out for methionine oxidation sites using SVM, neural network and random forest approaches with the latter achieving an accuracy of 0.75 [96]. Both of these studies are examples of how machine learning can be used to provide hypotheses and direct further experiments.

11.7.2 Machine Learning, Multiomics Data and FBA

Machine learning also has applications in the integration of multiomics data. The integration of omics data for machine learning purposes generally falls into three categories; concatenation-based, transformation-based and model-based [97]. Concatenation-based methods concatenate the matrices of multiple data types into a single matrix to which a machine learning algorithm can be applied. This method can lack reliability due to the noise and variance between datasets [98]. Transformation-based methods first transform the individual omics data into a uniform format such as a graph. Once transformed, the uniform data can be integrated into a final model to which machine learning methods can be applied. The model-based approach first applies machine learning methods to the data to create individual data-driven models. These models can then be integrated to produce a final model. Due to the stage at which the omics data is integrated these methods are also known as early, intermediate and late stage integration methods, respectively [97].

There are many ways in which machine learning can be applied to constraint-based modelling and FBA. These techniques including supervised fluxomic analysis, unsupervised fluxomic analysis, supervised multiomic analysis, unsupervised multiomic analysis and even for the generation of constraint-based models and fluxomic data have been thoroughly reviewed [97]. For example, in supervised fluxomic analysis, the output of the FBA model can be used to predict a biological result. One study was able to use machine learning to correlate growth conditions to simulated metabolic fluxes and ultimately allowed for the prediction of growth conditions using a given FBA solution [99]. Machine learning and constraint-based modelling can also be carried out in tandem by placing stoichiometric constraints in a learning task [97].

11.7.3 Machine Learning and Network Biology

Machine learning has been implemented in many areas of network biology. These applications have been thoroughly reviewed [94]. Of particular interest in the modelling of oxidative stress is the application of machine learning to disease biology. Interaction networks can be generated to represent a healthy individual and a machine learning approach applied to characterise what underpins a healthy state, this algorithm can then be fed data from patients to understand the dysregulation in disease states and this dysregulation can be explored further in the laboratory [94].

Other studies include the use PCA to determine the effectiveness of specific centrality measures at determining central nodes in a given network. The authors suggest that this analysis to be carried out on a network before choosing a suitable measure of centrality [55].

11.8 Concluding Remarks

It has been well established that oxidative stress underpins a wide range of disease states in human health. Many systems biology approaches are available for insights into the mechanisms of action of oxidative stress in these disease states and much progress has been made. The rise of high-throughput techniques has also provided research scientists with extremely large and complex datasets. The challenge moving forward is to move towards a more dynamic and integrated the view of this data to allow for a holistic view of the entire biological system in question to reveal unseen interactions between the datasets.

References

1. Reuter S, Gupta SC, Chaturvedi MM, Aggarwal BB (2010) Oxidative stress, inflammation, and cancer: how are they linked? Free Radic Biol Med 49:1603–1616
2. Sies H (2018) On the history of oxidative stress: concept and some aspects of current development. Curr Opin Toxicol 7:122–126
3. Poyton RO, Ball KA, Castello PR (2009) Mitochondrial generation of free radicals and hypoxic signaling. Trends Endocrinol Metab 20:332–340
4. Ray PD, Huang B-W, Tsuji Y (2012) Reactive oxygen species (ROS) homeostasis and redox regulation in cellular signaling. Cell Signal 24:981–990
5. Siauciunaite R, Foulkes NS, Calabrò V, Vallone D (2019) Evolution shapes the gene expression response to oxidative stress. Int J Mol Sci 20:3040
6. Scherz-Shouval R et al (2007) Reactive oxygen species are essential for autophagy and specifically regulate the activity of Atg4. EMBO J 26:1749–1760
7. West AP et al (2011) TLR signalling augments macrophage bactericidal activity through mitochondrial ROS. Nature 472:476–480
8. Perez-Torres I, Guarner-Lans V, Rubio-Ruiz ME (2017) Reductive stress in inflammation-associated diseases and the pro-oxidant effect of antioxidant agents. Int J Mol Sci 18
9. Lloret A, Fuchsberger T, Giraldo E, Vina J (2016) Reductive stress: a new concept in Alzheimer's disease. Curr Alzheimer Res 13:206–211
10. Jesmin J, Rashid MS, Jamil H, Hontecillas R, Bassaganya-Riera J (2010) Gene regulatory network reveals oxidative stress as the underlying molecular mechanism of type 2 diabetes and hypertension. BMC Med Genomics 3:45
11. De Las Rivas J, Fontanillo C (2012) Protein–protein interaction networks: unraveling the wiring of molecular machines within the cell. Brief Funct Genomics 11:489–496
12. De Las Rivas J, Fontanillo C (2010) Protein-protein interactions essentials: key concepts to building and analyzing interactome networks. PLoS Comput Biol 6:e1000807
13. Stark C et al (2006) BioGRID: a general repository for interaction datasets. Nucleic Acids Res 34:D535–D539
14. Hermjakob H et al (2004) IntAct: an open source molecular interaction database. Nucleic Acids Res 32:D452–D455

15. Launay G, Salza R, Multedo D, Thierry-Mieg N, Ricard-Blum S (2015) MatrixDB, the extracellular matrix interaction database: updated content, a new navigator and expanded functionalities. Nucleic Acids Res 43:D321–D327
16. Calderone A, Cesareni G (2012) Mentha: the interactome browser. EMBnet J 18:128
17. Alonso-López D et al (2016) APID interactomes: providing proteome-based interactomes with controlled quality for multiple species and derived networks. Nucleic Acids Res 44:W529–W535
18. Szklarczyk D et al (2019) STRING v11: protein-protein association networks with increased coverage, supporting functional discovery in genome-wide experimental datasets. Nucleic Acids Res 47:D607–D613
19. Shoemaker BA, Panchenko AR (2007) Deciphering protein–protein interactions. Part II. Computational methods to predict protein and domain interaction partners. PLOS Comput Biol 3:e43
20. Dong S, Provart NJ (2018) Analyses of protein interaction networks using computational tools. Methods Mol Biol 1794:97–117
21. Salwinski L et al (2009) Recurated protein interaction datasets. Nat Methods 6:860–861
22. Keseler IM et al (2014) Curation accuracy of model organism databases. Database (Oxford) 2014:bau058
23. Xing S, Wallmeroth N, Berendzen KW, Grefen C (2016) Techniques for the analysis of protein-protein interactions in vivo. Plant Physiol 171:727–758
24. Snider J et al (2015) Fundamentals of protein interaction network mapping. Mol Syst Biol 11:848
25. Shannon P et al (2003) Cytoscape: a software environment for integrated models of biomolecular interaction networks. Genome Res 13:2498–2504
26. Hagberg AA, Schult DA, Swart PJ (2008) Exploring network structure, dynamics, and function using NetworkX. In: Varoquaux G, Vaught T, Millman J (eds) Proceedings of the 7th python in science conference, pp 11–15
27. Kalaev M, Smoot M, Ideker T, Sharan R (2008) NetworkBLAST: comparative analysis of protein networks. Bioinformatics 24:594–596
28. Chiang SM, Schellhorn HE (2012) Regulators of oxidative stress response genes in Escherichia coli and their functional conservation in bacteria. Arch Biochem Biophys 525:161–169
29. Kentner D, Sourjik V (2009) Dynamic map of protein interactions in the Escherichia coli chemotaxis pathway. Mol Syst Biol 5:238
30. Ashburner M et al (2000) Gene ontology: tool for the unification of biology. The gene ontology consortium. Nat Genet 25:25–29
31. Dong S, Provart NJ (2018) Analyses of protein interaction networks using computational tools BT—two-hybrid systems: methods and protocols. In Oñate-Sánchez L (ed). Springer New York, pp 97–117. https://doi.org/10.1007/978-1-4939-7871-7_7
32. Palluzzi F et al (2017) A novel network analysis approach reveals DNA damage, oxidative stress and calcium/cAMP homeostasis-associated biomarkers in frontotemporal dementia. PLoS ONE 12:e0185797–e0185797
33. Chouchani ET et al (2014) Ischaemic accumulation of succinate controls reperfusion injury through mitochondrial ROS. Nature 515:431–435
34. Feist AM et al (2007) A genome-scale metabolic reconstruction for Escherichia coli K-12 MG1655 that accounts for 1260 ORFs and thermodynamic information. Mol Syst Biol 3:121
35. Duarte NC et al (2007) Global reconstruction of the human metabolic network based on genomic and bibliomic data. Proc Natl Acad Sci 104:1777–1782
36. Förster J, Famili I, Fu P, Palsson BØ, Nielsen J (2003) Genome-scale reconstruction of the Saccharomyces cerevisiae metabolic network. Genome Res 13:244–253
37. Mendoza SN, Olivier BG, Molenaar D, Teusink B (2019) A systematic assessment of current genome-scale metabolic reconstruction tools. Genome Biol 20:158
38. Karlsen E, Schulz C, Almaas E (2018) Automated generation of genome-scale metabolic draft reconstructions based on KEGG. BMC Bioinform 19:467

39. Heirendt L et al (2019) Creation and analysis of biochemical constraint-based models using the COBRA Toolbox v.3.0. Nat Protoc 14:639–702
40. Thiele I, Palsson BØ (2010) A protocol for generating a high-quality genome-scale metabolic reconstruction. Nat Protoc 5:93–121
41. Feist AM, Herrgård MJ, Thiele I, Reed JL, Palsson BØ (2009) Reconstruction of biochemical networks in microorganisms. Nat Rev Microbiol 7:129–143
42. Hoppe A, Hoffmann S, Holzhütter H-G (2007) Including metabolite concentrations into flux balance analysis: thermodynamic realizability as a constraint on flux distributions in metabolic networks. BMC Syst Biol 1:23
43. Edwards JS, Palsson BO (2000) The Escherichia coli MG1655 in silico metabolic genotype: Its definition, characteristics, and capabilities. Proc Natl Acad Sci 97:5528 LP–5533
44. Guo W, Feng X (2016) OM-FBA: integrate transcriptomics data with flux balance analysis to decipher the cell metabolism. PLoS ONE 11:e0154188
45. Klamt S, Saez-Rodriguez J, Gilles ED (2007) Structural and functional analysis of cellular networks with cell net analyzer. BMC Syst Biol 1:2
46. Ebrahim A, Lerman JA, Palsson BO, Hyduke DR (2013) COBRApy: COnstraints-Based Reconstruction and Analysis for Python. BMC Syst Biol 7:74
47. Wynn ML, Consul N, Merajver SD, Schnell S (2012) Logic-based models in systems biology: a predictive and parameter-free network analysis method. Integr Biol (Camb) 4:1323–1337
48. Kauffman SA (1969) Metabolic stability and epigenesis in randomly constructed genetic nets. J Theor Biol 22:437–467
49. Bornholdt S (2008) Boolean network models of cellular regulation: prospects and limitations. J R Soc Interface 5(Suppl 1):S85–S94
50. Bloomingdale P, Nguyen VA, Niu J, Mager DE (2018) Boolean network modeling in systems pharmacology. J Pharmacokinet Pharmacodyn 45:159–180
51. Sridharan S, Layek R, Datta A, Venkatraj J (2012) Boolean modeling and fault diagnosis in oxidative stress response. BMC Genom 13(Suppl 6):S4–S4
52. Nam D, Seo S, Kim S (2006) An efficient top-down search algorithm for learning Boolean networks of gene expression. Mach Learn 65:229–245
53. Krumsiek J, Pölsterl S, Wittmann DM, Theis FJ (2010) Odefy—from discrete to continuous models. BMC Bioinform 11:233
54. Albert I, Thakar J, Li S, Zhang R, Albert R (2008) Boolean network simulations for life scientists. Source Code Biol Med 3:16
55. Ashtiani M et al (2018) A systematic survey of centrality measures for protein-protein interaction networks. BMC Syst Biol 12:80
56. Freeman LC (1978) Centrality in social networks conceptual clarification. Soc Netw 1:215–239
57. Batool K, Niazi MA (2014) Towards a methodology for validation of centrality measures in complex networks. PLoS ONE 9:e90283
58. Bouttier J, Di Francesco P, Guitter E (2003) Geodesic distance in planar graphs. Nucl Phys B 663:535–567
59. Bonacich P (1972) Factoring and weighting approaches to status scores and clique identification. J Math Sociol 2:113–120
60. Barabási A-L, Oltvai ZN (2004) Network biology: understanding the cell's functional organization. Nat Rev Genet 5:101–113
61. Fernandes J et al (2016) 379—integration of multi-omics data reveal dynamic oxidative stress responses to manganese in human SH-SY5Y neuroblastoma cells. Free Radic Biol Med 100:S160
62. Lowe R, Shirley N, Bleackley M, Dolan S, Shafee T (2017) Transcriptomics technologies. PLoS Comput Biol 13:e1005457–e1005457
63. Stark R, Grzelak M, Hadfield J (2019) RNA sequencing: the teenage years. Nat Rev Genet. https://doi.org/10.1038/s41576-019-0150-2
64. Yu M et al (2019) RNA-Seq analysis and functional characterization revealed lncRNA NONRATT007560.2 regulated cardiomyocytes oxidative stress and apoptosis induced by high glucose. J Cell Biochem 120:18278–18287

65. Damiani C et al (2019) Integration of single-cell RNA-seq data into population models to characterize cancer metabolism. PLoS Comput Biol 15:e1006733–e1006733
66. Alber AB, Paquet ER, Biserni M, Naef F, Suter DM (2018) Single live cell monitoring of protein turnover reveals intercellular variability and cell-cycle dependence of degradation rates. Mol Cell 71:1079–1091.e9
67. Pajares M et al (2015) Redox control of protein degradation. Redox Biol 6:409–420
68. Schwanhäusser B et al (2011) Global quantification of mammalian gene expression control. Nature 473:337–342
69. Go Y-M, Jones DP (2013) The redox proteome. J Biol Chem 288:26512–26520
70. Forman HJ, Ursini F, Maiorino M (2014) An overview of mechanisms of redox signaling. J Mol Cell Cardiol 73:2–9
71. Choi H-J et al (2001) Structural basis of the redox switch in the OxyR transcription factor. Cell 105:103–113
72. Kansanen E, Kuosmanen SM, Leinonen H, Levonen A-L (2013) The Keap1-Nrf2 pathway: mechanisms of activation and dysregulation in cancer. Redox Biol 1:45–49
73. Butterfield DA, Perluigi M, Sultana R (2006) Oxidative stress in Alzheimer's disease brain: new insights from redox proteomics. Eur J Pharmacol 545:39–50
74. Zhang Z, Wu S, Stenoien DL, Paša-Tolić L (2014) High-throughput proteomics. Annu Rev. Anal Chem 7:427–454
75. Aslam B, Basit M, Nisar MA, Khurshid M, Rasool MH (2017) Proteomics: technologies and their applications. J Chromatogr Sci 55:182–196
76. Lamond AI et al (2012) Advancing cell biology through proteomics in space and time (PROSPECTS). Mol Cell Proteomics 11:O112.017731
77. Holmes WE, Angel TE, Li KW, Hellerstein MK (2015) Chapter seven—dynamic proteomics: in vivo proteome-wide measurement of protein kinetics using metabolic labeling. In: Metallo CMBT-M (ed) Metabolic analysis using stable isotopes, vol 561. Academic Press, pp 219–276
78. Altelaar AFM, Munoz J, Heck AJR (2013) Next-generation proteomics: towards an integrative view of proteome dynamics. Nat Rev Genet 14:35–48
79. Pavlou MP, Diamandis EP (2010) The cancer cell secretome: a good source for discovering biomarkers? J Proteomics 73:1896–1906
80. Song P, Kwon Y, Joo J-Y, Kim D-G, Yoon JH (2019) Secretomics to discover regulators in diseases. Int J Mol Sci 20:3893
81. Liao DF et al (2000) Purification and identification of secreted oxidative stress-induced factors from vascular smooth muscle cells. J Biol Chem 275:189–196
82. Caccia D, Dugo M, Callari M, Bongarzone I (2013) Bioinformatics tools for secretome analysis. Biochim Biophys Acta Proteins Proteomics 1834:2442–2453
83. Hu C, Wang M, Han X (2017) Shotgun lipidomics in substantiating lipid peroxidation in redox biology: methods and applications. Redox Biol 12:946–955
84. Halliwell B (2000) Lipid peroxidation, antioxidants and cardiovascular disease: how should we move forward? Cardiovasc Res 47:410–418
85. Spickett CM, Pitt AR (2015) Oxidative lipidomics coming of age: advances in analysis of oxidized phospholipids in physiology and pathology. Antioxid Redox Signal 22:1646–1666
86. Harayama T, Riezman H (2018) Understanding the diversity of membrane lipid composition. Nat Rev Mol Cell Biol 19:281–296
87. Yang K, Han X (2016) Lipidomics: techniques, applications, and outcomes related to biomedical sciences. Trends Biochem Sci 41:954–969
88. Johnson CH, Ivanisevic J, Siuzdak G (2016) Metabolomics: beyond biomarkers and towards mechanisms. Nat Rev Mol Cell Biol 17:451–459
89. Wang N et al (2016) Discovery of biomarkers for oxidative stress based on cellular metabolomics. Biomarkers 21:449–457
90. Soga T et al (2006) Differential metabolomics reveals ophthalmic acid as an oxidative stress biomarker indicating hepatic glutathione consumption. J Biol Chem 281:16768–16776
91. Lu Y et al (2019) Mass spectrometry-based metabolomics reveals occupational exposure to per- and polyfluoroalkyl substances relates to oxidative stress, fatty acid β-oxidation disorder, and kidney injury in a manufactory in China. Environ Sci Technol 53:9800–9809

92. Chicco D (2017) Ten quick tips for machine learning in computational biology. BioData Min 10:35
93. Xu C, Jackson SA (2019) Machine learning and complex biological data. Genome Biol 20:76
94. Camacho DM, Collins KM, Powers RK, Costello JC, Collins JJ (2018) Next-generation machine learning for biological networks. Cell 173:1581–1592
95. Sun M-A, Zhang Q, Wang Y, Ge W, Guo D (2016) Prediction of redox-sensitive cysteines using sequential distance and other sequence-based features. BMC Bioinform 17:316
96. Aledo JC, Cantón FR, Veredas FJ (2017) A machine learning approach for predicting methionine oxidation sites. BMC Bioinform 18:430
97. Zampieri G, Vijayakumar S, Yaneske E, Angione C (2019) Machine and deep learning meet genome-scale metabolic modeling. PLoS Comput Biol 15:e1007084–e1007084
98. Cavill R, Jennen D, Kleinjans J, Briedé JJ (2015) Transcriptomic and metabolomic data integration. Brief Bioinform 17:891–901
99. Sridhara V et al (2014) Predicting growth conditions from internal metabolic fluxes in an in-silico model of E. coli. PLoS ONE 9:e114608
100. Bader GD et al (2001) BIND—the biomolecular interaction network database. Nucleic Acids Res 29:242–245
101. Huttlin EL et al (2015) The BioPlex network: a systematic exploration of the human interactome. Cell 162:425–440
102. Salwinski L et al (2004) The database of interacting proteins: 2004 update. Nucleic Acids Res 32:D449–451
103. Kotlyar M, Pastrello C, Sheahan N, Jurisica I (2016) Integrated interactions database: tissue-specific view of the human and model organism interactomes. Nucleic Acids Res 44:D536–541
104. Breuer K et al (2013) InnateDB: systems biology of innate immunity and beyond–recent updates and continuing curation. Nucleic Acids Res 41:D1228–1233
105. Turner B et al (2010) iRefWeb: interactive analysis of consolidated protein interaction data and their supporting evidence. Database (Oxford) 2010:baq023–baq023
106. Licata L et al (2012) MINT, the molecular interaction database: 2012 update. Nucleic Acids Res 40:D857–861
107. Goll J et al (2008) MPIDB: the microbial protein interaction database. Bioinformatics 24:1743–1744

Chapter 12
Computational Modeling in Virus Infections and Virtual Screening, Docking, and Molecular Dynamics in Drug Design

Rachel Siqueira de Queiroz Simões, Mariana Simões Ferreira, Nathalia Dumas de Paula, Thamires Rocco Machado, and Pedro Geraldo Pascutti

Abstract Computer modeling is an area of broad multidisciplinary knowledge that includes the study of various biological systems. This chapter will describe the molecular aspects of viral infections and molecular modeling techniques applied to drug discovery with examples of applications in protein activity inhibition in several pathologies. The first part will cover topics of computational chemistry methods, DNA technologies, structural modeling of virus proteins, molecular biology, viral vectors, virus-like particles, and pharmaceutical bioprocess with application in some specific viruses such as papillomavirus, hepatitis B virus, hepatitis C virus, Coronavirus, and Zika Virus. The second part will deal with methods in Virtual Screening for the drug design based on ligands and on the structure of target macromolecules. Molecular docking in drug design, its search algorithms, and scoring functions will be covered in the third part. Finally, a protocol of the Molecular Dynamics technique for studies of protein-ligand complexes and analysis of free energy of binding will be exposed in the last part.

R. S. de Queiroz Simões
Center of Technology Development in Health and Laboratory of Interdisciplinary Medical Research, Oswaldo Cruz Foundation, Rio de Janeiro, Brazil
e-mail: rachelsqsimoes@gmail.com

M. S. Ferreira · N. Dumas de Paula · T. R. Machado · P. G. Pascutti (✉)
Molecular Modeling and Dynamics Laboratory, Institute of Biophysics Carlos Chagas Filho, Federal University of Rio de Janeiro, UFRJ, Rio de Janeiro, Brazil
e-mail: pascutti@biof.ufrj.br

M. S. Ferreira
e-mail: marianasimoesf@gmail.com

N. Dumas de Paula
e-mail: nathalia.dumas@gmail.com

T. R. Machado
e-mail: thamiresroccomachado@hotmail.com

F. A. B. da Silva et al. (eds.), *Networks in Systems Biology*, Computational Biology 32,
https://doi.org/10.1007/978-3-030-51862-2_12

Keywords Computational modeling · Docking · Drug design · Molecular dynamics · Virus · Virtual screening

The emergence of computers in the twentieth century allowed new advances in the field of life sciences because it allowed scientists to overcome limitations when exploring molecules in their experimental investigations. However, the term that represents this approach, involving simulations and the manipulation of a large amount of data, computational biology, was only used after the human DNA sequencing project in the 1990s. Since then, several approaches have used computational technology to solve problems in biology, which include, among others, the treatment of nucleic acid and protein sequences, description of biological systems networks, and modeling and simulation of the interactions of macromolecules and their ligands [1].

Advancements in computational technologies have also enabled improvements in the computational chemistry field of knowledge, such as in the development of new drugs. This way, computer software have become present in the areas of medicine, pharmacy, biology, chemistry, among others related to the development of biological processes or systems, investigating more complex processes at atomic-molecular scale each day at a much faster pace [2].

The investigation of molecular interactions is of great importance for biological understanding and allows observing and studying various processes involved in infectious diseases such as enzymatic action, binding of DNA repair proteins, and the formation of viral capsids. Moreover, molecular interactions also play a central role in the design of new drugs in the industry and a large part of the efforts are currently carried out using computational biology techniques, known as computer-aided drug design approaches. Thus, research has been developed through computer simulations in the investigation of molecular structures, protein-protein and protein-ligand complexes, biological membranes, and protein folding, with some of them being addressed throughout this chapter.

12.1 Computational Modeling in Viral Infections

In general, numerous conformations in protein folding have been documented by simulated annealing techniques through the application of Molecular Dynamics and Statistical Mechanics methods. These methods can simulate, for example, conformational changes in the antiviral proteins present in host cells, called viral restriction factors such as IFITM, SAMHD1, APOBEC3G and BST-2, when activated by the innate immune system during a viral infection and hinder the replication of viruses such as Influenza, Western Nile, Zika, HIV, and Hepatitis C virus [3].

Innate immune response, also known as natural or nonspecific response, acts mainly during viral infections in dendric cells (CD) and Natural Killer cells (NK), which have the ability to destroy virus-infected cells given its cytotoxic activity. They contribute to an antiviral defense action by secreting cytokines such as IFN capable of interfering with viral replication like IFN-α and IFN-β and by killing inhibitory receptors. In the acquired immune response, T helper lymphocytes (Th) and T cytotoxic lymphocytes (Tc) generate a specific cellular response while the antibodies produced by B lymphocytes generate a humoral response. The three-dimensional representation of protein structures by conformational analysis is obtained by Molecular Modeling, a branch of Computational Chemistry using mathematical models. Several computer programs are used, as described in this chapter, which allows the construction and identification of small molecules linked to proteins and nucleotides by analyzing protein databases such as Protein Data Bank (PDB) (Table 12.2), For example, the Osiris Property Explorer program is capable of signaling the molecule's behavior through two measures: Drug likeness and Drug score [2].

Computational Chemistry strategies are divided into the methods of Molecular Mechanics (MM) and Quantum Mechanics (QM). The MM approaches are based on Newtonian mechanics, in which the molecule is interpreted as a set of spheres connected by springs and can be used in calculations of very large molecules. In this method, the parameters associated with the atoms of the molecule are reasonably constant between different structures, as long as the hybridization of the atoms is maintained [2].

The QM methods, on the other hand, consider the movement of electrons, assuming fixed the nucleus, and represent molecular orbitals as a linear combination of atomic orbitals. They can be divided into strategies semi-empirical or ab initio. The semi-empirical approaches are developed based on empirical parameters or parameters already calculated from Schrodinger's equation, whereas the ab initio ones are based on the laws of quantum mechanics applying approximations for directly solving Schrodinger's equation that enable solutions for a wide range of molecules [2].

In Computational Chemistry, nevertheless, studies of new compounds have been applied regarding many infectious and viral diseases. An example is to analyze the activity of DNA polymerase of cytomegalovirus (CMV) against the physical-chemical properties of molecules in order to find potential candidates such as anti-CMV agents, in strategies such as Virtual Screening (VS), later described in this chapter [2]. It is an important investigation since in recent years there has been an increase in the number of opportunistic infections to human CMV, occurring mainly in transplant patients and in immunocompromised individuals.

Another example applies the molecular modeling of the interaction between human Respiratory Syncytial Virus (RSV) Fusion Protein (hRSV F protein) and Viral Action Inhibitors. RSV is responsible for outbreaks causing inflammatory disease in the lower respiratory tract, pneumonia in children, and cardiopulmonary disease in adults and immunodeficient patients. Currently, through the Molecular Docking technique, explained in detail further ahead in this chapter, possible sites of interaction of hRSV F protein with flavonoids have been proposed [2].

12.1.1 Viral Vectors

Experimental approaches in gene therapy commonly use viruses as vectors for inserting the genetic material into the cell. It applies the knowledge on the complex interaction between a virus and a host cell, known to include several steps. They include interactions with cell membrane receptors (adsorption), viral entry into the cell (penetration), partial or total denudation of the viral capsid, and release of viral nucleic acid for replication inside the cell, as well as early and late proteins synthesis, assembly and release by exocytosis, viral budding, cell lysis or even syncytia formation. Each step will be exemplified throughout the chapter and viral vector designs and applications will also be addressed as it progresses in Biotechnology. Currently used vectors in human gene therapy suffer from a number of limitations with respect to safety and reproducibility through the use of animal models, toxicology, and biodistribution studies. Therefore, other strategies such as nonviral and hybrid vectors are being explored trying to overcome these issues [4].

When it comes to the introduction of genes into mammalian cells, methodologies require the use of chemical, physical, and biological tools. Briefly, chemical methods may include DNA-calcium phosphate, DNA-DEAE dextran, DNA-lipide (liposomes not integrated into the genome of the host cell), DNA-protein and HCAs (artificial chromosomes). Even so, there are physical methods with high transfection rates such as direct microinjection, electroporation, high-pressure plasmid injection, and DNA ballistic injection [4].

Viral vectors for gene therapy can be enveloped as retrovirus, lentivirus and baculovirus, and non-enveloped, like adenovirus. The key point of human gene therapy clinical trials is the production of a therapeutic vector, which can be viruses of the *Retroviridae* family, mainly gammaretroviruses and lentiviruses, such as the Moloney-murine leukemia virus (Mo-MLV) and the chimeric Moloney-Human lentiviral (prototype HIV-1). Moreover, adenoviruses, adeno-associated virus (AAV), chimeric-AAV, and vaccinia virus are other types of biological vectors possible. There are a number of additional viral vectors based on Epstein–Barr virus, herpes simplex virus type 1 (HSV-1), simian virus 40, hepatitis virus, and papillomavirus, which are still being studied for application in clinical gene therapy [4].

12.1.2 Virus-like Particles

Virus-like particles (VLPs), structures that resemble viruses without the infectious capability of them since there is no genetic material present, show differences according to the presence of envelopment or not. In the case of non-enveloped VLPs, examples are the single protein hepatitis B core antigen and the two proteins of papillomavirus L1 and L2. As for the enveloped VLPs, there is Influenza virus with proteins hemagglutinin and neuraminidase. Some viral particles have the ability to self-assemble into VLPs regardless of the viral genome. HBV subviral particles

(HBsAg) have been used as a vector for the presentation of several antigen epitopes. The epitopes presented have included poliovirus capsid proteins, HCV envelope proteins, and human immunodeficiency virus (retrovirus) envelope proteins, among others [5].

VLPs can stimulate both innate and adaptive immune responses. Therefore, the particulate nature of VLPs generally induces an immune response that is more efficient than that generated by soluble proteins, and can trigger both a humoral and cellular response, without the need for an adjuvant. In addition, they are produced in large quantities, can be easily purified, and are not infectious. Currently, VLPs are a powerful tool in the presentation of immunogenic epitopes and the generation of recombinant VLPs, carrying relevant antigens, opening the way for the development of bivalent vaccines [6].

The incorporation of heterologous sequences of several viral structures into VLP particles has caused the development of chimeric VLPs. The use of this system as a platform for expression and presentation of exogenous epitopes favors the optimization of the immunogenicity of these peptide sequences. Numerous characteristics of VLPs allow the use of these particles as carriers of epitopes, the main one being that these particles are not infectious, since they do not carry the genetic material of the infectious agent, and still present the epitope expressed on its surface in a multi-sequential form [5, 7].

Several recombinant protein-based virus-like protein vaccines have been used as Human papillomavirus (HPV), Hepatitis B virus (HBV), and Hepatitis E vaccine. Others under development are the Influenza vaccine candidates, Ebola and Marburg virus vaccine, Hepatitis C virusHepatitis C vírus, and Human Immunodeficiency Virus (HIV) [5].

12.1.3 Pharmaceutical Bioprocess

Clustering algorithms and multivariate analysis are statistical data mining, classified in nonparametric methods. In a large scale, Bayesian networks and stoichiometric modeling for metabolic pathway analysis are also applied in Systems Biology. In addition, differential equations and several estimation algorithms are parametric modeling methods. So, pharmaceutical products are very different in size and complexity as small molecules, antibodies, viruses and viral vectors, and cells [8].

Molecular associations as gene expressions, analysis by DNA microarrays, PCR techniques, and kinetic experiments on molecular states are some examples of experimental methods. Modeling requires knowledge of mathematics, physics, biology, and information technology to develop processes for manufacturing therapeutic proteins in the biopharmaceutical field. It can be used, for example, to establish a cell culture process with higher productivity and lower costs [8].

12.1.4 Papillomavirus

The advancement of stem cell research provides tools to model diseases, test new drugs, and develop effective therapies for use in regenerative medicine and cell therapy. The mapping of the stem cells can generate biomedical implications with prospects for production in vitro of induced pluripotent stem cells (iPS). These pluripotent stem cells are reprogrammed from somatic cells using retroviral vectors containing many genes [8].

In the near future, the keratinocyte reprogramming will be possible in tissue engineering laboratories as creating new models for the study of tumors associated with HPV as anogenital cancers and oropharyngeal carcinoma. So, the expression of genes and the role of proteins involved in DNA damage repair pathways in cell lines as primary human keratinocytes (PHK), HPV-positive (SiHa—HPV-16 and HeLa—HPV-18), and HPV-negative (C33A) human cervical carcinoma cell lines, as also in immortalized keratinocyte cell lines (HaCaT, not tumor control), have been described as possible prognostic markers of cervical cancer [9].

Some studies have investigated the ability of the cytokine to inhibit the proliferation in vitro of normal and HPV infected keratinocytes, as well as the expression of E6 and E7 oncogenes. L1 capsids can trigger innate immune responses in some types of antigen-presenting cells, including dendritic cells. Thus, the dendritic cell control tissue immunity is a field with important implications in both basic and clinical immunology. So, the cytokine inhibition ability has been demonstrated [9].

Research has been carried out on the Molecular Modeling and Dynamics of the E6 oncoprotein of human papillomavirus type 18. This molecule is an essential protein for the formation of cervical cancer, since it is responsible for preventing cell apoptosis by binding in the tumor suppressor p53, promoting cell immortalization, and inducing carcinogenesis. In order to refine the generated E6 model, the N terminal version was modeled ab initio and part of the E6 model was done by homology based on the 4 GIZ protein [10].

New biotechnologies in Molecular Biology have been applied in papillomavirus research and are still being investigated for the application in gene therapy. For analysis of the transcriptome of genomes with high resolving power, RNA sequencing (RNA-seq) monitors the cell expression and allows quantifying the transcript levels using the Illumina platform for sequencing, which is a useful tool for gene mapping and identification of transcribed regions. Furthermore, examples of new techniques include the multiply primed rolling-circle amplification (RCA) method, which has been used to identify novel viruses, and the DNA sequencing strategy, based on real-time pyrophosphate.

The constant development of new biotechnologies is followed by the description of new therapeutic targets, for example, regarding HPV vaccination. They include DNA-based vaccines, recombinant proteins, nanoparticles, synthetic peptides, viral and nonviral vectors, and expressed chimeric proteins self-assembled into virus like particles (VLP) from L1 major capsid proteins to finally produce the HPV vaccines [7, 11].

12.1.5 Hepatitis B Virus (HBV)

Two billion people are infected and more than 350 million people are chronic carriers of the hepatitis B virus worldwide, according to the World Health Organization (WHO). Moreover, there are approximately six hundred thousand cases of HBV-related deaths, with 4.5 million of new infections around the world each year. It is estimated that 25% of this new infection will result in the development of liver diseases, meaning HBV is a major causative agent of chronic liver disease, presenting high morbidity and mortality rates. Therefore, it is necessary to study such virus at epidemiological and molecular levels, through the scope of the genomic variability analysis of HBV. Currently, it brings the classification of the genome into eight different genotypes named A–H that diverge by less than 8% from all the nucleotide sequence. Lastly, this virus is classified as being part of the family *Hepadnaviridae*, divided into the two genera *Orthohepadnavirus* (include viruses responsible for infections in mammals) and *Avihepadnavirus* (comprises bird infective viruses) [12–15].

The genome of this virus is composed of a double-stranded DNA with 3.200 base pairs (bp) having a partially incomplete smaller strand with positive polarity and a larger strand with negative polarity and complementary to the viral produce during replication. Four open reading frames (ORFs) are present in this genome: Pre-S/S, Pre-C/C, P, and X. The Pre-S/S ORF corresponds to the HBV surface gene (HBsAg). The Pre-S1, Pre-S2, and S regions have three initiation codons in the same reading phase which, after being translated, give rise to L "Large" (400 aa), M "Middle" (281 aa), and S "Small" proteins (226 aa), respectively [13, 16]. The HBsAg protein is present in the non-glycosylated (p24) and glycosylated (p27) forms in the S region, a glycoprotein (gp32) or two glycoproteins (gp36) in the M region, and the non-glycosylated (p37) and glycosylated (p39) in the region L. HBsAg hybrid particles, also called chimeras, have been shown in different immunization experiments to be very efficient proteins in the presentation of viral epitopes [12, 13].

Complete particles (42 nm), spherical, infectious, and containing the HBV genome, as well as subviral particles (22 nm), spherical or filamentous, noninfectious and composed exclusively of HBsAg, are produced during HBV infection. Subviral particles are in large quantities during infection [16].

HBV S protein is a glycoprotein, which is anchored in the endoplasmic reticulum (ER) membrane probably through four transmembrane (TM) segments. The first segment (TM1), located at the amino-terminal end of the protein (amino acids 4–24), is followed by a cytosolic loop (amino acids 24–80) comprising epitopes for T cells, a TM2 segment (amino acids 81–100), and an antigenic loop (amino acids 101–164), located in the luminal portion of the ER, which includes the main epitope for B cells (the determinant a, amino acids 124–147). The topology of the carboxy-terminal domain (amino acids 165–226) is not precisely known. It has been suggested that this domain contains two transmembrane segments (TM3 and TM4) at positions 173–193 and 202–222, respectively, being separated by a second "loop" (amino acids 194–201) [12, 15, 16].

The HBV S protein has the property of self-organizing into VLPs and carries the information necessary for its secretion mediated by mammalian cells. Each HBsAg particle is composed of 100–150 subunits of the S protein. Due to the high immunogenic potential, hybrid HBsAg particles, also called chimeras, have been shown in different immunization experiments to be very efficient proteins in the presentation of viral epitopes and antigenic sequences of various infectious agents, such as *Clostridium tetani*, Herpes simplex glycoprotein D, poliovirus capsid protein antigens derived from the malaria parasite, dengue virus, HCV, and HIV, which have been successfully fused to HBsAg, maintaining the antigenicity and immunogenicity of their regions after immunization in experimental animals [17, 18].

HBsAg has an antigenic region called a determinant (residues 124–147), which has the ability to induce a protective immune response against HBV. Neutralizing antibodies are directed to this region. Mutations in the determinant justify the development of new vaccines against HBV, among other factors that influence the vaccine response. New technologies such as DNA vaccines and bifunctional vaccines (for immunization against two vaccine targets simultaneously) have been a reality in the study and development of new vaccines [17, 19].

The first HBV vaccines, produced between 1981 and 1982, were created from the HBV surface antigen, HBsAg, from the plasma of chronic carriers of the virus. In the mid-1980s, with the introduction of recombinant DNA technology, this antigen started to be expressed in yeasts transfected by the HBsAg gene. This new technology offered unlimited production potential, which allowed the HBV vaccine to become more widely used worldwide. The development of recombinant vaccines composed exclusively of HBsAg (Engerix-B, SmithKline and Recombivax, Merck & Co) was possible with the advance of genetic engineering. So, the HBV vaccine was the first recombinant vaccine licensed and produced from yeast expression [15, 19].

12.1.6 Hepatitis C Virus (HCV)

Hepatitis C virus (HCV) is a chronic infection that affects more than 2% of the global population and is a leading cause of hepatocellular carcinoma and end-stage liver diseases. The progression from the viral infection to hepatocarcinoma is a worldwide public health problem that grows each year, supported by the lack of a vaccine against it to date. This virus belongs to the order *Nidovirales*, family *Flaviviridae*, genus *Hepacivirus* [14], and it is classified into seven genetically distinct genotypes, which differ approximately 30% at the nucleotide level. It is an enveloped virus, strand of positive polarity RNA, containing approximately 9,400 nucleotides. HCV is the virus that most infects humans among the viruses of clinical importance, being one of those with the highest mutation rate [20, 21].

The envelope glycoproteins (E1 and E2) have highly variable domains in their sequences, a large N-terminal ectodomain, and a C-terminal transmembrane portion. At the N-terminal end of the E2 protein, there is a region characterized by a high degree of variability called the hypervariable region 1 (HVR1), important in the neutralization of HCV. However, the high variability of this antigenic fragment plays a fundamental role in the viral escape mechanism of the host's immune response and represents the greatest obstacle in the development of an HCV vaccine [22, 23]. Despite this high variability, there are some residues in the HVR1 region that are conserved. Studies have shown that a HCV clone defective for HVR1 proved to be infectious in assays performed on chimpanzees, but with a highly attenuated infectivity, supporting the functional role of this domain in the entry of the virus.

E1 and E2 have 6 and 11 glycosylation sites, respectively, many of which are extremely conserved among HCV strains. It has been shown that these regions can play an important role in the entry of the virus into the cell and because of this, envelope glycoproteins have been the subject of extensive studies for the development of antiviral molecules, which prevent the entry of the virus into the hepatocyte [18, 24, 25].

The development of a chimeric vaccine, a prototype of heterologous antigens based on HBV envelope proteins, including HCV antigen, based on immunogenicity and antigenicity tests obtained from the collection and analysis of sera from animals laboratory immunized with synthetic peptides was assayed [26]. Production of specific antibodies in rabbits against conserved and potentially immunogenic peptides of the HCV envelope glycoprotein E2 was investigated. Possible cross-reactivity between the peptides and pre-immune sera of five rabbits that were further immunized was tested by *in house* Elisa. The sera at high concentrations showed a high level of cross-reactivity with the exception of anti-HVR1 antibody, which shows beyond the specific reaction, a nonspecific reaction with BSA. Thus, all the designed peptides were able to generate anti-E2 specific antibodies in rabbits at relatively high titles indicating candidate vaccine against HCV [15, 18, 25, 26].

12.1.7 Coronavirus

In the context of emerging diseases, a useful tool to investigate evolutionary history and viral origin from animals to humans is the phylogenetic analysis of molecular sequences of the viral genome sharing a common ancestor, since it may show the evolutionary relationships of novel viruses. The current outbreak of a coronavirus associated acute respiratory disease called coronavirus disease 19 (COVID-19), for example, has been designated as severe acute respiratory syndrome coronavirus 2 (SARS-CoV-2 or 2019-ncov) based on phylogeny and in the viral taxonomy among clusters to the prototype human and bat viruses [27, 28].

From that, it was possible to identify the differences between SARS-CoV-2 and two other zoonotic coronaviruses, the SARS-CoV and the MERS-CoV, classified in the Coronaviridae family. As it occurs with other RNA viruses, high genetic variability leads to gene recombination in coronaviruses, because they have similar and nonidentical genome sequences generating variants of the same virus. So, enveloped positive-sense RNA viruses with a 27 kb genome have the region of the most conserved replicative proteins encoded in the open reading frames 1a and 1b (ORF1a/1b) of the coronavirus genome [27–29].

In the envelope of SARS-CoV-2, there is a protein called spike (S), which has a receptor-binding domain (RBD) that binds to the cell surface and has the appearance of a crown when a viral particle is seen in great magnification in electron microscopy [27]. The three-dimensional structure of the spike protein of this new coronavirus has been recently obtained by computational modeling. A 3D map of the protein or blueprint was created using the cryoelectron microscopy technique. The discovery of this structure was only possible thanks to the recent complete sequencing of the genome of this coronavirus strain. The Pasteur Institute in France, responsible for monitoring the respiratory virus, sequenced the entire genome of the new coronavirus [28]. The coronavirus S glycoprotein is surface-exposed and mediates entry into host cells; for this reason, it is the main target of neutralizing antibodies (Abs) upon infection and the focus of therapeutic and vaccine design.

The pandemic COVID-19 has accelerated molecular studies using genomic tools with Next Generation Sequencing (NGS) for the characterization of viral samples and genetic engineering for the development of vaccines. In this regard, Israeli researchers are developing an oral vaccine based on one that targets infectious bronchitis caused by other coronavirus which affects birds. GSK announced a collaboration with the Chinese firm Clover Biopharmaceuticals to develop a COVID-19 subunit vaccine consisting of a trimerized SARS-CoV-2 S protein using their patented Trimer-Tag® technology. Other countries such as the USA, Brazil, England, and China are also developing possible candidates for vaccines against SARS-CoV-2 but with different strategies, using mRNA synthetic molecules and virus-like particles (VLP), for example.

With the global emergence of this new virus, several diagnostic tests such as real-time PCR and rapid kit have been developed for its early detection. Moreover, the implementation of stem cell therapy in the treatment of patients infected with COVID-19 capable of repairing the damage caused to the lungs, liver, and other organs has been successful, as well as the use of blood plasma from patients recovered from the infection. To date, Remdesivir, a drug that was also used in Ebola and Marburg virus treatment, has demonstrated antiviral activity against SARS-Cov-2. In addition, this drug has also been found to show antiviral activity against other coronaviruses including MERS and SARS. Other drugs are also being studied against the 2019-ncov, such as favilavir and chloroquine. More effective treatment has been achieved by association with antibiotic therapy [29]. Furthermore, new studies are constantly being carried out. An example of it is the action of the antiparasitic called Ivermectin, widely used in veterinary medicine, that was tested in cell culture, and showed activity on reducing the replication of the virus by decreasing the concentration of viral proteins present in the nucleus as well as other therapeutic targets investigated.

12.1.8 Zika Virus

The Zika virus (ZIKV) belongs to the family *Flaviviridae*, genus *Flavivirus,* which includes more than 70 viruses, as the Dengue virus (DENV), Yellow fever virus (YFV), West Nile virus (WNV), and Japanese encephalitis virus (JEV). In 1947, Zika fever was first diagnosed in a sentinel rhesus monkey in Africa, the Republic of Uganda, in the forest called Zika, which gave the name for the virus [4, 27, 30].

Regardless of the viral family, but because they have the RNA genome, both the coronavirus and the Zika has been detected by molecular biology using reverse transcription of polymerase chain reaction (RT-PCR) in several biological specimens, such as samples of amniotic fluid, blood, tissues (brain, liver, spleen, kidney, lung, and heart), saliva, urine, and semen [27]. It has already been confirmed on the relationship between Zika virus infection and the occurrence of microcephaly (brain malformation affecting babies at birth), including neurological disorders, the Guillain-Barré syndrome, causing muscle paralysis with special attention to respiratory muscles [27]. Another similarity is that both, to date, have vaccines in development and clinical trials in progress to find the least toxic antiviral drugs [30]. It has been tested that a DNA vaccine called pcTPANS1 against the dengue virus NS1 protein (N-linked glycoprotein conserved) induces the production of antibodies and activation of CD4+ lymphocytes acting in the infected cells tested in animal models as target studies for flavivirus infections [30].

Studies have performed computational modeling of the NS5 protein of the Zika virus complexed with the SAH cofactor. Viable alternative treatments have been investigated targeting proteins that are essential for viral replication. Nonstructural protein 5 (NS5) is an important pharmacological target, being the most conserved among flaviviruses because it contains a methyltransferase (MTase) domain in the N-terminal region and an RNA polymerase-dependent RNA domain (RdRp) in the C portion terminal. The inhibition of methyltransferase is lethal for virus replication, thus, it is considered an excellent target for drug design [31, 32]. Finally, with computational methods, such as Molecular Dynamics (MD) simulations, technique thoroughly exemplified in the last part of this chapter, there are studies that were able to map binding sites and search for inhibitors based on the interaction of NS5 in its complex with SAH [32].

12.2 Virtual Screening in Drug Design

Virtual Screening (VS) is an in silico technique that consists in applying algorithms to filter from a library of small compounds, the ones with the potential to bind a target molecule. The surging of this methodology represented a major increase in the capability of designing and developing new drugs. It has allowed researchers to find, among millions of molecules in a dataset, lead compounds for binding a biological target in a very short amount of time, while saving costs, in comparison to experimental approaches such as High Throughput Screening (HTS), in which hundreds of compounds are experimentally tested. This is possible through the constant development and optimization of faster computers and robust algorithms that continuously diminish the time such analysis takes, while increasing the accuracy of the interactions predicted. The VS method can be combined with other techniques described in this chapter, such as MD and Docking and, more recently, with machine learning algorithms. With them, it is possible to predict the binding and check the profile of the protein-ligand interaction through time and molecular flexibility, representing the selection of a more assertive set of lead compound candidates. Currently, VS is the first step in a drug development process, narrowing down the investigation on effective binding molecules before continuing exploration through in vitro assays (Fig. 12.1) [33, 34].

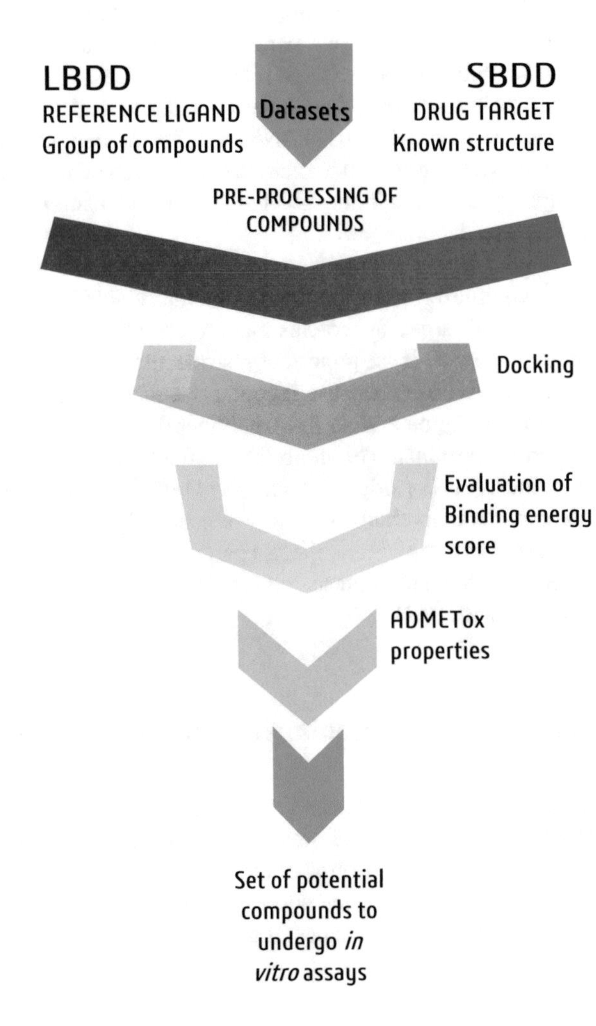

Fig. 12.1 Virtual screening workflow. From different perspectives, LBDD or SBDD, it is possible to obtain compounds for in vitro tests

12.2.1 Computer-Aided Drug Design (CADD)

Structure-based drug design (SBDD) and ligand-based drug design (LBDD) are two types of computer-aided drug design (CADD) approaches. SBDD methods analyze three-dimensional structural information from the macromolecular target, usually from proteins or RNA, to identify important locations and interactions that are fundamental to their respective biological functions. LBDD methods concentrate on known ligands in order to establish a relationship between their physical-chemical properties and biological activity, known as the structure-activity relationship (SAR). This information can be used to optimize drugs or guide the design of new ones [35].

The virtual screening technique can be approached through these two general knowledge-driven strategies. They both fight the problem from different, but complementary, perspectives. While the LBDD takes a group of compounds and searches for similarity to a known ligand of the biological target, the SBDD explores a specific site or the whole 3D structure of the target searching for compounds that are able to bind it [35].

12.2.1.1 Ligand-Based Drug Design

In the lack of reliable biological target structure, or in the presence of ligands known to have activity against the target, a better strategy is the ligand-based approach. It consists in taking a known ligand, such as a natural substrate of an enzyme, and screening in a given dataset for other small molecules that share similarity at structural or physicochemical level, which must present lower binding energy [35].

In this regard, LBDD is related to the ligand structure-activity relationship (SAR) studies, for example, the chemical group position, binding affinity, logP, number of rotatable chemical bonds, and number of donors and acceptors of hydrogen bonds. A strategy for calculation uses quantitative analysis of structure-activity relationship (QSAR), in which the molecule descriptors can be quantified. It is also possible for the association with the exploration of the tridimensional structure of the compounds (3D-QSAR). Results when applying such strategies are more energy accurate by taking the geometry of the compounds in the complex into account. Beyond tridimensional descriptors, many strategies involve ligand exploration in many more variables, from multiple conformations to solvation profile. An example is 4D-QSAR, that, by exploring molecular mechanics (MM) or even quantum mechanics (QM), adds the representation of the multiple conformations of the same ligand as another variable. The selection of descriptors in either SAR methodology must be carefully done based on each investigation, excluding highly correlated ones, for avoiding overlapping data, that can be identified by multiple linear regression (MLR) analysis, or even principal component analysis (PCA) [35–37].

Another assessment in LBDD is proposing a pharmacophore model to find other effective ligands. The modeling requires analyzing the functional groups, from hydrogen bond donors and acceptors to aromatic or hydrophobic groups in a known ligand or set of ligands superimposed, and proposing the minimum of electrostatic energy and spatial distribution of some groups as required for effective activity against the aimed target (Fig. 12.2). Closely related to the 3D-QSAR technique, investigating a pharmacophore model, is an extremely powerful way to screen for new active compounds as well as a way to optimize the known ones. It is important to address that from analysis of a certain complex, there is not a unique answer, which means many different pharmacophore models can be proposed, and will be valid as well. Superimposing of compounds structures to the pharmacophore model comprehends the next steps on this process, since it will be able to identify the ones that present the desirable structure. Software to be used in such strategy can be Discovery Studio Visualizer, LigandScout, and the visualizing platforms for analyzing the structures are in Table 12.1 [37].

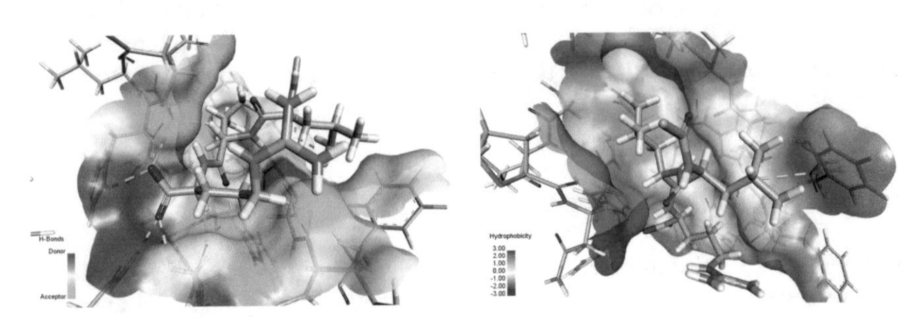

Fig. 12.2 Hydrogen bonds (on the left, dashed lines) and hydrophobic contacts between receptor and ligand (on the right, dashed line) in the enzyme catalytic site. Visualizing the interactions between the ligand and protein residues helps the researcher to propose pharmacophoric groups. Images obtained by Discovery Studio Visualizer

Table 12.1 Platforms for structural visualization

Softwares	Web link
Free	
Avogadro	https://avogadro.cc/
PyMOL	https://pymol.org/2/
VMD	https://www.ks.uiuc.edu/Research/vmd/
Mercury	https://www.ccdc.cam.ac.uk/Community/csd-community/freemercury/
Discovery Studio Visualizer	https://www.3dsbiovia.com/products/collaborative-science/biovia-discovery-studio/
Commercial	
Molecular Operating Environment (MOE)	https://www.chemcomp.com/Products.htm
OpenEye	https://www.eyesopen.com/
Spartan Pro	https://www.wavefun.com/spartan

12.2.1.2 Structure-Based Drug Design

In contrast to LBDD, structure-based drug design requires the knowledge of the macromolecular structure in order to predict the binding of a library of compounds. Once defined the biological target, information on its structure must be gathered at several different repositories (Table 12.2), the most important one being the Protein Data Bank (PDB), where macromolecular structures solved experimentally by X-ray crystallography diffraction, nuclear magnetic resonance (NMR), or cryoelectron microscopy (Cryo-EM) are found. In the absence of the structure solved by experimental techniques, the alternative is to get an already proposed model or to create one based on homology using software such as the robust Modeler (https://salilab.org/modeller/). Once the structure of the biological target is obtained, the definition of the binding sites to have the screening compounds docked there should be the next step. If well-established, the binding site may even be identified with structures in the presence of ligands on repositories (Fig. 12.3). But, when exploring a new biological target or when studying other possible allosteric sites in the structure, investigation can occur through software such as MDpocket [38], where it can discover transitional pockets and potential target sites, in combination. Once the binding site of the biological target is defined and clustered, the following step is the docking of the compounds in it [36, 39].

12.2.2 The Virtual Screening Process

After defining the starting strategy (either LBDD or SBDD) to undergo virtual screening, the next steps are as follows. The first and mandatory one is the preprocessing of the compounds database (Table 12.3). It comprises especially identifying and obtaining the tridimensional geometry of such structures, if it is not available in the database chosen. Several software (Schrodinger's LigPrep, OpenBabel) can be used to convert the 2D into the 3D structures. More confident structures may be obtained by Cambridge Structure Database (CSD) that stores the crystal structures of compounds. Before heading to the next steps, checking the protonation and tautomeric state is also mandatory to allow confident and accurate interactions in docking them into the target. Moreover, regarding the biological target preparation, beyond correcting the structure for a given pH, an interesting optional step can be to cluster the binding site among an ensemble of structures, either from different

Table 12.2 Free repository of protein structures

Repositories	Data	Web link
Protein Data Bank	Experimental	https://www.rcsb.org/
SWISS-MODEL	Model	https://swissmodel.expasy.org/repository
ModBase	Model	https://modbase.compbio.ucsf.edu/modbase-cgi/index.cgi

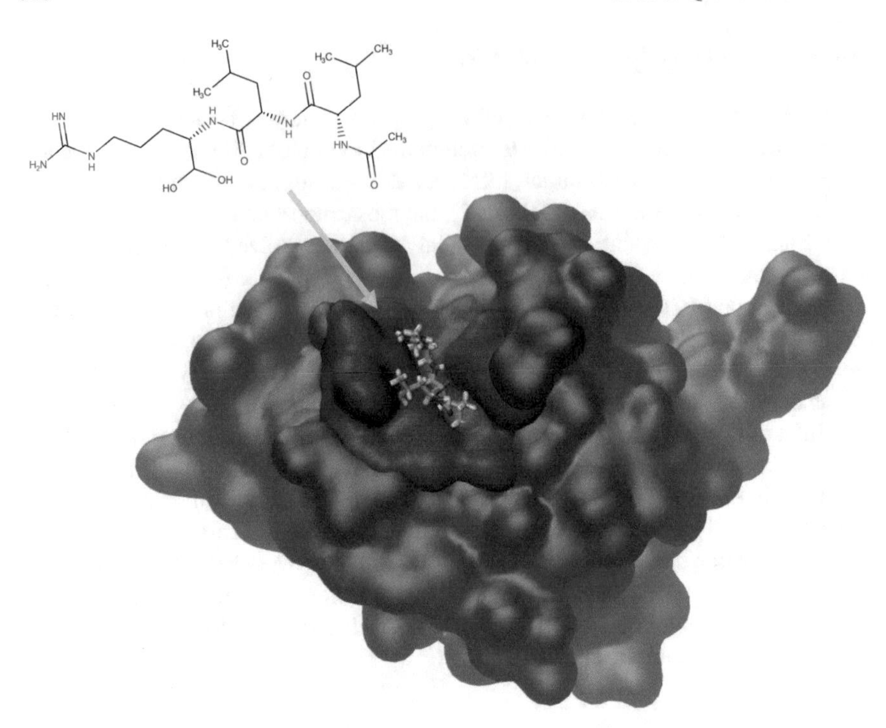

Fig. 12.3 Example of Virtual Screening preparation. In the absence of experimental structure, the protein vivapain-4 (surface representation in gray and its catalytic site in red) of *Plasmodium vivax* was modeled from homology toward the PDB structure ID:3BPM (falcipain-3 from *Plasmodium falciparum*) with the presence of Leupeptin (represented in 2D and 3D at the catalytic site of vivapain-4), a known inhibitor for cysteine-proteases

Table 12.3 Free database of compounds

Repositories	Data	Web link
ZINC	Commercially available compounds	https://zinc.docking.org/
Cambridge Structure Database (CSD)	Crystal structures of compounds (3D)	https://www.ccdc.cam.ac.uk/
ChEMBL Database	Bioactive molecules	https://www.ebi.ac.uk/chembl/
ChemSpider	Chemical structures	http://www.chemspider.com/
DrugBank	Drugs library (2D and 3D)	https://www.drugbank.ca/
PubChem	Small compounds library (2D)	https://pubchem.ncbi.nlm.nih.gov/
DEKOIS 2.0	Decoys	http://www.dekois.com/
DUD-E	Decoys	http://dude.docking.org/

models or structures found in repositories or from MD simulations, for finding a more accurate pose in docking. Therefore, clustering can be done using software of MD such as Amber's cpptraj or GROMACS's clustering function [33].

Compounds in investigation are usually docked in a grid assigned at one of the many docking software (Table 12.4). During the docking process, carefully explained in the next section of this chapter, the degrees of freedom of the molecules are explored, and software do that using different strategies, from systematic methods such as conformational search, to stochastic ones such as Monte Carlo. This step gives scores based on Van der Waals and electrostatic interactions at the binding site for the predicted poses, which will be used to rank and choose the best potential ligands further along. It is an extremely important step that helps us reduce the number of molecules analyzed, through scoring and observing the compounds that are capable of binding and the ones that do not interact in a stable manner at the biological binding site [40]. After evaluating the energy of the complex protein-ligand, the evaluation of pharmacokinetics properties is the last step before entering in vitro assays with the set of compounds left [37].

Scoring functions for ranking the poses from docking, however, are still subject to improvement. That is, the predicted binding affinity of the compounds does not necessarily represent that the active compounds are the ones located at the top of the rank exclusively. Errors in the energy of the systems are one of the issues compromising the analysis. To overcome that, the most effective techniques developed so

Table 12.4 Docking softwares to be used in a VS process

Software	Algorithm	Web
Free		
AutoDock4.2	Genetic Algorithm	http://autodock.scripps.edu/
AutoDock Vina	Broyden–Fletcher–Goldfarb–Shanno (BFGS) Method	http://vina.scripps.edu/
DOCK 6	Geometric Algorithm	http://dock.compbio.ucsf.edu/DOCK_6/index.htm
DockThor (Server)	Genetic Algorithm	https://dockthor.lncc.br/v2/
PLANTS	Stochastic search algorithm	https://uni-tuebingen.de/fakultaeten/mathematisch-naturwissenschaftliche-fakultaet/fachbereiche/pharmazie-und-biochemie/pharmazie/pharmazeutische-chemie/pd-dr-t-exner/research/plants/
Commercial		
Glide	Systematic search algorithm	https://www.schrodinger.com/glide
GOLD	Genetic Algorithm	https://www.ccdc.cam.ac.uk/solutions/csd-discovery/Components/Gold/

far are machine learning and deep learning. Machine learning consists of an algorithm that is able to recognize patterns (chemical descriptors) in a dataset. It can be a training dataset and applied to the list of compounds in the test, or it can be the list of compounds itself, in which the algorithm tries to identify a pattern without previous exposure. Examples of algorithms available for such assessment can be the K-Nearest Neighbor (KNN), Naive Bayes (NB), and Random Forest (RF). Deep learning, on the other hand, is considered a class of machine learning approaches which works with neural networks. It comprises layers of multiple nonlinear arrangements of processing units to solve hierarchically the issue. Even though it represents a powerful way to process many descriptors quickly by presenting intrinsic flexibility in its connections, it still requires reproducibility improvements [34, 41].

Any model should be first tested on a training set of complexes, to be certain of the prediction scores on your own system. The selection is tested using algorithms in a training set of compounds, with known activity against a specific target. If no *in house* library of reference compounds is available for training the VS model developed, useful dataset training for different types of the target can be obtained at Database of Useful Decoys—Enhanced (DUD-E) repository, for example, where already known active compounds and their targets are found. Training evaluation may be accessed by specificity (*SP*) for identifying true non-active compounds, selectivity (*SE*) for correctly identification of true active compounds and through a combined prediction (*Q*), in equations as follows:

$$\text{(a) } SP = \frac{TN}{TN + FP} \qquad \text{(b) } SE = \frac{TP}{TP + FN} \qquad \text{(c) } Q = \frac{TP + TN}{TP + TN + FP + FN}$$

Equations for validation scoring: (a) Specificity (*SP*), (b) selectivity (*SE*), and (c) total prediction of the system (*Q*). Compounds in the test will be scored as true positive (*TP*), true negative (*TN*), false positive (*FP*), or false negative (*FN*) [42].

The question "does my model truly separate active from inactive compounds?" can be solved by analysis of the receiver operating characteristics (ROC) curve. It helps visualize the profile of prediction throughout the compound list scores, from descriptors selected. Area under the curve (AUC) must be closer to one in order of the model in question to be considered effective, which means the model is effective in distinguishing an active compound as true positive and a compound not active as true negative [43].

12.2.3 Repurposing of Drugs

In a new disease outbreak, or in fighting the constant new drug resistant-diseases, it is extremely desirable to find compounds that have passed tests and are already in the market, that present activity against the new problem, in a strategy called drug repurposing. It is important because when redirecting an already commercial compound, several in vitro or in vivo tests can be skipped, since data on their toxicity, pharmacokinetic, and pharmacodynamics profiles are determined and they are, to a certain extent, safe for human administration [44].

This is of particular interest and use in the recent Ebola, Zika, and Covid-19 viral outbreaks, since developing a new drug takes a very long time and money to be available for the public. The repurposing process involves the assessment of databases of already commercial compounds, such as ZINC, followed by the standard steps in a VS process. Urgency is required in an outbreak event to reduce the spread of the disease and its consecutive deaths, and potent hardware and algorithms are, therefore, necessary for this combat. The most recent example of it is the pandemic Covid-19, which pushed scientists to their limits on finding the fastest alternatives to contain its spread, that is, by developing, in a very short amount of time, the screening of 1000 hits among 1.3 billion already regulated compounds against the main protease of such virus [44–47].

12.3 Molecular Docking in Drug Design

Molecular Docking, together with VS, is one of the initial stages of investigating molecules as potential drugs and aims to predict the conformation and orientation of ligands at the target macromolecule binding site [48]. It is based on the premise that the energy of the receptor-ligand complex is lower than that of the separate parts, being, therefore, favorable to the complex formation [35]. Complexes formed by protein-protein and protein-ligand are the most studied currently in drug discovery.

Molecular Docking can be divided into two main parts:

I. the investigation of the orientation and conformation of a ligand at the receptor binding site carried out by search methods;
II. and the prediction of the binding free energy, or affinity, of a receptor-ligand complex through a scoring function.

In order to perform Molecular Docking, the macromolecule receptor and the ligand molecule structures are needed. To apply the technique, some steps for preparing the system components are necessary (Fig. 12.4).

Initially, one should obtain the three-dimensional structure of the receptor, usually a protein, ideally from experimental data that can be acquired in libraries such as the Protein Data Bank (PDB). PDB has the coordinates of the atoms of individual molecules and resolved complexes by experimental methods such as X-ray diffraction crystallography and nuclear magnetic resonance (NMR). For proteins that have not yet been solved, some computational techniques can be used as alternatives. One of these is the Comparative Modeling, in which the protein to be discovered has its amino acid sequence aligned with the sequences of other proteins with previously known structures [49]. Ab initio modeling can also be performed using theoretical calculation for structural prediction [50].

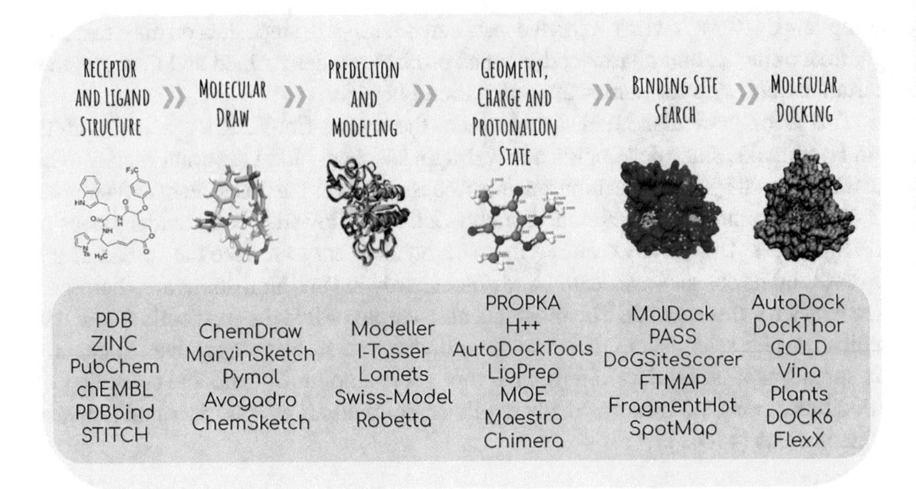

Fig. 12.4 Overview of Molecular Docking process and their respective software examples. Not all steps are always required [77–103]

The binding site at the receptor to be analyzed must be selected manually and delimited by specifying the coordinates or selected automatically using coordinates of ligands already bonded. In cases where the site is not known, prediction algorithms can be used to find possible cavities. Different algorithms are used for this purpose, from the ones that use small probes to map possible interaction sites on the protein surface [51] to algorithms in which druggability scores are predicted after training and testing a target library [52].

Computational mapping methods of co-solvent based on Molecular Dynamics (MD) are also being employed with the objective of searching at the entire surface of the protein, encompassing orthosteric and allosteric sites. In this mapping, a set of conformations of the target proteins, sampled by MD and fragments of small molecules, are used to identify and characterize binding sites [53, 54]. In addition, most software used for Docking have the option of "Blind Docking", where the search for potential interaction sites is done over the entire surface of the receptor structure, thus presenting a high computational cost compared to those previously described [55].

For the standard molecular docking calculation, to reduce computational costs, the preparation of the receptor involves the construction of a grid, which can be understood as a box containing a cubic network of points where the interaction energy is calculated, delimiting the binding site [55]. The receptor potential energy is pre-calculated for each point of the grid and, subsequently, the interaction electrostatic and Lennard-Jones energies of the ligand are calculated with each point in this grid [56]. The choice of box size influences the correct results and virtual screening time. For the Vina software, higher precision was observed when the dimensions of the search space are about 3 times greater than the ligand's rotation radius [57]. Some software perform the calculation of the grid previously, as occurs in AutoDock, and others perform in real time producing a faster calculation, such as Vina [55].

In addition, it is important to investigate the protonation states of the amino acids that perform interactions at the receptor binding site. Some residues, such as Aspartate, Glutamate, Histidine, and Cysteine, are not commonly found in their usual protonation states due to the electrostatic environment in which they are inserted. In this case, it is possible to evaluate these amino acids in other complexes of the target protein and proteins homologous to several solved ligands, as well as using programs that determine the pka of each residue, such as the PROPKA software and the H++ server.

The ligands can have their structures obtained from public databases such as PubChem and Zinc [58, 59], virtual compounds, or organic synthesis. To build the input files with the three-dimensional coordinates, specific programs for the molecular draw can be used, such as MarvinSketch, ChemDraw, and Avogadro [60]. Some points must be taken into account when preparing the ligand file. These include charges, stereochemical geometry, and just like the receptor, the protonation states of chemical groups [54]. To determine the protonation states of the ligand groups, not only the pH should be considered, but also the interaction with the macromolecule binding site, which makes it a slightly more complex task. However, in some cases Docking can be performed with several protonation states of the ligand [61]. The charges are usually distributed as partial charges to the constituent atoms of the molecule, from the calculated net charge. Most Docking programs assume that both charges and protonation states are fixed and do not vary between on and off states [56].

Some important conformational changes can occur in the molecular recognition of ligands that contain nonaromatic cyclic structures. Software such as LigPrep [62] can generate representative structures of these changes that can be used in Molecular Docking with the receptor. The type of Docking to be calculated must be defined, which directly influences the results obtained. The ligand and the receptor can be considered rigid or flexible in the calculations. In the case of the flexible ligand, some additional terms for energy must be included, which means that the potential is "smoothed out" to allow the flexibility of the residue and its interaction with the ligand [55]. The degree of complexity in the calculation of flexible receptors and ligands is higher in relation to rigid ones, since the number of degrees of freedom to be evaluated is higher. Some software that perform docking with flexible receptors and ligands are based on the induced-fit theory proposed by Koshland, where both must suffer small changes to the binding that may occur [63].

12.3.1 Theory of Molecular Docking

As described before, the Molecular Docking method is divided into two parts: Search Algorithm and scoring functions, which involves a conformational search and the selection of best pose by the scoring functions (Fig. 12.5).

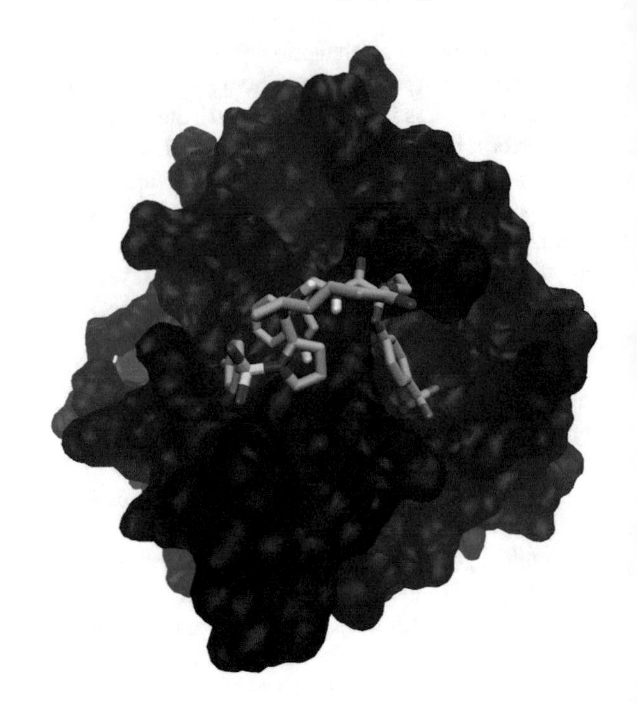

Fig. 12.5 Best pose found by Molecular Docking of Endonuclease of apurinic/apyrimidinic sites 1 (APE1) surface structure (black) and a macrocycle inhibitor (colored by atom type)

12.3.1.1 Search Algorithm

The exploration of different orientations and possible conformations for a ligand into a receptor binding site by the Docking software should find the best solution, which means to find the global minimum of energy. The function of energy needs to reproduce the entropic and enthalpic effects associated with the free energy of the system so that the global minimum of energy corresponds to the mode of binding described experimentally. However, due to the approximations introduced in the molecular interaction model, the global minimum does not always achieve this correspondence [61]. The selection of Search Algorithm is extremely important to reliable results of conformation and placement of the ligand at the receptor binding site. This search can be systematic or stochastic.

Systematic Search comprises a large sampling of conformations and a combination of structural parameters exploring each ligands' degree of freedom, including the rotational, translational, and conformational (rotatable dihedral angles) movements, which can generate a combinatorial explosion in the number of attempts. This search method uses more resources and spends significantly more time to generate the conformation poses and assesses them individually. By building the ligand from different fragments, selecting one as an anchor and sequentially adding combinations of remaining fragments, this problem is avoided. Some Docking algorithms also apply pharmacophoric data from proteins and ligands and try to match the distances between each point of them. Stochastic search is made randomly, developing different conformations based on bond rotations as degrees of freedom. Some implementations are necessary to guarantee the convergence to the best solution. Genetic Algorithm, Monte Carlo, Swarm Optimization, and Tabu Search are commonly implemented methods [55].

12.3.1.2 Scoring Functions

A large range of conformations is generated by the searching algorithms, although just a moiety of them are biologically relevant. To evaluate the quality of these conformations and to order according to the binding affinity to the receptor, it's necessary to apply scoring functions. Scoring functions are mathematical models, usually linear functions, composed of terms related to physicochemical properties as intermolecular interactions, desolvation, electrostatic, and entropic effects, involved at interactions of the small molecule and the binding site of the receptor.

There are different functions of scoring, that may vary mainly at the number and types of terms, mathematical complexity, and the form of parametrization. These functions can be applied according to the objective and the stage of the study of molecular docking. In the first stages of the docking, simpler functions can be used to evaluate the conformations generated by the search method, to reduce computational costs. At the final stages, more sophisticated and complex functions are employed to obtain greater accuracy in predicting the correct binding mode and affinity for the receptor [63]. Scoring functions can be classified as force field-based, empirical, and knowledge-based.

Force field-based functions are a sum of terms from a classic molecular force field, as MMFF94, CHARMM, and AMBER, including experimental data or quantum calculations as parameters. The terms are divided by bonding interactions, as energy associated with the torsional chemical bonds, and non-bonding interactions, as associated with Van der Waals interactions, electrostatics and hydrogen bonds. To include additional effects as solvation effects and hydrophobic contacts, some terms are incorporated.

Empirical scoring functions are based on the idea that the binding free energy can be related through the sum of uncorrelated variables. This function is developed from tridimensional structures and affinities of known receptor-ligand complexes, on which the terms are adjusted to reproduce experimental data more accurately possible. Therefore, these functions are limited, as they are derived from heterogeneous data in training sets. Compared to the force field-based functions, the empirical ones are computed much faster.

Knowledge-based functions are based on the statistical mechanics of simple fluid systems, which employ the potentials of mean force (PMF). They are built from the statistical analysis between the atom pairs of experimentally solved receptor-ligand complexes. Thereby, like the empirical functions, these functions are differentiated by the size of the training set and the type of receptor-ligand interactions considered in parameterization. A combined approach using more than one type of scoring function can be also applied to improve accuracy.

12.3.2 *Challenges of Molecular Docking*

One of the best ways to evaluate a protein-ligand complex is to use a dynamic system approach. However, the protein flexibility remains a challenge in Molecular Docking and most current flexible ligand Docking programs treat the receptor as rigid. Some methods apply different tools to solve that. The Induced-fit method considers the neighboring residues of the binding site as flexible. The backbone movement affects the side chains, which leads to a higher order of magnitude in terms of the number of degrees of freedom [63]. These flexible docking algorithms predict the binding mode of a molecule, and also its binding affinity relative to other compounds, more accurately than rigid body ones.

Protein Energy Landscape Exploration (PELE) is a Monte Carlo method to sample flexibility combining protein and ligand perturbations. This technique is based on a steered localized perturbation followed by side-chain sampling and minimization cycles [64]. It acts as a useful refinement, in particular, if the binding site of a protein is not known. Compared to Molecular Dynamics simulations, PELE has a smaller computational cost associated with the exploration. Machine Learning (ML) functions are also recently applied to simulate the flexibility of the receptor in the scoring functions for Molecular Docking.

An interesting strategy to improve the performance with respect to compound scoring and pose prediction is to combine two or more Docking methods, as a Consensus Method [65]. These methods showed better results than the single ones that compose it. They can vary in how conformations are obtained, the scoring functions selected, and the algorithm to achieve the consensus. Even though Molecular Docking has been developed for about 10 years, it remains being updated and new features are incorporated. Currently, the Virtual Screening of thousands of compounds is performed by using the background of the Molecular Docking technique.

12.4 Molecular Dynamics (MD)

Molecular Dynamics (MD) is an in silico technique based on Newton's laws that describes the variation of atomic positions in a molecule as a function of time. Therefore, it is different from the docking technique, which is mainly stochastic. With the use of MD, it is possible to evaluate characteristics such as content of secondary structure, orientation of side chains, conformation of loops, and the energy of interaction between different molecules, such as protein and ligands, over time. For this reason, it presents similar results to experimental methods such as Nuclear Magnetic Resonance (NMR), but with the advantage of more reduced costs, MD can be applied to bigger molecular systems, has better reproduction of biological systems, and costs less time [61].

MD is part of the so-called classical methods, or methods of Molecular Mechanics (MM), and it is based on the solution of Newton's second law, according to the equation:

$$F_{x_i} = \frac{d^2 x_i}{dt^2} m_i = \frac{\Delta v_i}{\Delta t} m_i = a_i m_i$$

where F_{xi} is the force applied to atom i at position x, t is time, v the velocity, and a_i the acceleration of atom i.

12.4.1 Force Fields

In Molecular Mechanics (MM), molecules are described as a set of connected atoms, instead of nuclei and electrons as in quantum methods. Since these atoms are linked to other atoms, they are subject to intermolecular and intramolecular forces. A complete set of interaction potentials between particles is referred to as a "force field". An empirical force field, as it is known, is a mathematical function that allows the calculation of total potential energy of a system, $V(r)$, from the coordinates of its particles. In molecular systems, $V(r)$ is described as the sum of several energy terms, including terms for bonded atoms (lengths and angles of bonding, dihedral angles) and terms for nonbonded atoms (Van der Waals and Coulomb interactions) [66]. In most force fields, the potential energy of the molecules can be represented by the equation:

$$V(r) = \Sigma El + \Sigma Ea + \Sigma Et + \Sigma Enl$$

where Σ is a sum of the energy terms as a function of the bonding length *El*, the bonding angles *Ea*, the torsion of the angles *Et*, and the interactions of nonbonded atoms *Enl* (Coulombian and Van der Waals).

The existing force fields have been developed independently and with all specific parameter sets. Some include other terms to specifically describe hydrogen bonds or to couple oscillations between bond angles and lengths, in order to achieve better agreement with vibrational spectra. The reliability of the results is based on the elaboration of a force field with well-defined parameters. The choice of the force field depends, to a large extent, on the system to be studied and on the properties that will be investigated. For example, while a type of force field can describe proteins with high fidelity, it can be quite limited in reproducing the geometry of carbohydrates or nucleic acids. In the case of biomolecular systems, the most used force fields are CHARMM, GROMOS, AMBER, OPLS, among others [61, 66].

There are different levels of simplification in the description of atoms in a force field depending on the type of system to be studied. Force field can describe all atoms in the system (all-atom), which requires a higher computational cost and may be unfeasible for large systems depending on the computational power available. The first simplification would be to join the hydrogen atoms attached to carbon atoms as a pseudo-atom, representing the properties of CH, CH2, or CH3 groups. Another possibility for simplification is through the method called coarse-grained (CG). In this force field, several atoms can be aggregated into a single particle, analogous to the pseudo-atom of the joined atom model. This model is useful for systems such as lipid membranes, allowing long simulation times, since they are much faster to perform. This model is not indicated in studies where changes in secondary structure content are expected and for studies involving systems of proteins and organic molecules, such as drugs [61].

The simulations can also consider the solvent explicitly, cases wherein the water molecules are physically included in the simulations, or implicitly, in which the water molecules are not included in the simulations, only the dielectric properties of the solvent are represented, considerably reducing the time of the simulation. Another difference between the force fields is the description of the water molecules, an example of a widely used model is the TIP3P, used in the force field of AMBER, OPLS, and GROMOS [61].

The programs currently available for performing MD differ in relation to the force field used, computational cost and access; some of the most used are AMBER, GROMACS, and NAMD. These differences need to be evaluated by the researcher according to the system to be studied.

12.4.2 MD Simulations

MD simulation involves several steps, from obtaining the protein structure either by accessing the repositories of experimental or modeled structures (Table 12.2) to parameterization, minimization, equilibration, simulation, and analysis. The diagram below demonstrates the basic steps for a protein MD simulation (Fig. 12.6) [67].

12.4.2.1 Energy Minimization and Equilibration

Before performing the molecular dynamics simulation, it is necessary to perform an energy minimization of the system, which aims to eliminate bad contacts between atoms. During this calculation, the geometry of the system is optimized causing a reduction of its overall energy, reaching a more stable conformation, which is a state of energy minimum. This conformation will be the starting point for molecular dynamics. A widely used method is the steepest descent, which uses the first derivative to determine the direction to the minimum. It is a robust technique used to initially optimize a structure that is far from a minimum point [66]. A new structure optimization is then performed with a second derivative minimization of energy method to improve the result.

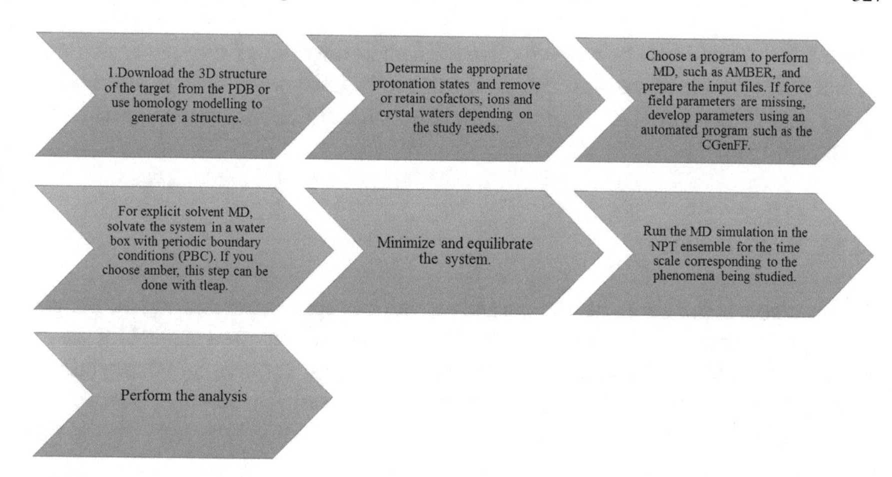

Fig. 12.6 Steps to perform an MD simulation

After minimizing the energy, it is necessary to heat the system at constant volume, bringing it to a temperature of interest, which is usually 300 K. After heating, equilibration is performed, with the function of controlling the system pressure (protein + water + ions), and stabilizing the system properties. In this stage, it is common to restrict the movement of the atoms of the protein and crystallographic water molecules from PDB, which allows the free water molecules and ions to organize and tend to come into balance, with the position restriction gradually decreasing until all the components of the system have obtained total freedom of movement. The equilibration period is variable and depends on the system under study. Generally, it is considered finished when the thermodynamic balance is reached. From that point, you can then generate the MD trajectories and calculate the different properties for the system of interest.

12.4.2.2 Periodic Boundary Conditions

The use of the explicit solvent in Molecular Dynamics is what demands the most computational cost, therefore, it is necessary to optimize the number of water molecules in order to obtain a faster simulation. For this reason, boxes are used, which are the three-dimensional space where the biomolecule and solvent are placed; the size and shape of these boxes are important in defining how many water molecules will be used in the simulation. The most common forms are cubic, octahedral, and dodecahedral. The form to be used will depend on the system and the computer power available for such simulation, but in the case of proteins, the cubic form is widely used [61].

Unlike biological conditions, in MD we have a vacuum around the simulation box, which could cause the solvent to evaporate. To prevent the system's molecules from exceeding the box's limits, a strategy called periodic boundary conditions was created. In this strategy, the box is replicated in all its directions from space, making it so that if a molecule leaves the central box, one of its images enters from the opposite side, representing a continuity of the simulation, in a closer way to the experimental conditions [61].

12.4.2.3 Ensemble

The ensemble is the group of settings and properties kept constant during the simulation, representing the state of the system. Examples of widely used ensembles are the canonical or NVT (with a constant number of particles, volume, and temperature), generally used for the equilibration stage, and the isothermal-isobaric or NPT (with a constant number of particles, pressure, and temperature), normally used during MD simulation.

12.4.2.4 Sampling

The ability of a simulation to describe the expected behavior of a system is related to sampling. It is necessary that the simulation has enough time to observe the phenomena of interest, which is a great challenge in the simulation of biomolecules, since the computational cost to obtain a sample close to the ideal is very high. The longer the MD simulation, the greater the sampling, however, the number of atoms in the system has a great influence on how many conformations can be adopted. An all-atom simulation would need a larger time scale to obtain the same sampling as a coarse-grained simulation, for example. The type of problem to be analyzed also has a great influence on sampling and, consequently, on the necessary simulation time. The analysis of domain movement, for example, requires a longer simulation time, on the microseconds scale, whereas in the analysis of protein-ligand interaction, usually, the nanoseconds scale is sufficient.

12.4.3 Analysis of MD Simulations

Analyzing MD simulations is considered by many to be the most difficult part of MD. The first and most basic analysis to be performed is the stability of properties such as temperature, volume, pressure, density, and total energy throughout the simulation. If these were stable, you can move on to the rest of the analyses.

Among the most common analyses to be carried out, mainly for proteins, there is the root mean square deviation (RMSD), which allows evaluating how much the protein structure varied along the MD simulation in comparison with an initial structure, which is usually the crystallographic structure or the minimized structure before MD. In this analysis, the RMSD starts from 0 and rises until it reaches a point of equilibrium, where it remains until the end of the simulation (Fig. 12.7). Root

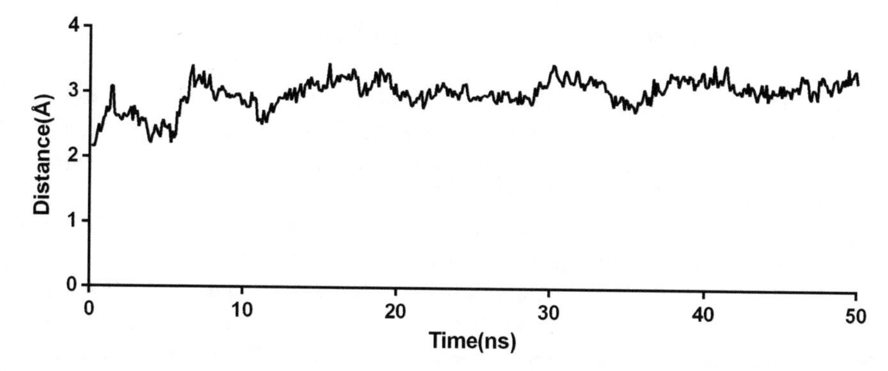

Fig. 12.7 RMSD of the p38 MAPK protein in complex with an inhibitor candidate in a 50 ns MD simulation. It is possible to observe greater stability of the protein at the end of the simulation

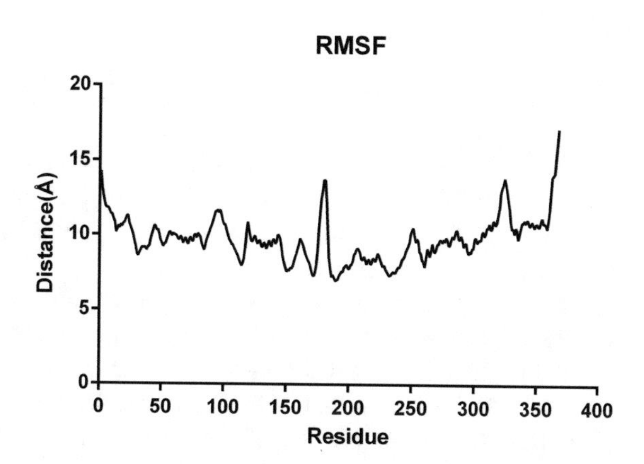

Fig. 12.8 RMSF of the p38 MAPK protein in complex with an inhibitor candidate in a 50 ns MD simulation. The activation loop consists of residues 172–183, where it is possible to observe a greater variation

mean square fluctuation (RMSF) evaluates the variation of each protein residue in relation to an average structure along the molecular dynamics. More flexible regions, such as loops, will show greater variation (Fig. 12.8).

12.4.4 MD Applications

Molecular Dynamics has numerous applications, such as the prediction of protein structures, analysis of protein-protein, protein-lipid, and protein-ligand interactions, virus studies, and drug design for the most diverse diseases.

12.4.4.1 Drug Design and Free Energy Calculation

Among a variety of approaches used for structure-based drug design, MD simulation combined with a free energy calculation method is a great tool, as it can provide detailed information about protein-ligand interactions and consider the environment with solvent and protein flexibility [68].

The calculation of binding free energy can be performed using the Molecular Mechanics energies combined method with the Poisson–Boltzmann or Generalized Born and surface area continuum solvation (MM/PBSA and MM/GBSA). The purpose of this method is to calculate the difference in free energy between the bound and unbound states of two solvated molecules. However, in this calculation, most of the energy contributions would come from solvent-solvent interactions, making the calculation take a long time. Thus, in the AMBER program, an effective method is used, which is the division of the calculation according to the following equation [69]:

$$\Delta G^0_{\text{bind,solv}} = \Delta G^0_{\text{bind,vacum}} + \Delta G^0_{\text{solv,complex}} - (\Delta G^0_{\text{solv,ligand}} + \Delta G^0_{\text{solv,receptor}})$$

where $\Delta G^0_{\text{bind, solv}}$ is the free energy of binding in aqueous medium, $\Delta G^0_{\text{bind, vacum}}$ is the free energy of binding in vacuum, $\Delta G^0_{\text{solv, complex}}$ is the free energy of the solvated complex, $\Delta G^0_{\text{solv, ligand}}$ is the free energy of the solvated ligand, and $\Delta G^0_{\text{solv, receptor}}$ is the free energy of the solvated protein [68].

The analysis of free energy is important mainly to compare the interaction of different ligands with the same protein performing an MD simulation, as more negative values indicate that that compound could be more promising, as it performed more relevant intermolecular interactions with the target. Programs such as AMBER also allow the decomposition of the binding free energy by aminoacid residue, which is very important to evaluate which residues have the best contribution to protein-ligand interaction.

The analysis of the prevalence of hydrogen bonds over time is another important application of MD for drug design, as it allows to evaluate which residues made stable hydrogen bonds with the ligand and to compare with the existing data in the literature for the protein under study, allowing greater reliability of the results obtained (Table 12.5).

Table 12.5 Prevalence of hydrogen bonds between p38 MAPK and an inhibitor candidate in a 50 ns MD simulation

Ligand	Protein	Prevalence (%)
N1H1	GLU71_OE2	53.8
O1	ASP168_NH	65.4
N2H2	GLU71_OE2	22.5
O3	MET109_H	43.7
N1H1	GLU71_OE1	37.0

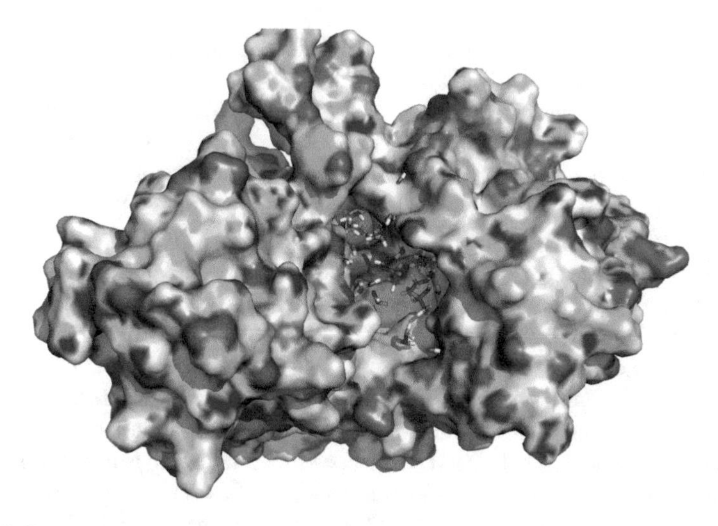

Fig. 12.9 Representative structure of p38 MAPK with water (red) and ethanol (purple) molecules at the ligand binding site extracted from a 100 ns MD simulation

Another interesting approach to drug design in Molecular Dynamics is the solvent mapping technique, which allows the simulation of proteins in different solvents, such as ethanol/water, to identify hotspots that could be important in the interaction with possible inhibitors, through observations of intermolecular interactions between residues of the studied protein and solvent molecules. This technique is an in silico version of the multiple solvent crystal structures (MSCS) approach, but with the advantage of being much faster and cheaper. Water represents polar interactions well, while the ethanol molecule is highly miscible in water and represents polar interactions, through hydroxyl, and nonpolar interactions, through the methyl group [70]. Structural clustering algorithms can be used to extract representative conformations from MD trajectories to understand different interaction patterns between the ligand, in this case water and ethanol molecules, and the protein that contributes to binding (Fig. 12.9).

12.4.4.2 Study of Viruses

Another interesting application of Molecular Dynamics is the study of viruses. In 2006, Schulten and co-workers reported the first all-atom simulation of a complete non-enveloped virus, called the satellite tobacco mosaic virus (STMV). Regarding enveloped viruses, the first capsid simulation of an immature HIV virus was carried out in 2010 by Ayton and Voth. However, as the viral particles are large for all-atom simulations, most studies are directed to simulations of viral proteins and capsids. The limitation for this area is the study of more complex processes such as replication, membrane fusion, virus maturation, and nuclear entry. However, the study of viral proteins provides a lot of important information, also allowing the drug design for viral diseases through MD [71, 72].

Among the most recent examples of MD applications in viruses is the study of the Ebola virus by Pappalardo and co-workers, wherein they determined that VP24 protein of the Ebola virus is a key protein for pathogenicity, through the observation of its interactions with the human protein KPNA5 [73]. Nasution and co-workers also conducted a drug design study to combat the Ebola virus through Molecular Dynamics, finding promising compounds [74]. Another example is the study of influenza B virus, by Zhang and Zheng, wherein they provided structural and dynamic details of the effects of serine triad on proton conduction in the tetrameric channel of influenza B channel M2 (BM2), important in the virus life cycle, at the atomic level [75]. Studies of ZIKA virus are also being carried out, such as by Bowen and co-workers, wherein a collection of more than 7 million commercially and freely available compounds from the ZINC15 database were subjected to a virtual screening procedure consisting of consensus-based docking followed by MD simulation and binding energy calculations in order to identify promising potential inhibitors of the Zika NS2B-NS3 [76].

References

1. Hagen JB (2000) The origins of bioinformatics. Nat Rev Genet 1(3):231–236
2. Mesquita APR (2014) Modelagem molecular de compostos anti-citomegalovírus. Trabalhos de Conclusão de Curso (Universidade Federal Fluminense) 1–52
3. Vale G, Silva T, Ferreira A, Bou-Habib D, Siqueira M, Lopes TM, Miranda M (2020) Inibição da replicação do influenza através da modulação de fatores restritivos pelos ligantes dos receptores CCR5 e CXCR4. Resumos Caderno Simpósio de Virologia (Universidade Federal do Rio de Janeiro) 53
4. Simões RSQ, Barth OM (2015) Papillomavirus: viral vectors in the gene therapy and new therapeutic targets. Int J Biomed Res 6(10):763–768
5. Jain NK, Sahni N, Kumru OS, Joshi SB, Volkin DB, Middaugh CR (2015) Formulation and stabilization of recombinant protein based virus-like particles vaccine. Adv Drug Deliv Rev 93(1):42–45
6. Huber B, Schellenbacher C, Shafti-Keramat S, Jindra C, Christensen N, Kirnbauer R (2017) Chimeric L2-based virus-like particle (VLP) vaccines targeting cutaneous human papillomaviruses (HPV). Plos One 1–27
7. Simões RSQ, Barth OM (2017) Immunological and structural analysis of HPV-positive cervical carcinoma cell lines and bovine papillomavirus virus-like particles (BPV-VLP). Int J Adv Res 5:1003–1009
8. Lavine BK, Mirjankar N (2012) Clustering and classification of analytical data. Encycl Anal Chem
9. Lenz P, Day PM, Pany YYS, Frye SA, Jensen PN, Lowy DR, Schiller JT (2020) Papillomavirus-like particles induce acute activation of dendritic cells. J Immunol 166:5346–5355
10. Nagib NRC (2017) Modelagem e dinâmica molecular da oncoproteína E6 do vírus do papiloma humano (HPV) tipo 18. Trabalho de Conclusão de Curso (Universidade Federal de Uberlândia) 1–37

11. Simões RSQ, Barth OM (2018) Papillomavirus (PV)—associated skin diseases in domestic and wild animals: animal nucleotide sequence identity of PV types to their closest related PV and HPV sequences deposited in the gen bank. Int J Curr Microbiol Appl Sci 6:938–951
12. Dehesa-Violante M, Nunez-Nateras R (2007) Epidemiology of hepatitis virus B and C. Arch Med Res 38(6):606–611
13. Heermann KH, Goldmann U, Schwartz W, Seyffarth T, Baumgarten H, Gerlich WH (1984) Large surface proteins of hepatitis B virus containing the pre-S sequence. J Virol 52:396–402
14. Arauz-Ruiz P, Norder H, Robertson BH, Magnius LO (2002) Genotype H: a new Ameridian genotype of hepatitis B virus revealed in Central America. J Gen Virol 83:2059–2073
15. Vieira MB (2010) Estudos de antigenicidade e imunogenicidade de vetores HBsAg carreadores de epítopos do HCV. Dissertação de Mestrado (Fundação Oswaldo Cruz) 1–145
16. Seeger C, Mason W (2000) Hepatitis B virus biology. Microbiol Mol Biol Rev 64:51–68
17. Delpeyroux F, Chenciner N, Lim A, Malpiece Y, Blondel B, Crainic R et al (1986) A poliovírus neutralizing epitope expressed on hybrid hepatitis B surface antigen particles. Science 233:472–475
18. Netter HJ, Macnaughton TB, Woo W, Tindle R, Gowans E (2001) Antigenicity and immunogenicity of novel chimeric hepatitis B surface antigen particles with exposed hepatitis C virus epitopes. J Virol 75:2130–2141
19. Patient R, Hourioux C, Vaudin P, Pages JC, Roingeard P (2009) Chimeric hepatitis B and C viruses envelope proteins can form subviral particles: implications for the design of new vaccine strategies. New Biotechnol 25(4):226–234
20. Forns X, Bukh J, Purcell RH (2002) The challenge of developing a vaccine against hepatitis C virus. J Hepatol 37:684–695
21. Martell M, Esteban JI, Quer J, Genesca J, Weiner A, Esteban R (1992) Hepatitis C virus (HCV) circulates as a population of different but closely related genomes: quasispecies nature of HCV genome distribution. J Virol 66:3225–3229
22. Bukh J, Emerson SU, Purcell RH (1997) Genetic heterogeneity of hepatitis C virus and related viruses. In: Rizzeto M, Purcell RH, Gerin JL, Verme G (eds) Viral hepatitis and liver disease. Minerva Médica, Turin, pp 167–175
23. Major MM, Vivitski L, Mink MA, Schleef M, Whalen RG, Trepo C (1995) DNA-based immunization with chimeric vectors for the induction of immune responses against the hepatitis C virus nucleocapsid. J Virol 69:5798–5805
24. Geissler M, Gesein A, Tokushige K, Wands JR (1997) Enhancement of cellular and humoral immune responses to hepatitis C virus core protein using DNA-based vaccines augmented with cytokine-expressing plasmids. J Immunol 158:1231–1237
25. Drazan KE (2000) Molecular biology of hepatitis C infection. Liver Transpl 6:396–406
26. Geissler M, Tokushige K, Wakita T, Zurawski VR, Wands JR (1998) Differential cellular and humoral responses to HCV core and HBV envelope proteins after genetic immunizations using chimeric constructs. Vaccine. 16:857–867
27. Simões RSQ, Barth OM (2019) Emerging and reemerging virus. In: Human and veterinary virology, vol 1, pp 317–24
28. Yan R, Zhang Y, Li Y, Xia L, Guo Y, Zhou Q (2020) Structural basis for the recognition of the SARS-CoV-2 by full-length human ACE-2. Science
29. Liu J, Cao R, Xu M, Wang X, Zhang H, Hu H, Li Y, Hu Z, Zhong W, Wang M (2020) Hidroxychoroquine, a less toxic derivate of chloroquine is effective in inhibiting SARS-CoV-2 infection in vitro. Cell Discov 6:16
30. Simões RSQ, Barth OM (2016) Historical and epidemiological aspects of some human diseases just to Zika virus, a short review. Int J Res Stud Biosci 4:46–54
31. Magnani DM et al (2017) Neutralizing human monoclonal antibodies prevent Zikavirus infection in macaques. Sci Transl Med 9:8184
32. Oliveira LMA, Pascutti PG, Souza RC, Gomes PSFCG, Gomes DEB (2020) Modelagem computacional da proteína NS5 do zika vírus ao cofator SAH. Resumos Caderno Simpósio de Virologia (Universidade Federal do Rio de Janeiro) 58

33. Sliwoski G, Kothiwale S, Meiler J, Lowe EW Jr (2013) Computational methods in drug discovery. Pharmacol Rev 66:334–395
34. D'Souza S, Prema KV, Balaji S (2020) Machine learning in drug–target interaction prediction: current state and future directions. Drug Discov Today
35. Yu W, Mackerel AD Jr (2017) Computer-aided drug design methods. Methods Mol Biol Antibiot 1520:85–106
36. Macalino SJY, Gosu V, Hong S, Choi S (2015) Role of computer-aided drug design in modern drug discovery. Arch Pharmacal Res 38:1686–1701
37. Shim J, Mackerell AD Jr (2011) Computational ligand-based rational design: role of conformational sampling and force fields in model development. MedChemComm 2:356–70
38. Schmidtke P, Bidon-Chanal A, Luque FJ, Barril X (2011) MDpocket: open-source cavity detection and characterization on molecular dynamics trajectories. Bioinformatics 27:3276–3285
39. Anderson AC (2003) The Process of Structure-Based Design. Cell Chem Biol 10:787–797
40. Kitchen DB, Decornez H, Furr JR, Bajorath J (2004) Docking and scoring in virtual screening for drug discovery: methods and applications. Nat Rev Drug Discovery 3:935–949
41. Wang D, Cui C, Ding X, Xiong Z, Zheng M, Luo X, Jiang H, Chen K (2019) Improving the virtual screening ability of target-specific scoring functions using deep learning methods. Front Pharmacol 10:924
42. Ren J-X, Zhang R-T, Zhang H (2020) Identifying novel ATX inhibitors via combinatory virtual screening using crystallography-derived pharmacophore modelling, docking study, and QSAR analysis. Molecules 25:1107
43. Swift RV, Jusoh SA, Offutt TL, Li ES, Amaro RE (2016) Knowledge-Based methods to train and optimize virtual screening ensembles. J Chem Inf Model 56:830–842
44. Zheng W, Sun W, Simeonov A (2018) Drug repurposing screens and synergistic drug-combinations for infectious diseases. Br J Pharmacol 175:181–191
45. Schuler J, Hudson ML, Schwartz D, Samudrala R (2017) A systematic review of computational drug discovery, development, and repurposing for ebola virus disease treatment. Molecules 22:1777
46. Santos F, de Nunes DAF, Lima WG, Davyt D, Santos LL, Taranto AG, Maria Siqueira Ferreira J (2019) Identification of Zika virus NS2B-NS3 protease inhibitors by structure-based virtual screening and drug repurposing approaches. J Chem Inf Model
47. Ton AT, Gentile F, Hsing M, Ban F, Cherkasov A (2020) Rapid identification of potential inhibitors of SARS-CoV-2 main protease by deep docking of 1.3 billion compounds. Mol Inf
48. Morris GM, Lim-Wilby M (2008) Molecular docking
49. Webb B, Sali A (2016) Comparative protein structure modeling using MODELLER. Curr Protoc Protein Sci 86:2.9.1–2.9.37
50. Barreiro EJ, Rodrigues CR (1997) Modelagem molecular: Uma Ferramenta Para O Planejamento Racional De Fármacos Em Química Medicinal. Química Nova 20(1)
51. Kozakov D, Grove LE, Hall DR, Bohnuud T, Mottarella SE, Luo L, Xia B, Beglov D, Vajda S (2015) The FTMap family of web servers for determining and characterizing ligand-binding hot spots of proteins. Nat Protoc 10(5):733–755
52. Volkamer A, Kuhn D, Grombacher T, Rippmann F, Rarey M (2012) Combining global and local measures for structure-based druggability predictions. J Chem Inf Model 52(2):360–372
53. Sayyed-Ahmad A (2018) Hotspot identification on protein surfaces using probe-based MD simulations: successes and challenges. Curr Top Med Chem 18(27):2278–2283
54. Arcon JP, Defelipe LA, Modenutti CP, Lopez ED, Alvarez-Garcia D, Barril X, Turjanski AG, Martí MA (2017) Analyzing the molecular basis of enzyme stability in ethanol/water mixtures using molecular dynamics simulations. J Chem Inf Model 57:846–863
55. Prieto-Martínez FD, Arciniega A, Medina-Franco JL (2018) Molecular docking: current advances and challenges. TIP Rev Esp Cienc Quím Biol 21:1–23
56. Torres PHM, Sodero ACR, Jofily P, Silva FP Jr (2019) Key topics in molecular docking for drug design. Int J Mol Sci 20:4572019

57. Feinstein WP, Brylinski M (2015) Calculating an optimal box size for ligand docking and virtual screening against experimental and predicted binding pockets. J Cheminform 7:18
58. Kim S, Chen J, Cheng T, Gindulyte A, He J, He S, Li Q, Shoemaker BA, Thiessen PA, Yu B, Zaslavsky L, Zhang J, Bolton EE (2019) PubChem 2019 update: improved access to chemical data. Nucleic Acids Res 47
59. Sterling T, Irwin JJ (2015) ZINC 15—ligand discovery for everyone. J Chem Inf Model 55(11):2324–2337
60. Hanwell MD, Curtis DE, Lonie DC, Vandermeersch T, Zurek E, Hutchison GR (2012) Avogadro: an advanced semantic chemical editor, visualization, and analysis platform. J Cheminform 4:17
61. Verli H (2014) Bioinformática: da Biologia à Flexibilidade Molecular. Sociedade Brasileira de Bioquímica e Biologia Molecular
62. LigPrep, Schrödinger, LLC, New York, NY, 2020. https://www.schrodinger.com/ligprep. Accessed 29 March 2020
63. Pagadala NS, Syed K, Tuszynski J (2017) Software for molecular docking: a review. Biophys Rev. 9:91–102
64. Borrelli KW, Vitalis A, Alcantara R, Guallar V (2005) PELE: Protein Energy Landscape Exploration. a novel monte carlo based technique. J Chem Theory Comput 1(6):1304–1311
65. Ren X, Shi YS, Zhang Y, Liu B, Zhang LH, Peng YB, Zeng R (2018) Novel consensus docking strategy to improve ligand pose prediction. J Chem Inf Model 58(8):1662–1668
66. Namba AM, Da Silva VB, Da Silva CHTP (2008) Dinâmica molecular: Teoria e aplicações em planejamento de fármacos. Eclet Quim 33(4):13–24
67. Yu W, Jr ADM (2016) Computer-aided drug design methods. Methods Mol Bio Antibiot 85–106
68. Yang Y, Shen Y, Liu H, Yao X (2011) Molecular dynamics simulation and free energy calculation studies of the binding mechanism of allosteric inhibitors with p38 α MAP Kinase. J Chem Inf Model 3235–3246
69. Walker R (2020) Amber advanced tutorials–tutorial 3–MM-PBSA–introduction. http://ambermd.org/tutorials/advanced/tutorial3/. Accessed 27 March 2020
70. Arcon JP, Defelipe LA, Modenutti CP, López ED, Alvarez-Garcia D, Barril X, Martí MA (2017) Molecular dynamics in mixed solvents reveals protein-ligand interactions, improves docking, and allows accurate binding free energy predictions. J Chem Inf Model 57(4):846–863
71. Huber RG, Marzinek JK, Holdbrook DA, Bond PJ (2017) Multiscale molecular dynamics simulation approaches to the structure and dynamics of viruses. Prog Biophys Mol Biol 128:121–132
72. Perilla JR, Goh BC, Cassidy CK, Liu B, Bernardi RC, Rudack T, Schulten K (2015) Molecular dynamics simulations of large macromolecular complexes. Curr Opin Struct Biol 31:64–74
73. Pappalardo M, Collu F, Macpherson J, Michaelis M, Fraternali F, Wass MN (2017) Investigating Ebola virus pathogenicity using molecular dynamics. BMC Genomics 18(Suppl 5)
74. Nasution MAF, Toepak EP, Alkaff AH, Tambunan USF (2018) Flexible docking-based molecular dynamics simulation of natural product compounds and Ebola virus Nucleocapsid (EBOV NP): a computational approach to discover new drug for combating Ebola. BMC Bioinform 19(Suppl 14)
75. Zhang Y, Zheng QC (2019) What are the effects of the serine triad on proton conduction of an influenza B M2 channel? an investigation by molecular dynamics simulations. Phys Chem Chem Phys 21(17):8820–8826
76. Bowen LR, Li DJ, Nola DT, Anderson MO, Heying M, Groves AT, Eagon S (2019) Identification of potential Zika virus NS2B-NS3 protease inhibitors via docking, molecular dynamics and consensus scoring-based virtual screening. J Mol Model 25(7)
77. Brady GP Jr, Stouten PF (2000) Fast prediction and visualization of protein binding pockets with PASS. J Comput Aided Mol Des 14:383–401

78. Berman HM, Westbrook J, Feng Z, Gilliland G, Bhat TN, Weissig H, Shindyalov IN, Bourne PE (2000) The Protein Data Bank. Nucleic Acids Res 28:235–242
79. Wang R, Fang X, Lu Y, Yang CY, Wang S (2005) The PDBbind database: methodologies and updates. J Med Chem 48(12):4111–4119
80. Szklarczyk D, Santos A, von Mering C, Jensen LJ, Bork P, Kuhn M (2016) STITCH 5: augmenting protein-chemical interaction networks with tissue and affinity data. Nucleic Acids Res 44(D1):D380–D384
81. ChemDraw (2020) PerkinElmer Informatics. https://www.perkinelmer.com/category/chemdraw
82. MarvinSketch (2020) (version 20.9, calculation module developed by ChemAxon). http://www.chemaxon.com/products/marvin/marvinsketch
83. ACD/ChemSketch (2019) Advanced Chemistry Development, Inc., Toronto, On, Canada. www.acdlabs.com
84. Pymol (2020) Schrödinger, New York, NY. Version 2.3. https://pymol.org/2/
85. Maestro (2020) Schrödinger, LLC, New York, NY. https://www.schrodinger.com/maestro
86. Sali A, Blundell TL (1993) Comparative protein modelling by satisfaction of spatial restraints. J Mol Biol 234:779–815
87. Yang J, Yan R, Roy A, Xu D, Poisson J, Zhang Y (2015) The I-TASSER suite: protein structure and function prediction. Nat Methods 12(1):7–8
88. Zheng W, Zhang C, Wuyun Q, Pearce R, Li Y, Zhang Y (2019) LOMETS2: improved meta-threading server for fold-recognition and structure-based function annotation for distant-homology proteins. Nucleic Acids Res 47:W429–W436
89. Waterhouse A, Bertoni M, Bienert S, Studer G, Tauriello G, Gumienny R, Heer FT, de Beer TAP, Rempfer C, Bordoli L, Lepore R, Schwede T (2018) SWISS-MODEL: homology modelling of protein structures and complexes. Nucleic Acids Res 46:W296–W303
90. Kim DE, Chivian D, Baker D (2004) Protein structure prediction and analysis using the Robetta server. Nucleic Acids Res 32:W526–W531
91. Olsson MHM, Sondergaard CR, Rostkowski M, Jensen JH (2011) PROPKA3: consistent treatment of internal and surface residues in empirical pKa predictions. J Chem Theory Comput 525–537
92. Ananda Krishnan R, Aguilar B, Onufriev AV (2012) H++ 3.0: automating pK prediction and the preparation of biomolecular structures for atomistic molecular modeling and simulation. Nucleic Acids Res 40(W1):W537–W541
93. Morris GM, Huey R, Lindstrom W, Sanner MF, Belew RK, Goodsell DS, Olson AJ (2009) Autodock4 and AutoDockTools4: automated docking with selective receptor flexibility. J Comput Chem 16:2785–2791
94. Pettersen EF, Goddard TD, Huang CC, Couch GS, Greenblatt DM, Meng EC, Ferrin TE (2004) UCSF Chimera–a visualization system for exploratory research and analysis. J Comput Chem 25(13):1605–1612
95. Vilar S, Cozza G, Moro S (2008) Medicinal chemistry and the molecular operating environment (MOE): application of QSAR and molecular docking to drug discovery. Curr Top Med Chem 8(18):1555–1572
96. Radoux CJ, Olsson TSG, Pitt WR, Groom CR, Blundell TL (2016) Identifying interactions that determine fragment binding at protein hotspots. J Med Chem 59:4314–4325
97. dos Santos KB, Guedes IA, Karl ALM, Dardenne L (2020) Highly flexible ligand docking: benchmarking of the DockThor program on the LEADS-PEP protein-peptide dataset. J Chem Inf, Model
98. Jones G, Willett P, Glen RC, Leach AR, Taylor R (1997) Development and validation of a genetic algorithm for flexible docking. J Mol Biol 267:727–748
99. Trott O, Olson AJ (2010) AutoDock Vina: improving the speed and accuracy of docking with a new scoring function, efficient optimization and multithreading. J Comput Chem 31:455–461
100. Korb O, Stützle T, Exner TE (2009) Empirical scoring functions for advanced protein-ligand docking with PLANTS. J Chem Inf Model 49(1):84–96

101. Allen WJ, Balius TE, Mukherjee S, Brozell SR, Moustakas DT, Therese Lang P, Case DA, Kuntz ID, Rizzo RC (2015) DOCK 6: impact of new features and current docking performance. J Comput Chem 36(15):1132–1156
102. Kramer B1, Rarey M, Lengauer T (1999) Evaluation of the FLEXX incremental construction algorithm for protein-ligand docking. Proteins 37(2):228–241
103. Ferreira LG, Dos Santos RN, Oliva G, Andricopulo AD (2015) Molecular docking and structure-based drug design strategies. Molecules 20

Chapter 13
Cellular Regulatory Network Modeling Applied to Breast Cancer

Luiz Henrique Oliveira Ferreira, Maria Clicia Stelling de Castro, Alessandra Jordano Conforte, Nicolas Carels, and Fabrício Alves Barbosa da Silva

Abstract Cells in an organism interact with each other and with the environment through a complex set of signals, which triggers responses and activates cellular regulation mechanisms. Models obtained by computational mathematics for cellular signaling dynamics are used to understand factors and causes of deregulation of internal biological processes, which is a relevant knowledge in a disease such as cancer. Gene regulatory networks describe gene interactions and how these relationships control cellular processes such as growth and cell division, which relate this disease to the regulatory network. Despite its simplicity, Boolean networks may accurately model some biological phenomena, such as gene regulatory network dynamics. Indeed, several reports in the literature show that they are accurate enough to build models of regulatory networks of cell lines related to breast cancer. In this chapter, we present a methodology for building cellular regulatory networks based on the Boolean paradigm, which uses entropy as a criterion for selecting genes that are included in the network. The main objective is to understand dynamical behaviors related to situations that cause breast cancer and tumor lineages and to suggest experimentations to verify the outcome of interventions in networks, in order to support the identification of new therapeutic targets.

L. H. O. Ferreira · M. C. S. de Castro
State University of Rio de Janeiro - UERJ - Rua São Francisco Xavier, 524 Maracanã, Rio de Janeiro, Brazil
e-mail: lhrique@gmail.com

M. C. S. de Castro
e-mail: mariaclicia@gmail.com

A. J. Conforte · N. Carels
Plataforma de Modelagem de Sistemas Biológicos, Center of Technology Development in Health, Oswaldo Cruz Foundation, Rio de Janeiro, Brazil
e-mail: conforteaj@gmail.com

N. Carels
e-mail: nicolas.carels@gmail.com; nicolas.carels@cdts.fiocruz.br

A. J. Conforte · F. A. B. da Silva (✉)
Laboratório de Modelagem Computacional de Sistemas Biológicos, Scientific Computing Program (PROCC), Oswaldo Cruz Foundation, Rio de Janeiro, Brazil
e-mail: fabs.fiocruz@gmail.com; fabricio.silva@fiocruz.br

F. A. B. da Silva et al. (eds.), *Networks in Systems Biology*, Computational Biology 32,
https://doi.org/10.1007/978-3-030-51862-2_13

Keywords Cellular signaling dynamic · Cellular regulatory network · Boolean network · Breast cancer

13.1 Introduction

The understanding of biological phenomena has made great progress in the last decades. The knowledge of component structures of living beings, the relationship between theses structures, and the function they have, make it possible to explore computational models that can predict how their systems develop.

Ancestry and characteristics of living beings that are transmitted from generation to generation suggest the existence of similar biological control mechanisms.

Cells in an organism interact with each other and with the environment through a complex set of signals [1], which triggers responses and activate cellular regulatory mechanisms (growth, division, differentiation, death) [2, 3].

Understanding these mechanisms and building models that represent them enables to set up experiments and hypothesis that have the potential of speeding up pre-clinical trials. Therefore, it is possible to reduce invasive exploration techniques for saving patient suffering.

Models obtained by computational mathematics for cellular signaling dynamics are used to understand factors and causes of deregulation of their functioning, which is relevant in a disease like cancer. Deregulation of cancer-affected cells provides uncontrolled and exaggerated proliferation and resistance to programmed cell death [4].

Gene regulatory networks describe gene interactions and how these relationships control cellular processes such as growth and cell division. Several of these cellular processes are related to distinctive cancer hallmarks, which relate the regulatory network to this disease.

There are different cellular regulatory network models [5–7]. Nevertheless, there are not many models describing networks whose disorders give rise to breast cancer.

We did not find evidence of a Boolean network in the literature for describing breast cancer in cell lines with a huge number of genes. Similarly, no satisfactory method using available data could be found for modeling cell lines with Boolean inferences. Thus, in this chapter, we present a methodology for building a cellular regulatory network based on the Boolean paradigm related to breast cancer.

With this contribution, we aim at the prediction, prognosis, and specialized treatment of network disorders that may promote the malignant condition. Despite its simplicity, Boolean networks may model accurately some biological phenomena [8–10]. Moreover, they are accurate enough to build models of gene regulatory networks of cell lines related to breast cancer.

In this chapter, we describe a methodology for network simplification based on the selection of genes using entropy as a criterion. Interactome's information is used to build a network, where each node represents a gene and each edge matches the physical interaction between genes. The inference of Boolean functions occurs considering interaction information and selected gene state.

We considered data from eight cell lines: one non-tumoral (MCF10A) used as a control (reference), and seven malignant ones (BT20, BT474, MDA-MB-231, MB468, MCF7, T47D, and ZR741).

As the number of genes is very large, the corresponding regulatory network is also huge. Nevertheless, not every gene needs to be considered. So, we selected genes according to Carels et al. [11], i.e., we only considered the genes that were up-regulated in the malignant cell line under consideration relative to the reference (MCF10A) and assess their connection potential by reference to the human interactome from EMBL-EBI (IntAct). The potential connectivity of each gene, as obtained from the interactome allows the calculation of the local degree entropy corresponding to each up-regulated gene. These data together with the relationships between them (activation and inhibition) enable discovery of the rules that govern these genes.

Using Boolean networks, we can reproduce the treatment described in [12] and [13], which uses the model of ordinary differential equations that has a greater precision potential. Another possibility is to use the control algorithm, described in [14], to study the disturbances in the regulatory network. The idea of this algorithm is to identify the specific treatments that enable to bring the network back to its healthy state using control theory, which is however impossible in practice due to the irreversible trajectory taken by malignant cells. Nevertheless, control theory can be used to forward malignant cells toward death. In addition, the network can be controlled through its Boolean model, as proposed by [15].

The methodology described below allows: (i) the modeling of cellular regulatory networks based on Boolean representation using entropy as a criterion [11, 16]; (ii) the implementation of different models to better understand how cellular regulatory networks might evolve from normal to malignant cell lines; and (iii) the inference of bench experiments for validating the outcome of network interventions for supporting drug development.

The network model that we describe below behaves in different ways, depending on the chosen parameters for network building. The methodology allows exploring the effect of different simplification criteria (expression levels, reaction to the inhibition of molecular targets induced by given substances).

13.1.1 Background in Gene Regulatory Networks Modeling

Boolean networks representing cellular regulatory network models need the inference of several individual functions, one for each node in the network. The construction process takes effort and time. In addition, it requires biological knowledge. There have been many studies to represent the dynamic behavior of cellular regulatory networks in the literature.

Genetic regulatory networks can be modeled through Bayesian, Boolean, and neural networks, or even differential and/or linear equations [17]. More details about other models can be found in [17–22].

Trairatphisan et al. [23] highlight that modeling through ordinary differential equations has greater precision potential, but little information and experimental data are available to feed them, which restricts these models to smaller networks. On the other hand, models based on Boolean networks support larger networks. Smolen et al. [24] believe Boolean networks are a good option for modeling large gene regulatory networks.

Shmulevich et al. [25] quote studies indicating that, despite their simplicity, Boolean networks are capable of modeling various biological functions. Other works reinforce this position [26, 27]. Krumsiek et al. [28] propose a combination of two models: a Boolean model that is then converted to ordinary differential equations by using the Odefy tool [29].

Von der Heyde et al. [5] presented an inference method of Boolean networks based on three cancer cell lines (BT474, SKBR3, and HCC1954). From a biological standpoint, Rodriguez et al. [6] described rules that allow the characterization, simulation, and interpretation of the FA/BRCA network. On the other hand, Mohanty et al. [7] focused on building a model that describes heterogeneity in tumors.

Campbell and Albert [15] designed Boolean Net, which is a repository of models with Boolean networks tools. Perturbations (node alteration) can be introduced directly through adjustments in the Boolean network itself. This strategy is different from that of Cornelius et al. [14], who first proposed the transformation of the Boolean model into differential equations prior to the introduction of perturbations.

Krumsiek et al. [13], Naldi et al. [30] and, Krumsiek et al. [10] indicated repositories of Boolean network information for gene interactions. Pujana et al. [31], Zhu et al. [32] and Ruz et al. [33] reported on Boolean modeling, but without specifying the used networks. Martin et al. [34] developed a method for Boolean inference based on an expression data time series.

Reverse Engineering Algorithm (REVEAL) [20] extracts the relationship between inputs and outputs from state transitions analysis, entropy, and mutual information of information sets. The logic functions inference is made consulting pre-existing logic rules.

Akutsu et al. [35–37] developed an inference algorithm called BOOL-1 whose inference method consists of an exhaustive search for all logical combinations that satisfy the truth tables. Next, these authors [37] developed BOOL-2 for addressing the noise issue in gene expression data.

Lahdesmaki et al. [9] offer a method for Boolean networks inference using consistency and the best allocation problems (best-fit extension problem) as theoretical paradigms. The experimental results were performed only with small networks.

Koyuturk [1] and Benso et al. [38] provided an understanding of factors that affect regulatory networks. Extended BN Toolkit (EBNT) [39] is a tool for simulations with Boolean networks, but it does not have inference functions. D'haeseleer et al. [19] mentioned the Discrete Dynamics Lab (DDLAB) [40]. Margolin et al. [41] presented ARACNE: Algorithm for the Reconstruction of Accurate Cellular Networks [42].

Karlebach and Shamir [43] published ModEnt [44]. The use of all these solutions has not proved to be viable for the huge number of genes composing the regulatory network that we addressed here.

We would like to emphasize that it does not matter how good a model is; it is still a simulation of a real phenomenon. Therefore, certain premises and situations are simplified to allow the more appropriate treatment or to make the system treatable.

This chapter addresses the modeling of gene regulatory networks using Boolean networks. We propose a methodology that considers the data of static gene expression and the relationship between genes without restrictions on the number of genes that may compose the network.

13.1.2 Cellular Activity Regulation and Cancer Biology Aspects

Cells that compose living beings have a complex set of functions and can generate tissues and organs with completely different organizations. Thereby, it is necessary to decode the function of proteins and nucleic acids in cells.

Proteins can act for signal transmission, process regulation, structural purpose, or as catalysts for chemical processes (enzymes). They provide structural support for cellular organelles, which are responsible for the recognition and transmission of external signals. They also act inhibiting or activating certain cellular functions.

Nucleic acids store and transmit genetic information. They are polymers composed of the four nucleotides Cytosine, Guanine, Adenine, and Thymine or Uracil. A Deoxyribonucleic Acid (DNA) molecule is the basic unit of chromosomes and genes.

Genes contain all the organism's information, it means how they develop, what features they transmit to the progeny among others. A gene set is triggered by cells according to the function type needed in response to an event, or even to various functions that occur in the normal progression of cellular life [3].

The cell cycle is basically composed of multiple phases (lifespan), which prepare for the reproduction process (duplication). Resulting cells can enter a new cell cycle, suffer apoptosis, or enter a phase where cells remain with low metabolic activity and no response to stimuli [45].

Cells do not replicate infinitely; after several divisions, they enter in a programmed death stage. This control mechanism prevents errors occurring during the replication process. Errors become more frequent as the cell ages and cause changes in DNA [2].

There are different mechanisms to control the stability and accuracy of cell replication. The biochemical nature of this process allows some inaccuracy degree, which can generate changes in DNA of the newly generated cells, with positive, negative, or neutral impacts on cell functioning.

Other aspects related to cell functioning is differentiation. A single cell may evolve in different tissues with different cell functions by sharing the same basic methods for translating the information encoded in genes into proteins.

Concerning *Gene Regulatory Networks* (GRNs), many authors emphasize the importance of interactions of various factors in cellular behavior [18, 38, 46]. Factors vary from environmental conditions to cell's molecular components (enzymes, proteins). Other authors express the relation of life concept to cellular regulatory network in a similar way [3, 47].

What determines how the cell develops and responds to stimuli is encoded in the genetic material. Thus, it is also important to know what regulation signals influence gene expression.

The genetic information contained in DNA is first transcribed into a special type of RNA (mRNA) and then translated into proteins. Some genes do not generate proteins as their final product. However, they are also transcribed into non-coding or regulating RNA, which perform specific functions in cells.

Genes can be expressed or not expressed resulting in the activation or deactivation of regulatory functions [25]. Considering that all cells of an organism share the same genetic code, the differences between them come from gene expression as a consequence of their regulation [3].

Products of gene expression can be activators or inhibitors in regulatory networks. Understanding the phenomena behind activation or inhibition can promote the development of therapeutic interventions and treatments for physiologic diseases such as cancer.

There is a consensus regarding the states to which cellular regulatory networks may converge. It is known that, in the absence of abnormal factors, biological functions occur in such a way that cells converge to specific steady states. These states represent cellular phenotypes, which are biophysical conformations derived from genetic characteristics, known as genotype, and from environment variables. Under normal conditions, those conformations are normal results of cellular differentiation.

Cancer resilience is related to an accumulation of genetic mutations and evolutionary selection of new cell lines [47]. Cancer is a genomic disease that causes deregulation in systems that control various cell functions. The most important gene types for cancer are the oncogenes, whose alteration triggers mutations in cell proliferation. There are also tumor suppressor genes whose mutations can lead to functional loss. Therefore, we can consider cancer as a disease that causes deregulation in the regulatory network [2] .

It is not a trivial task to know how the most diverse cancer types acquire malignant features. Hanahan and Weinberg [4, 48] described ten attributes necessary for cancer development, that are considered as hallmarks: (i) sustaining proliferative signaling; (ii) deregulating cellular energetics; (iii) resisting to cell death; (iv) promoting genome instability and mutation; (v) inducing angiogenesis; (vi) activating invasion and metastasis; (vii) promoting tumor and inflammation; (viii) enabling replicative immortality; (ix) avoiding immune destruction; (x) evading growth suppressors.

These hallmarks pointed to the cell functions that are affected by the deregulation in gene expression. Huang et al. [49] did not consider mutations as a primary cancer cause but rather reinforced the importance of gene regulatory network deregulation. Based on cell mobility, Mukherjee [50] stresses how tumors use cellular functions to their own benefit as is the case in tissue invasion and cell proliferation, which are strategies similar to that used by immune cells.

Cancer features can be explained using the hallmarks proposed by Hanahan and Weinberg [4], and the classification of Barillot et al. [2]. These characteristics are (i) cell cycle control failures; (ii) failures to control cell death, and (iii) potential for invasion and metastasis.

The addition of new hallmarks, proposed by Hanahan and Weinberg [48], also enabled to include the following emerging and facilitating features.

Emerging characteristics provide energy for the growth processes and cellular division, ensuring sustained proliferation. They prevent cancer cells from being detected and attacked by the immune system. Hallmarks associated with emerging characteristics are changing in cellular metabolism and resistance to the immune system.

Facilitating characteristics favor the deregulation of cellular activity and the acquisition of evolutionary advantages via natural selection. The inflammatory process, although part of the immune response, is used by tumors to promote their growth, taking advantage of the biological material from their microenvironment. Hallmarks associated with the facilitating characteristics are genetic instability, mutation, and inflammation, which induce tumor growth.

Causes of endogenous cancer are more directly related to regulatory networks. However, external carcinogenic factors can also occur or even facilitate internal causes, as exposure to radiations (X-rays, gamma rays, ultraviolet), chemical elements (asbestos, tobacco, food conservatives) and infection by viruses (HPV) or bacteria (*Helicobacter pylori*).

The differences between tumors occur in their morphology, clinical manifestations (evolution type, response to treatments), molecular levels, and in the tumor's microenvironment that are related to different cell lines found in the same tumor (tumor heterogeneity) [2].

The main breast cancer subtypes are classified according to the expression of protein markers, indicating the presence of hormone receptors, such as estrogen (ER) and progesterone (PR), and *human epidermal growth factor receptor 2* (HER2). Table 13.1 shows the classification according to the presence (+) or absence (−) of

Table 13.1 Breast cancer subtypes (adapted from [2] and [11])

Subtype	Markers			Lines
	ER	PR	HER2	
Luminal A	+	+	−	MCF7, T47D, ZR51
Luminal B	+	+	+	BT474
HER2+	−	−	+	
Triple negative	−	−	−	BT20, MB231, MB468

markers together with the classification of malignant cell lines considered in this study.

The presence of Luminal A, Luminal B, and HER2 markers induces specific therapeutic targets while the triple-negative subtype is more difficult to treat due to the absence of specific targets and for this reason it is considered the most aggressive subtype of breast cancer. Personalized cancer treatment, according to specific patient characteristics and cancerous manifestation, involves the intervention on particular protein targets.

13.2 Methodology for Building Gene Regulatory Networks Models

Figure 13.1 shows a simplified scheme of the proposed methodology for building cellular regulatory networks based on the Boolean paradigm and the control of regulatory networks.

13.2.1 Network Characterization

In this work, the proposed model is based on experiments and data obtained from previous work [11, 16]. All data analyses were performed with the network built from the interactome used by these authors, considering the relationship between genes, expression data, and entropy as discussed below.

The complete interactome is described through 2,133 interactions between the up-regulated genes belonging to seven breast cancer tumor cell lines (BT20, BT474,

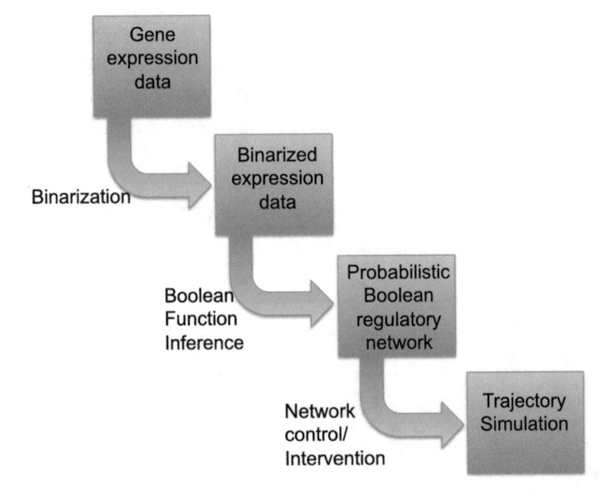

Fig. 13.1 Methodology for building gene regulatory network models

MB231, MB468, MCF7, T47D, ZR751) and the non-tumoral control cell line (MCF10A).

An interactome subset considering all cell lines, called reduced interactome, contains only interactions where at least one of the genes is overexpressed or under-expressed. The reduced interactome has 819 genes, where interactions between a gene and itself are maintained because they have a relevant function in the model of biological phenomenon.

Originally each gene is identified with its UniProtKB accession number. Later, we adopted the usual names (gene symbol), related to biological functions. For genes where it was not possible to establish disambiguation by gene symbol, we used gene symbol plus the number UniProtKB as an identifier (see [51] for more detailed information on the interactome and construction of the model).

The reduced interactome considers gene expression data in each cell lines [51]. Since gene entropy data, although not needed for construction of the initial model, plays a key role in the target-network generation, we focused on the genes with the highest entropy [51] because their inhibition has a greater potential to affect the network, which is a powerful mean for evaluating drug therapeutic potential [11].

The construction of the reduced interactome graph uses expression and entropy data. However, to develop a model based on Boolean networks, we also needed the Boolean function that defines each network node depending on the state of incident nodes. Therefore, the network was constructed considering two issues: the binary state of each incident node (input node) and the direction of interaction between genes, which identifies the gene that regulates the state of another one. Information about directionality between genes was obtained by searching the Metacore (Clarivate) database [52] and the type of interaction documented was relative to gene inhibition as well as activation or information transmission through the signaling network.

For building the network model, we also took the molecular weights of each gene into account.

13.2.2 Reference Model to Construct Boolean Networks

Berestovsky and Nakhleh [53] proposed a general scheme to build Boolean models of gene regulatory network in two steps: (i) expression data discretization or binarization and (ii) logical functions inference at each node in the network. This scheme was later complemented in Trairatphisan et al. [23] for including probabilistic Boolean network aspects.

We propose a different and simplified approach. Reduced interactome data is used to define the network topology, and gene expression data is used to infer Boolean functions. There are divergent positions, in the literature, that consider whether gene regulatory network should be the same for all cell lines or if it should be unique for each one [16, 21]. We chose the option of simplifying the model by building a single network for representing all cell lines. This approach is consistent with [46, 49, 54,

55], whose authors claim that the differentiation is responsible for different types of cell lines that share the same genome. Another argument in favor of the decision to use a single network for all cell lines is related to the limitation of available information (static expression data).

The proposed model based on Boolean networks can be used to generate target-networks with a high number of nodes. It allows networks larger than those used by Von der Heyde et al. [5], which varied between 16 and 26 nodes only for the BT474 cell line.

Target-network generation with a large degree of connectivity can result in a huge amount of misinterpreted information and consequently has the inference of less accurate logical functions resulting in inconsistent nodes. Thus, the choice of connectivity degree affects the target-network generation. It is important to note that the reduced interactome (819 genes) has genes with a degree of connectivity greater than 200, which raises the need for simplification.

13.2.3 Reduced Network Topology

The graph representing the reduced network is given in Fig. 13.2. In Fig. 13.2, node names were omitted for better viewing. The difficulty of dealing with a large number of interactions is obvious and highlights the need for further network simplification.

The first version of the algorithm for network simplification was described in [56] and was improved as described below.

13.2.4 Building a Simplified Target-Network

13.2.4.1 Parameters

Here we called, as target-network, a simplified network that was derived from the reduced network of 819 genes by removing elements according to criteria chosen for reflecting specific interest or properties.

The pruning process was based on a predefined list of criteria or an establishment of rules to choose genes from the reduced network. Of course, a threshold must be imposed to stop the process of target-network pruning at some reasonable level.

This threshold was be chosen according to parameters such as amplitude (m), degree (n) and depth (p), where $m, n, p \in N^*$. Amplitude (m) is related to the number of vertices of the generated target-network, i.e., the number of genes of the target-network. New genes (vertices) and interactions (edges) can be added to m. Degree (n) refers to the number of connections of each vertex. Depth (p) considers the geodesic distance (i.e., number of edges in a minimum path between two vertices) between initial and final vertex when assembling the target-network.

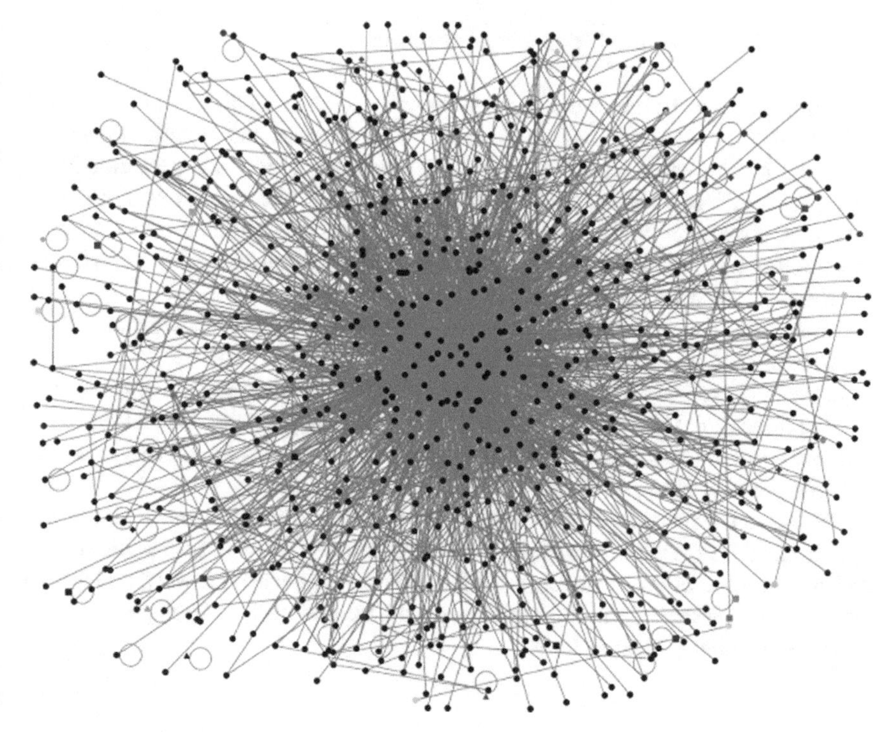

Fig. 13.2 Topology of the network built from the reduced interactome

We established a ranking for selecting genes based on their entropy [16]. Entropy values were calculated according to the n value of each gene in the target-network. The score associated with each gene is the average of entropy values for all cell lines whose genes were expressed (not null since).

In addition, we added the molecular weight to differentiate genes scores, since molecular weight allows the elimination of ambiguity of genes that would be associated with the same entropy score. Intuitively, information carried by small proteins should pass through signaling and regulation networks faster than that carried by larger proteins. So, if genes have equal entropies, we attributed a higher score to genes with lower molecular weight. In the simplified network, genes with the largest scores were selected preferentially in relation to those with the lowest scores.

The edges of the graph from Fig. 13.2 are not oriented, consequently, the direction of interactions between genes are undetermined. Gene interactions and inhibition or activation relationships were then added in the target-network construction according to the information available in Metacore.

13.2.4.2 Target-Network Topology Definition

As outlined above, the target-network was built based on the entropies and molecular weights of reduced networks of up-regulated genes in seven malignant cell lines and on simplification parameters (m, n, and p). The construction was performed in p layers with m initial genes, starting with those having the highest ranking score and, then, connecting their neighbors (n). The process of simplified target-network construction was based on the simplification algorithm proposed by Ferreira, Castro and Silva [56], but adding the interaction directionalities.

We assessed three different constructions of target-networks that we called POS, FWD, and REV as defined below.

In the POS version, the network simplification was done without prior knowledge of edge direction. This network topology was based only on the connection of top connected n genes (top ranking). The direction of each edge was assigned by reference to Metacore after network simplification.

FWD and REV versions considered edge orientation before network simplification. In the FWD version, the target-network was constructed in forwarding mode; starting from the m initial genes, we choose the top-n genes that were under their influence. In the REV from the initial m genes, we choose the top-n genes among the neighbors that influence them.

13.2.4.3 Expression Data Discretization

When building a Boolean model, it is necessary to transform the expression data into binary data. We use Binarize to binarize expression data [57].

Binarization vector was obtained by considering expression data of all genes per cell line (cell line vectors) or by averaging expression data for each gene across cell lines (gene vectors).

Cell line vectors have a large sample of data for discretization. However, after binarization, we obtained an enormous amount of '0' logical values, due to the great variation of expression values and bias for those of lower values. The distribution of gene expressions follows a power-law distribution, with a larger proportion of genes associated with zero expression and a smaller proportion associated with the maximum of gene expression. Using the K-means algorithm as a binarization algorithm in cell line vectors, we found that only 5% of logical values were assigned to '1' and 95% of values to '0'. By contrast, even if binarized gene vectors can be considered less precise because the number of samples is small (only eight values), it carried more information with 32% of '1' and 68% of '0'.

Hopfenstiz et al. [58] compared different binarization methods and showed the benefits of the BASC-B algorithm for binarizing gene expression data in models based on Boolean networks. We also used the BASC-B algorithm for data binarization.

13.2.4.4 Truth Tables Construction Related to Genes of Simplified Target-Network

Truth table construction for each gene in the target-network was based on the topology and input states at each node of cell lines.

Each row of a truth table is related to the binary expression data in all of the eight cell lines. Each row corresponds to genes states for each cell line. Two different situations need to be addressed before Boolean inference (Table 13.2): (i) two lines of the truth table may have equal inputs, which is the case of ZR751 and MCF7 (blue) as well as BT474 and MCF10A (green). The MDA-MB-231 (abbreviated as MB231) and T47D have entries (in red) identical to that of MCF10A, but their outputs are different.

Rows corresponding to MDA-MB-231 and T47D lines are logically inconsistent with that of MCF10A. These rows should not be used simultaneously in the logical inference. This situation, anticipated in [25], is treated in [18] as an open problem. Inconsistencies are due to errors in gene expressions reading or incorrect attribution of logical values to genes (bias induced by the binarization process).

To avoid disregarding relevant information, we built more than one truth table for each gene. A gene can be represented by several truth tables according to the number of inconsistencies.

Shmulevich et al. [25] considered several possible Boolean functions to define a node; there is no reason to consider only one. We adopted the model of a *probabilistic Boolean network* (PBN). According to this model, each function associated with a network node must be assigned with an occurrence probability. The probability impacts the choice of tools to analyze the dynamic behavior of the target-network.

Table 13.2 Truth table of the GRP78 gene (node output)

Lines	Inputs					
	CDC37	EGFR	GABARAPL2	GRB2	TP53	GRP78
MCF10A	0	0	0	0	0	0
BT20	0	1	0	0	0	0
MB231	0	0	0	0	0	1
MB468	1	1	0	0	0	0
MCF7	0	0	0	1	0	0
T47D	0	0	0	0	0	1
ZR751	0	0	0	1	0	0
BT474	0	0	0	0	0	0

We assumed that the functions related to all genes have the same occurrence probability (p). For each gene v with q inconsistent rows, the probability of occurrence of each function is $p = 1/2^q$.

Gene interactions identified as inhibition should be taken into account because it represents a strong biological factor. Inhibition takes precedence over any conflict existing in the model. Priority in the inhibition treatment prevents the treatment of inconsistencies due to overlapping inhibition. Indeed, it can reduce the number of inconsistencies found, and consequently the number of truth tables.

The truth table construction algorithm is described in Fig. 13.3.

Basically, the fluxogram from Fig. 13.3 works as follow:

1. Build an incident nodes list (input list) related to each node (v) from the list of all edges of the target-network. If gene w inhibits gene v, include w in the incident list (or inhibition list) that will compose the truth table inputs.
2. Go through the nodes list and for each node v that has incident nodes (input nodes):

   a. Insert the node (v) and the number of incident nodes (q) as columns of the truth tables associated to the node;
   b. For all cell lines, create a vector of the node expression value v followed by the expression values of the q incident nodes in v, corresponding to a row in

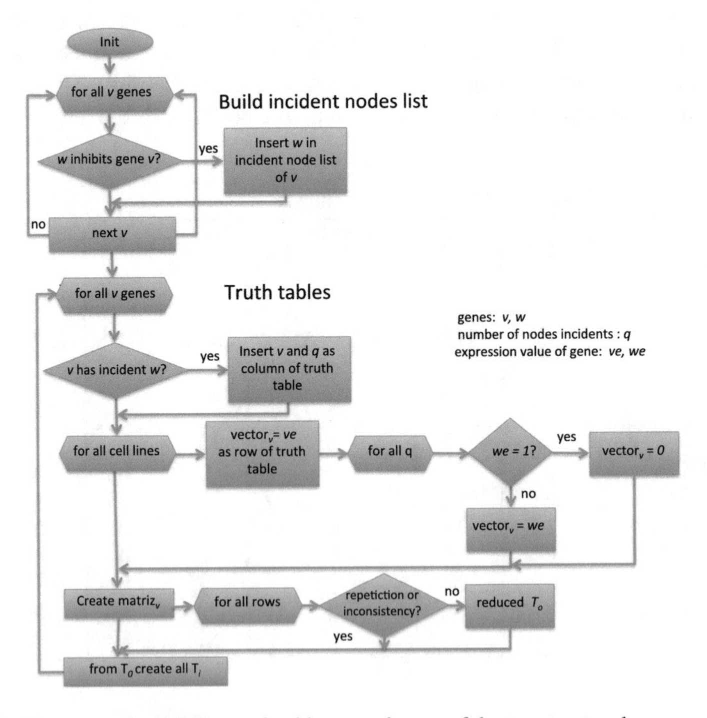

Fig. 13.3 Fluxogram for building truth tables at each gene of the target-network

the truth table. For every activated w_i ($w_i = 1$) included in the v inhibition table, v is inhibited ($v = 0$);

c. Create a matrix (v truth table) with the vectors;
d. For all rows in truth table, check for repetitions or inconsistencies. Rows without inconsistencies are preserved for construction of the reduced truth table (T_O);
e. Starting from the reduced table (T_O), create others (T_i) truth tables, considering all the inconsistencies.

13.2.4.5 Boolean Function Inference of Truth Tables

The BOOM-II: Boolean Minimizer [59] tool, described by Fišer and Kubátová [60], was used to infer the Boolean functions of target-networks.

The truth tables related to target-network nodes were used as input in the execution of the BOOM-II tool. The output was a file containing function inferred from the truth table. If the node had more than one function, the occurrence probability of each function was added to the output.

13.3 Results of Simplified Target-Network Construction

As binarization of gene expression data can be done by different methods, the simplification had three execution versions and, the inference of Boolean functions performed by BOOM-II was subjected to randomness [60, 61]. So, it was necessary to perform several experiments with different configurations to obtain the target-network.

The computational environments used in all experiments are described in Table 13.3. Binarization and simulations of trajectories were performed on a computer with 64 GB RAM running the Linux Ubuntu Server.

Table 13.3 Computational environments

Computer 1	Computer 2	Computer 3
VM Windows Server 2012 x64 DC 16 GB RAM 2 vCPU Intel Xeon 2.3 GHz	PC Windows Pro SP1 x64 8 GB RAM 1 CPU Intel Core i5 3.3 GHz	PC Linux Ubuntu Server 64 GB RAM

Table 13.4 Percentage of '0' logical values after binarization

Cell lines	Cell line vectors			Gene vectors		
	BASC-A	BASC-B	K-means	BASC-A	BASC-B	K-means
BT20	99.3	94.6	93.3	65.1	67.0	67.3
BT474	99.9	94.4	94.6	67.7	67.4	69.2
MB231	91.8	94.6	93.5	63.7	65.2	66.9
MB468	99.8	93.8	94.0	65.4	66.7	66.4
MCF10	99.8	92.4	95.0	69.1	71.2	70.3
MCF7	99.9	94.9	94.9	68.1	68.7	69.0
T47D	99.9	94.7	93.2	63.1	64.5	65.0
ZR751	99.8	93.4	93.7	65.1	66.4	67.8

13.3.1 Binarization of Gene Expression Data

A comparative analysis was performed using BASC-A, BASC-B and K-means algorithms for the binarization of expression data.

The time spent for binarizing the gene vectors (order of minutes) was shorter when compared to the time involved in the processing of cell line vectors (order of hours). We observed that all three algorithms provided similar results using gene vectors, when compared to each other, reaching over 93% coincidence.

Table 13.4 shows the percentage of '0' logical values after binarization with all algorithms considering cell line and gene vectors.

Table 13.4 shows a very large percentage of '0' logical values in cell line vectors regardless of the algorithm under consideration. Therefore, the best option was to consider gene vectors.

Binarization can be done using absolute or differential expression values (RNA-seq of a cell line minus that of the reference). Experiments performed using BASC-B showed no significant differences between absolute and differential expression. Thus, we used BASC-B for the binarization of absolute expression values.

13.3.2 Parameters Impact

The tuning of m, n, and p parameters allowed the generation of different target-networks. However, it was not trivial to find parameter values generating the target-network that match the biological phenomena under modeling. The original network metrics of the Reduced Interactome (RI) (Fig. 13.2) are given in Table 13.5.

The number of genes (m) first added to the target-network is only related to network size, and we assume $m = 10$.

We assigned a depth limit (p) of $p = 4$ to the reduced network graph (Table 13.5), whose average geodesic distance was 3.692 edges.

Table 13.5 Network metrics of the Reduced Interactome (RI)

Metrics	Values
Graph type	Not directed
Vertex number	819
Edge number (excluded loops)	1979
Duplicated edge number Loop number	10 144
Edge number (including loops)	2133
Diameter	12
Average geodesic distance	3.692

Parameter n defines how many interactions are included in target-network for each gene. Depending on n, interactions can be removed or maintained in target-network, which can be decisive in characterizing the dynamic behavior of the model and its adherence to the biological phenomenon under consideration. Parameter n varied from 3 to 20, and several experiments were performed to analyze how it might influence m and p parameters.

Table 13.6 shows the metrics used to compare the simplified versions of target-network with the reduced one.

Table 13.6 shows that simplifying the reduced network decreases the number of elements composing the networks regardless of the method (FWD, REV, POS) used.

The loops represent relevant information on the dynamic behavior of the network (self-regulation of gene expression). On 144 loops of the reduced interactome, 142 are reported as bidirectional. This situation has no sense in Boolean network modeling. Therefore, the loops were considered as unidirectional, which explains the larger number of loops observed in networks generated by the FWD, REV, and POS algorithms.

Table 13.6 Metrics of simplified target-networks

Metrics	Values			
	FWD	REV	POS	RI
Graph type	Directed	Directed	Directed	Not directed
Number of vertices	148	135	63	819
Number of edges (without loops)	330	318	183	1979
Number of duplicate edges	0	0	0	10
Number of loops	330	318	183	144
Number of edges (considering loops)	0	0	0	2133
Diameter	9	9	6	12
(Average geodesic distance)	3.917	3.886	2.871	3.692

We can conclude that it is useful to use a simplification method when the objective is to study a network region. Here the region matched the network space corresponding to the genes that were up-regulated in malignant cell lines compared to MCF10A.

The choice of FWD, REV, or POS methods depended on: (i) the study of the impact that the deregulation of certain genes may have on network dynamics, and on its equilibrium state; and (ii) how changing expression value of some genes might influence the network portion that is dependent on those genes. In the FWD algorithm, which proved to be more appropriate, the gene interactions that are used to build the topology of the target-network are influenced by the genes of the previous level.

13.3.3 Choice of the Boolean Function Used in Simulations

Execution of several experiments using the same set of parameters (m, n, p) gave rise to different instances of Boolean target-networks because of randomness in the inference of Boolean functions.

Different instances of Boolean target-networks are acceptable if they are purely logical, but different inputs regulating the same gene cause problems for the interpretation of biological models. The analysis of dynamic behavior must be done in just one instance. So, we defined the target-network using parameters $m = 10, p = 4$ and $n = 7$ and an objective criterion of choice for functions, i.e., the function that the best represents a gene under consideration.

We evaluated the results according to the following characteristics: (i) the number of generated inconsistencies in the truth tables; (ii) the number of non-adherences to the inhibition found in the truth tables, i.e., when there was an inhibition indication, but it did not appear in the truth tables; and (iii) the characteristics of Boolean functions inferred at each node.

By executing the simplification program with $m = 10$, $p = 4$ and varying n from 3 to 20, we obtained several target-networks. The best parameter n could be assessed through the inconsistencies percentage in the truth tables. Above a certain value n, there was no inconsistency reduction. The experiments carried out with FWD variation shown that the limit was for $n = 7$.

All loops were maintained, and we executed the simplification algorithm 50 times. The generated target-network showed similar results after 30 executions. The amount of executions seeks to mitigate the randomness impact in inference and in logical minimization of Boolean functions.

The chosen function for a gene depended on its occurrence in the sample. For instance, if a function occurred 32 times in 50 network instances for a gene; it was natural to choose that function to represent the gene. When we found the same frequency values between two and more functions, the one corresponding to the lowest order instance was chosen (occurred first).

The target-network used for trajectory simulations and attractors search is shown in Fig. 13.4. This target-network had 177 nodes, 452 edges and was organized as

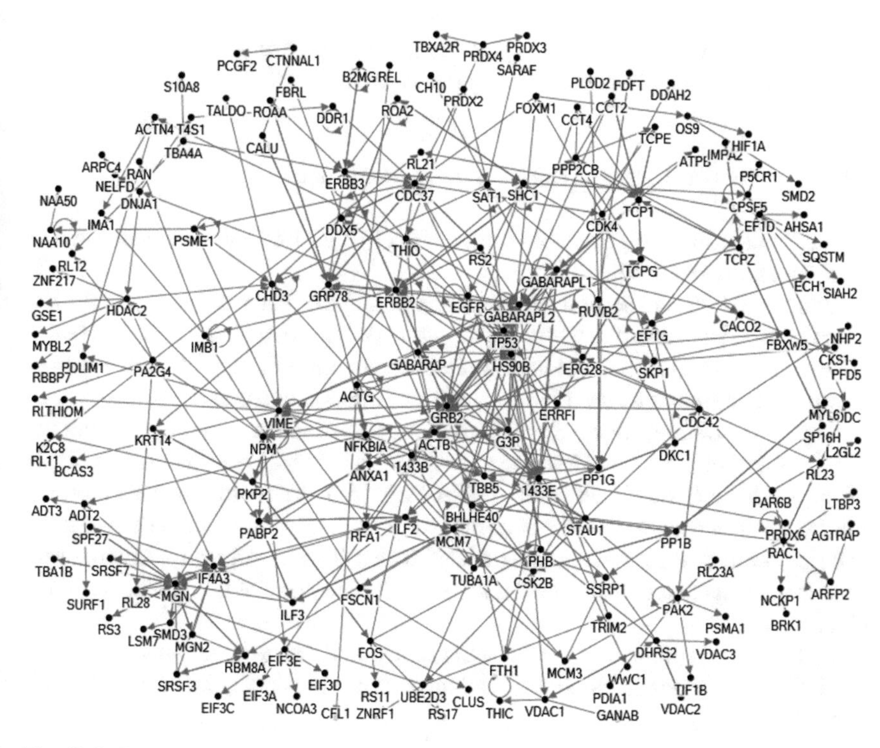

Fig. 13.4 Target-network used in simulations

a PBN. The states of its genes were defined by Boolean functions, which were individually inferred from 30 instances.

13.3.4 *Simulations and Search for Attractors and Steady States*

The dynamic behavior of a gene regulatory network can simulate how to bring malignant cells to a desirable state through controlled interventions [14].

Attractors identification, steady states, and basins of attraction are key to test target-network performance. Interventions in the states of some genes can converge to a desired steady state, which can identify new targets for drug development.

A trajectory shows the sequence of states in the evolution of the network. The trajectory simulations were performed starting at each characteristic state vector of the eight cell lines.

Trajectory analysis of network state transitions can contribute to the understanding of how a given state is reached. Such analysis allows speculation of which state alterations are needed to drive a regulatory network toward the malignant state as well as those that potentially may bring it toward cell death.

13.3.4.1 Tools to Search Attractors and Steady States

Public domain tools are available to find attractors on Boolean networks, such as BoolNet and ASSA-PBN.

Boolnet [62, 63] has several features for exploring Boolean networks, which allow simulation of state transitions, search for attractors and visualization of the results. However, the search for attractors is limited to networks of at most 29 nodes (small networks) and cannot be used for probabilistic Boolean networks.

The *Approximate Steady-State Analyzer of Probabilistic Boolean Networks* (ASSA-PBN) proposed by Mizera et al. [64] was used because we focus on a model of probabilistic Boolean networks. ASSA-PBN [65] is able to (i) analyze probabilistic Boolean networks with a huge number of nodes, (ii) simulate state transition, and (iii) compute the probability associated with a state set. Objective criteria were used to understand tool behavior and to assess the adequacy of functionalities to the objectives.

In the analysis of the network dynamic behavior, the state transitions were simulated aiming at reproducing known biological states previously mapped in the cell lines of this study.

13.3.4.2 State Transitions Simulation

In order to simulate state transitions, it was necessary to set the network initial states (initial conditions). The program used initial states as input parameters for the running simulation.

The simplified target-network had 177 nodes and a huge state vectors number ($1.9\text{x}10^{53}$). It was not computationally feasible to perform simulations starting from each possible state. Thus, it was essential to define vectors of initial states based on a biological criterion. An alternative was to use, as initial states, those that characterize the cell lines (MCF10A, BT20, MB231, MB468, MCF7, T47D, ZR751, and BT474).

13.3.4.3 Analysis of the State Transitions Simulation

It is mandatory to define criteria for enabling results interpretation. However, there was a huge volume of data for simulations for state transition combined with the absence of tools for analyzing trajectories in PBNs. A criterion for result interpretation was to use the occurrence of cycles in state transitions.

Here, we choose only one network to represent all cell lines. Several studies support this approach; Ribeiro and Kauffman [54] proposed that attractor sets are related to mechanisms of cellular differentiation with typically only one of these sets

Table 13.7 Percentage of coincidence comparing the 177 gene states of the target-network in eight reference cell lines

	MCF10A	BT20	MB231	MB468	MCF7	T47D	ZR751	BT474
MCF10A	100	56	60	60	52	55	61	54
BT20	56	100	57	59	56	53	50	55
MB231	60	57	100	60	58	66	54	55
MB468	60	59	60	100	43	47	58	58
MCF7	52	56	58	43	100	64	59	56
T47D	55	53	66	47	64	100	68	62
ZR751	61	50	54	58	59	68	100	**69**
BT474	54	55	65	58	56	62	**69**	100

per cell line. Huang et al. [49] and Serra et al. [55] argued that the same process that generates cellular differentiation can lead to cancer.

Since mutations in the genes controlling cell division and DNA repair may transform healthy cells in malignant ones, they help identifying factors, during the network evolution, that lead to malignant transformations. Based on this evidence, the following analyses were performed: (i) network states that represent cell lines are steady states in the network evolution; (ii) since malignant cell lines are derived from healthy cell lines due to changes in certain gene states, there is a network evolution from the state of the reference cell line to the state of malignant cell lines; and (iii) malignant cell lines may evolve to the death state by changing certain gene states.

During the simulation, only the mapped states, i.e., the state vectors that characterize the eight cell lines, were observed. Considering a network with 177 genes, there was a low probability of occurrence of trajectory states exactly equal to those of the eight reference vectors in trajectory simulations. Therefore, we adopted a threshold to compare the trajectories and the expression features of cell lines.

The threshold choice involved the percentage of coincidences between the states of the eight cell lines (Table 13.7). Table 13.7 gives the expression values of each cell line binarized using BASC-B applied on gene vectors. The table shows the percentages of coincidence comparing cell lines pair by pair. The objective was to identify the threshold below which coincidences uniquely identify each cell line, and it also shows that there is up to 69% coincidence between them. Therefore, a threshold higher than 69% reduces the possibility of identifying more than one cell line in the simulated trajectory. The threshold of 75% was used to identify mapped states in trajectory simulations.

13.3.4.4 Trajectory Analyses

A trajectory shows the sequence of states in the simulation of the network dynamics. Using ASSA-PBN, trajectory simulations were performed starting at each characteristic state vector of the eight cell lines. Because of the probabilistic nature of the

Table 13.8 Summary of simulation and analysis

Volume	Analysis of 480 trajectory simulations with 30,000 state transitions with 5,310,000 gene comparisons per simulation
Execution time	149 h 55 min
Simulation with coincidences	Coincidences occurring among the mapped states of 372 simulations (78%)
Coincidences number	1,034 state coincidences were identified, 99% of them in trajectory parts that have the same state at the origin and destination
Coincidence degree	On average, the coincidences have 77% of genes with values identical to the mapped states

network, we ran 60 simulations. As a result, an output file of approximately 10 MB per trajectory simulation was generated. Table 13.8 shows a summary analysis of the simulations that were performed.

As explained previously, the 75% coincidence threshold was sufficient to identify the mapped states whose vector characterized one of the cell lines during the simulated trajectories.

Table 13.9 shows how the cycles with the same initial and final states are distributed in the network evolution. They are the most frequent cycles seen in simulations.

State transition cycles are concentrated on the initial steps of trajectories with 82.6% of them occurring before step 100 and 99.7% before step 940. This observation was consistent with the typical human cell cycle during their lifespan suggesting that simulations with more than a thousand transition steps were unnecessary.

Cycles that lead the network from healthy to malignant states were rare. Empirical experiments confirm this conjecture [18, 25, 46, 47], which pledge that cell lines rarely evolve from one to another naturally, without intervention or artificial stimuli.

As stated before, most cycles occurred in the initial stages of network evolution, where the difference between the states of a trajectory was smaller. The results associated with the average lifespan of cells, both healthy and malignant, supported the conjecture that a thousand steps of simulations of state transitions were sufficient. Barbosa [45] mentions that the cell cycle in a human organism varies from 24 to 48 h in healthy cells and 72 to 120 h in malignant tumors. Kauffman [66] estimates that the time required to activate a gene varies from 5 to 90 s and quoted studies showing that the time of cell cycle, in human cells, is about 1,000 min (~16 h 30).

Table 13.9 Cycles with origin and destination equals to the network initial state

Step	BT20	BT474	MB231	MB468	MCF10A	MCF7	T47D	ZR751	%accum[a]
1–10	44	48	71	62	39	47	28	36	36.8
11–20	8	5	20	1	8	24	16	19	46.7
21–30	6	6	15	3	5	11	15	13	53.9
31–40	5	2	17		9	16	7	17	61.1
41–50	3	5	13	1	3	9	8	10	66.2
51–60	1	4	7	3	3	9	8	9	70.5
61–70	2	5	7		2	10	10	11	75.1
71–80	1	3	4	1	3	6	10	2	78.0
81–90	2	9	2				7	4	80.4
91–100	2	1	5		1	6		8	82.6
101–110	3	1	1		2	3	5	5	84.6
111–120	1	1	1		1	3	5		85.8
121–130	1				1	1	4	2	86.7
131–140		1			2	5	4	2	88.0
141–150					5	4	2	4	89.5
151–160	1					6	6	1	90.9
161–170	2				1	4	1	6	92.3
171–180	1	1	3			4	3	1	93.5
181–190	1					1	1	1	93.9
191–200		3				2	1	4	94.9
201–210			2			2		1	95.4
211–220					1	1			95.6
221–230			1		1	2	1		96.1
231–240		1	1			1	3	1	96.8
241–250		1				1			97.0
251–260							1	1	97.2
261–270		1				1		1	97.5
271–280						1	1		97.6
281–290							4		98.0
291–300						1		1	98.2
301–310						2			98.4
311–320						1			98.5
331–340								1	98.6
371–380								1	98.7
391–400								1	98.8
401–410								1	98.9
411–420								1	99.0

(continued)

Table 13.9 (continued)

Step	BT20	BT474	MB231	MB468	MCF10A	MCF7	T47D	ZR751	%accum[a]
431–440						1			99.1
441–450								1	99.2
511–520								1	99.3
541–550								1	99.4
861–870								1	99.5
921–930							1		99.6
931–940							1		99.7
4.171–4.180								1	99.8
10.851–10.860		1							99.9
27.721–27.730		1							100.0
Total	84	100	171	71	87	185	153	170	1,020

[a] Accumulated percentage

13.4 Conclusions

This chapter presents a methodology to build gene regulatory networks based on Boolean networks.

The proposed methodology is consistent with other ones found in the literature, as in [18]. Other authors, such as Huang et al. [49], relate attractors to steady states and cell differentiation. The use of static expression data in only one network to represent all cell lines is acceptable. Of course, the inference of Boolean functions using the time series of gene expression data would be more accurate.

Although there is no standard in the literature, the treatment of inhibitions and inconsistencies is supported by similar approaches [38, 46]. The use of public domain tools aided in the model construction and supported the relevance and consistency of the results presented in this chapter.

Garg et al. [46] claim that in a reliable model of gene regulatory network, the robustness of steady states must be confirmed by the robustness of the attractors from the model. Our model had such robustness since it had the tendency to return to its original state or stay in a steady state during the simulation of the dynamics of the network.

We were able to model several gene regulatory networks by choosing different parameters. If metrics different from entropy are to be considered, one could construct new networks and change the filtering criteria.

Experiments can also be performed by focusing on network parts, adapting amplitude (m), degree of connectivity (n), and depth (p) for network simplification, which would allow the evaluation of other relevant aspects in greater detail.

In addition, one can alternate between the execution of the REV and FWD methods for searching the causes and consequences of specific situations.

Control methods can be applied to the model of target-network described here. The idea is to drive the malignant cell to specific steady states (cell death) through intervention on certain genes (inhibition), simulating therapeutic treatments.

References

1. Koyutürk M (2010) Algorithmic and analytical methods in network biology. Wiley Interdiscip Rev Syst Biol Med 2(3):277–292
2. Barillot E, Calzone L, Hupé P, Vert JP, Zinovyev A (2012) Computational systems biology of cancer. Mathematical & computational biology, 461 Pages. Chapman & Hall/CRC
3. Hunter L (1993) Molecular biology for computer scientists. Artificial intelligence and molecular biology, pp 1–46
4. Hanahan D, Weinberg RA (2000) The hallmarks of cancer. Cell 100(1):57–70
5. Von der Heyde S, Bender C, Henjes F, Sonntag J, Korf U, Beissbarth T (2014) Boolean ErbB network reconstructions and perturbation simulations reveal individual drug response in different breast cancer cell lines. BMC Syst Biol 8(1):75
6. Rodriguez A, Sosa D, Torres L, Molina B, Frias S, Mendoza L (2012) A Boolean network model of the FA/BRCA pathway. Bioinformatics 28(6):858–866
7. Mohanty AK, Datta A, Venkatraj V (2013) A model for cancer tissue heterogeneity. IEEE Trans Biomed Eng 61(3):966–974
8. Akutsu T, Kuhara S, Maruyama O, Miyano S (2003) Identification of genetic networks by strategic gene disruptions and gene overexpressions under a boolean model. Theoret Comput Sci 298(1):235–251
9. Lähdesmäki H, Shmulevich I, Yli-Harja O (2003) On learning gene regulatory networks under the Boolean network model. Mach Learn 52(1–2):147–167
10. Wittmann DM, Blöchl F, Trümbach D, Wurst W, Prakash N, Theis FJ (2009) Spatial analysis of expression patterns predicts genetic interactions at the mid-hindbrain boundary. PLoS Comput Biol 5(11):e1000569
11. Carels N, Tilli TM, Tuszynski JA (2015) Optimization of combination chemotherapy based on the calculation of network entropy for protein-protein interactions in breast cancer cell lines. EPJ Nonlinear Biomed Phys 3(1):6
12. Krumsiek J, Pölsterl S, Wittmann DM, Theis FJ (2010) Odefy-from discrete to continuous models. BMC Bioinform 11(1):233
13. Krumsiek J, Wittmann DM, Theis FJ (2011) From discrete to continuous gene regulation models–a tutorial using the Odefy toolbox. Applications of MATLAB in science and engineering, p 35
14. Cornelius SP, Kath WL, Motter AE (2013) Realistic control of network dynamics. Nature Commun 4(1):1–9
15. Campbell C, Albert R (2014) Stabilization of perturbed Boolean network attractors through compensatory interactions. BMC Syst Biol 8(1):53
16. Carels N, Tilli T, Tuszynski JA (2015) A computational strategy to select optimized protein targets for drug development toward the control of cancer diseases. PloS One 10(1)
17. Kaderali L, Radde N (2008) Inferring gene regulatory networks from expression data. Computational intelligence in bioinformatics, pp 33–74. Springer, Berlin, Heidelberg
18. Karlebach G, Shamir R (2008) Modelling and analysis of gene regulatory networks. Nat Rev Mol Cell Biol 9(10):770–780
19. D'haeseleer P, Liang S, Somogyi R (2000) Genetic network inference: from co-expression clustering to reverse engineering. Bioinformatics 16(8):707–726
20. Liang S, Fuhrman S, Somogyi R (1998) Reveal, a general reverse engineering algorithm for inference of genetic network architectures. http://ntrs.nasa.gov/search.jsp?R=20010002317

21. Bansal M, Belcastro V, Ambesi-Impiombato A, Di Bernardo D (2007) How to infer gene networks from expression profiles. Mol Syst Biol 3(1)
22. Ristevski B (2013) A survey of models for inference of gene regulatory networks. Nonlinear Anal Model Control 18(4):444–465
23. Trairatphisan P, Mizera A, Pang J, Tantar AA, Schneider J, Sauter T (2013) Recent development and biomedical applications of probabilistic Boolean networks. Cell Commun Signal 11(1):46
24. Smolen P, Baxter DA, Byrne JH (2000) Mathematical modeling of gene networks. Neuron 26(3):567–580
25. Shmulevich I, Dougherty ER, Kim S, Zhang W (2002) Probabilistic Boolean networks: a rule-based uncertainty model for gene regulatory networks. Bioinformatics 18(2):261–274
26. Shmulevich I, Dougherty ER, Zhang W (2002) From Boolean to probabilistic Boolean networks as models of genetic regulatory networks. Proc IEEE 90(11):1778–1792
27. Shmulevich I, Wei Z (2002) Binary analysis and optimization-based normalization of gene expression data. Bioinformatics 18(4):555–565
28. Krumsiek J et al (2010) Odefy-from discrete to continuous models. BMC Bioinformatics 11(1):233
29. Krumsiek J et al (2011) From discrete to continuous gene regulation models-a tutorial using the Odefy toolbox. INTECH Open Access Publisher
30. Naldi A, Berenguier D, Fauré A, Lopez F, Thieffry D, Chaouiya C (2009) Logical modelling of regulatory networks with GINsim 2.3. Biosystems 97(2):134–139
31. Pujana MA, Han JDJ, Starita LM, Stevens KN, Tewari M, Ahn JS et al (2007) Network modeling links breast cancer susceptibility and centrosome dysfunction. Nat Genet 39(11):1338
32. Zhu P, Liang J, Han J (2014) Gene perturbation and intervention in context-sensitive stochastic Boolean Networks. BMC Syst Biol 8(1):60
33. Ruz GA, Timmermann T, Barrera J, Goles E (2014) Neutral space analysis for a Boolean network model of the fission yeast cell cycle network. Biol Res 47(1):64
34. Martin S, Zhang Z, Martino A, Faulon JL (2007) Boolean dynamics of genetic regulatory networks inferred from microarray time series data. Bioinformatics 23(7):866–874
35. Akutsu T, Miyano S, Kuhara S (1999) Identification of genetic networks from a small number of gene expression patterns under the Boolean network model. In: Proceedings of Biocomputing '99, pp 17–28
36. Akutsu T, Miyano S, Kuhara S (2000) Algorithms for inferring qualitative models of biological networks. Biocomputing 1999:293–304
37. Akutsu T, Miyano S, Kuhara S (2000) Inferring qualitative relations in genetic networks and metabolic pathways. Bioinformatics 16(8):727–734
38. Benso A, Di Carlo S, Politano G, Savino A, Vasciaveo A (2014) An extended gene protein/products boolean network model including post-transcriptional regulation. Theor Biol Med Model 11(1):S5
39. Benso A et al (2016) BNToolkit – SysBio Group. https://www.sysbio.polito.it/bntoolk
40. Wuensche A (2019) Tools for researching cellular automata, random Boolean networks, multi-value discrete dynamical networks, and beyond. Discrete Dynamics Lab. http://www.ddlab.com
41. Margolin AA, Nemenman I, Basso K, Wiggins C, Stolovitzky G, Dalla Favera R et al (2006) ARACNE: an algorithm for the reconstruction of gene regulatory networks in a mammalian cellular context. BMC Bioinform 7(Suppl 1):S7. BioMed Central
42. Califano A (2019). http://califano.c2b2.columbia.edu/aracne
43. Karlebach G, Shamir R (2012) Constructing logical models of gene regulatory networks by integrating transcription factor–DNA interactions with expression data: an entropy-based approach. J Comput Biol 19(1):30–41
44. Karlebach G (2019) ModEnt a tool for reconstructing gene regulatory networks. http://acgt.cs.tau.ac.il/modent
45. Instituto Nacional de Câncer (Brasil), Barbosa MBA (2008) *Ações de enfermagem para o controle do câncer: uma proposta de integração ensino-serviço*. INCA

46. Garg A, Mohanram K, Di Cara A, De Micheli G, Xenarios I (2009) Modeling stochasticity and robustness in gene regulatory networks. Bioinformatics 25(12):i101–i109
47. Bellomo N, Li NK, Maini PK (2008) On the foundations of cancer modelling: selected topics, speculations, and perspectives. Math Models Methods Appl Sci 18(04):593–646
48. Hanahan D, Weinberg RA (2011) Hallmarks of cancer: the next generation. Cell 144(5):646–674
49. Huang S, Ernberg I, Kauffman S (2009) Cancer attractors: a systems view of tumors from a gene network dynamics and developmental perspective. Semin Cell Dev Biol 20(7):869–876. Academic Press
50. Gomes RS (2012) O imperador de todos os males: uma biografia do câncer. In Mukherjee S (ed) Companhia das Letras, São Paulo, 634 pp
51. Ferreira LHO (2019) Modelagem de redes de regulação celular aplicada ao câncer de mama. Dissertação de Mestrado. Programa e Pós-graduação em Ciências Computacionais, Instituto de Matemática e Estatística, Universidade do Estado do Rio de Janeiro
52. Clarivate Analytics (2019) MetaCore. https://clarivate.com/products/metacore/
53. Berestovsky N, Nakhleh L (2013) An evaluation of methods for inferring boolean networks from time-series data. PloS One 8(6)
54. Ribeiro AS, Kauffman SA (2007) Noisy attractors and ergodic sets in models of gene regulatory networks. J Theor Biol 247(4):743–755
55. Serra R, Villani M, Barbieri A, Kauffman SA, Colacci A (2010) On the dynamics of random Boolean networks subject to noise: attractors, ergodic sets and cell types. J Theor Biol 265(2):185–193
56. Ferreira LHO, Castro MCS, Silva FA (2016) Modeling gene regulatory networks: a network simplification algorithm. AIP Conf Proc 1790(1):100003. AIP Publishing LLC
57. Mundus S, Müssel C, Schmid F, Lausser L, Blätte, TJ, Hopfensitz M et al (2015) Binarize: binarization of one-dimensional data
58. Hopfensitz M, Müssel C, Wawra C, Maucher M, Kuhl M, Neumann H, Kestler HA (2011) Multiscale binarization of gene expression data for reconstructing Boolean networks. IEEE/ACM Trans Comput Biol Bioinf 9(2):487–498
59. Fišer P (2006) BOOM-II: the PLA minimizer. https://ddd.fit.cvut.cz/prj/BOOM
60. Fišer P, Kubátová H (2004) Two-level boolean minimizer BOOM-II. In: Proceedings of 6th international workshop on Boolean problems (IWSBP'04), Freiberg, Germany, vol 23
61. Coudert O, Sasao T (2002) Two-level logic minimization. Logic synthesis and verification, pp 1–27. Springer, Boston, MA
62. Müssel C, Hopfensitz M, Kestler HA (2010) BoolNet—an R package for generation, reconstruction and analysis of Boolean networks. Bioinformatics 26(10):1378–1380
63. Müssel C, Hopfensitz M, Zhou D, Kestler HA, Biere Hanson DT (2018) BoolNet: construction, simulation and analysis of Boolean networks. https://cran.r-project.org/web/packages/BoolNet/index.html
64. Mizera A, Pang J, Yuan Q (2015) ASSA-PBN: an approximate steady-state analyser of probabilistic Boolean networks. In: International symposium on automated technology for verification and analysis. Springer, Cham, pp 214–220
65. Mizera A et al (2019) ASSA-PBN 3.0: a software tool for probabilistic Boolean networks (PBNs). https://satoss.uni.lu/software/ASSA-PBN
66. Kauffman SA (1969) Metabolic stability and epigenesis in randomly constructed genetic nets. J Theor Biol 22(3):437–467

Index

F. A. B. da Silva et al. (eds.), *Networks in Systems Biology*, Computational Biology 32,
https://doi.org/10.1007/978-3-030-51862-2

E

F

G

N

O

P

Q

R

Printed in the United States
by Baker & Taylor Publisher Services